Applications of Nanocomposites

Editor

Ahalapitiya H. Jayatissa

Mechanical, Industrial and Manufacturing Department
College of Engineering
The University of Toledo
Toledo, Ohio
USA

CRC Press
Taylor & Francis Group
Boca Raton   London   New York

CRC Press is an imprint of the
Taylor & Francis Group, an **informa** business

A SCIENCE PUBLISHERS BOOK

First edition published 2022
by CRC Press
6000 Broken Sound Parkway NW, Suite 300, Boca Raton, FL 33487-2742

and by CRC Press
4 Park Square, Milton Park, Abingdon, Oxon, OX14 4RN

© 2022 Ahalapitiya H. Jayatissa

CRC Press is an imprint of Taylor & Francis Group, LLC

Library of Congress Cataloging-in-Publication Data (applied for)

ISBN: 978-1-032-16096-2 (hbk)
ISBN: 978-1-032-16105-1 (pbk)
ISBN: 978-1-003-24707-4 (ebk)

DOI: 10.1201/9781003247074

Typeset in Times New Roman
by Shubham Creation

Preface

The synthesis of nanomaterials has attracted a great deal of attention in the past two decades. The applications of nanomaterials have been investigated as a part of the characterization processes. These topics have been evolved in many different directions because functionalities of nanomaterials can be tuned to obtain numerous advantages over existing bulk materials. Currently, these materials have been used in some manufacturing technologies. There are many technologies at the transition stage between synthesis and applications. In this book, the authors present the possible application of nanomaterials and related scientific and economic impacts.

Chapter-1 brings the science behind the incorporation of nanomaterials in composite matrixes by surface treatments and modifications. This chapter will provide a greater understanding of the modification of carbon-based nanomaterials for various applications. This chapter also provides key technologies necessary to understand the interface characteristics of nanomaterial and polymer matrixes.

Chapter-2 brings the application of nanomaterials in automobiles and related industries. The chapter provides a substantial review of nanotechnology applications related to their mechanical properties. This chapter reviews and summarizes many published research on the application of nanomaterials in automotive industries.

Chapter-3 provides catalytic application of nanomaterials related to enhancing the surface area of nanostructured materials. This chapter also provides substantial fundamental knowledge of optoelectronic properties and surface modifications of nanomaterials.

Chapter-4 is related to the possible application of nanomaterials in energy storage in lithium-ion batteries. This is a current topic that attracted global attention in next-generation rechargeable batteries.

Chapter-5 gives recent developments related to flexible electronic devices, which use two-dimensional carbon sheets with one atom thick planar packed in a honeycomb crystal lattice structure has recently attracted significant academic and industrial interest because of its exceptional electronic and optoelectronic properties and excellent performance in mechanical, electrical, and thermal applications. This chapter describes the fabrication and characterization of flexible sensors using pure graphene sheets and graphene-based composites.

Chapter-6 discusses the importance of nanocomposites in tribology applications. The desirable qualities that make it superior to conventional liquid lubrication; reasons for their widespread acceptance are covered. The challenges, drawbacks of these coatings are explained from a cost and reliability standpoint. A comprehensive review of the application of metal, ceramic and polymer-based nanocomposites is covered in the chapter.

Chapter-7 is about the technological development and practical application of carbon fiber reinforced plastics, natural fiber-reinforced plastics, and GFRP recycling in Japan are described. First, the Carbon Fiber Reinforced Polymer (CFRP) parts having high-strength and lightweight using carbon fiber and epoxy resin were developed, and its practical use has expanded. The CFRP has been used in airplanes and automobiles, improving safety and fuel efficiency by reducing weight. Second, to reduce the consumption of petroleum resources, Natural Fiber Reinforced Polymer (NFRP) using natural fiber of reinforcement and Polylactic Acid (PLA) of base polymer made from biomass are also discussed.

Synthesis of nanocomposites by Liquid Phase Deposition is discussed in Chapter-8. These materials in which nanoparticles are dispersed into the matrix material to obtain improved properties. In recent years, the demand for lightweight and strong materials in the automotive industry, which directly affects fuel efficiency, vehicle performance, and exhaust gases emissions, are increasing to manufacture various auto parts such as brakes, engine covers, glasses, and mirror coatings, seat and door trim, chassis, and tires. These materials offer excellent mechanical, electrical, and thermal properties, flame retardance, wear and corrosion resistance, and self-cleaning, which ensure safety, comfort, and durability of the vehicle.

Chapter-9 describes the use of layered silicates inorganic material as nanofillers in the thermoset material to improve the mechanical, physical, thermal, and barrier properties of the polymer. Montmorillonite (MMT) is one of the layered silicates that are commonly used as reinforcement nanofillers in polymeric matrices, including the epoxy owing to its excellent physicochemical characteristics. This chapter reviews and summarizes the findings collected from several published research on the epoxy composites and nanocomposites with single and hybrid fillers, emphasizing the use of layered silicate as co-filler in the hybrid nanocomposite system. Furthermore, the published works on the epoxy nanocomposite with hybrid MMT/geopolymer fillers for piping application are compiled and summarized at the end of this chapter. These findings indicate that the use of hybrid MMT/geopolymer fillers can significantly improve the mechanical properties of the epoxy, revealing the great potential of this hybrid filler system in reinforcing the epoxy for piping application.

In Chapter-10, the properties and characterizations of MCC and NCC will be clarified, and their toxicity effect will be discovered to make them compatible as a new alternative approach in the pharmaceutical industry, especially for safety consumption purposes. The increasing use of nanomaterials in biomedical applications such as drug delivery and targeted drug therapy has triggered the global research community to develop strategies to improve engineered materials performance by incorporating nanomaterials with biocompatible polymers. Microcrystalline cellulose (MCC) and nanocrystalline cellulose (NCC) from agricultural waste possess bio-composite characterizations, which offer significant sustainability and eco-efficiency in the development of biomedical products. The high crystallinity index and yield percentage from both MCC and NCC were proven they had the potential in the pharmaceutical industry in drug delivery, tablet formulation, and wound healing. They were non-toxic and did not show inflammation effects on both cell lines and animal modes when presented in low concentration. Possible applications of MCC and NCC from agricultural wastes are potential candidates are also discussed in this chapter.

Chapter 11 is devoted to describing recycling, landfilling, and incineration of waste containing nanoparticles (NPs) is handled. But there are still no specific methods or systems in place to efficiently capture and recycle the NPs released into the waste streams. To avoid landfilling non-recyclable nanomaterials, especially of the manmade engineered nanomaterials (ENMs), and to counteract their harmful effects on the environment, innovative approaches such as green synthesis of nanocomposites, manufacturing biodegradable nanocomposites using renewable sources as raw materials and magnetic nanoparticles have piqued the interest of researchers to help synthesize and manufacture sustainable nanocomposites. Further, the need for and importance of conducting life cycle analysis of nanocomposites along with the health, safety, and environmental hazards' assessment by considering the impact of all the stages throughout their manufacturing, service, consumption, and disposal are emphasized. Some of the challenges faced by the emerging sustainable technologies and potential future directions of research are also outlined.

As the editor of this book, I owe a debt of gratitude to all the authors who prepared these manuscripts. Their dedicated work, patience, understanding of the process, and timely submission of manuscripts and reviews are greatly appreciated. I want to thank my graduate students, Suren, Victor, and Bodhi, for their support in contact with authors and reviewers continuously.

Editor
Ahalapitiya H. Jayatissa
Mechanical, Industrial and Manufacturing Engineering
Department College of Engineering, The University of Toledo
Toledo, Ohio, USA

Contents

Chapter 1

Surface Functionalized Nanoparticles

Kabir Rishi

Chemical & Materials Engineering, University of Cincinnati, 560 Engineering Research Center, 2901 Woodside Drive, Cincinnati, OH 45219, 513 888 1879

National Institute for Occupational Safety & Health, Centers for Disease Control and Prevention, 5555 Ridge Avenue, Cincinnati, OH 45213, 513 458 7130
Krishi@cdc.gov; rishikr@mail.uc.edu

Gregory Beaucage*

Chemical & Materials Engineering, University of Cincinnati, 492 Rhodes Hall 2851 Woodside Drive, Cincinnati, OH 45219, 513 373 3454
beaucag@uc.edu

1   INTRODUCTION

Industrial nanoparticles are essential for modern times and have advanced technological applications such as in automotive tires, inks, lithium-ion batteries and consumer goods such as sun block lotions and cosmetics. The dispersion of nanoparticles in viscous polymers is dictated by kinetics, the interaction potential between particles and the interfacial compatibility between the matrix and dispersed phases. Industrial polymer nanocomposites can be produced by a variety of processes. A common method is to melt mix viscous polymers with nanofillers including shear mixing in a Banbury mixer, twin-screw or single-screw extrusion and calendering in a two-roll mill. Dispersion and subsequent distribution and re-aggregation/network formation of nanoparticles do not occur under equilibrium. However, a pseudo-thermodynamic approach can be used that takes advantage of an analogy between the kinetically mixed polymer compounds and thermally dispersed colloids. A pseudo-second-order virial coefficient can adequately delineate the systemic interactions while having the ability to quantify dispersion as a function of the accumulated strain during the mixing process (Jin et al. 2017; McGlasson et al. 2019). It also has the advantage of being used as an input for coarse-grained simulations thereby combining experiments with modeling (Gogia et al. 2021). The degree of filler dispersion is dictated by binary filler-polymer interactions which

---

*Corresponding Author

depend on the surface characteristics. Weakly interacting systems such as carbon black, surface-modified silica in low dielectric constant polymers can be described by a mean field. However, this approach is expected to fail for nanoparticles with inherent surface charges such as precipitated and pyrogenic silicas, especially in high dielectric constant polymers since the fillers interact strongly. Conversely, a mean-field model may be suitable for surface-charged nanoparticles in low dielectric polymers with filler particle spacings larger than the Debye screening length (low concentrations), while specific interactions may dominate at higher concentrations and for higher dielectric constant polymers (McGlasson et al. 2020; Rishi et al. 2020; Okoli et al. 2021).

Commercial polymer nanocomposites involve a multi-hierarchical structure that includes a material spanning filler network leading to many of the signature properties such as tear resistance, conductivity, strength, optical scattering for paints and pigments and cosmetics. The network emerges from a complex interplay between miscibility, immiscibility, kinetic dispersion, particle breakup and aggregation. The multi-hierarchical nanofiller structure includes primary nanoparticles and fractal nano to colloidal scale aggregates and larger agglomerates and clusters that grow to an emergent network. Surface functional groups, surface polarity and charge, the Debye screening length dependent on the dielectric constant of the polymer, and the processing kinetics often described by the mean accumulated strain (Rishi et al. 2019a) have a tremendous impact on the properties of nanocomposites (Rishi et al. 2019a; 2020; McGlasson et al. 2019). For example, in commercial tires carbon, black nanofillers form hierarchical filler networks at different size scales that impact the dynamic response at highway speeds (Rishi et al. 2018) as well as the tear strength (Hamed 1991). On the contrary, fumed and precipitated silica nanofillers tend to correlate into somewhat ordered structures (McGlasson et al. 2020) that could potentially impact the rolling resistance of the tire. This chapter discusses how to leverage small-angle X-ray scattering to quantify binary interactions in these systems. It also describes how these interactions dictate both the local as well as the global structural emergence at multiple percolation concentrations pertaining to different sizes of the multi-hierarchical structure. The impact of these multi-hierarchical architectures on the dynamic and dielectric response in polymer nanocomposites is explored.

2   INDUSTRIAL NANOPARTICLES: IMAGING THE FRACTAL STRUCTURE AND HIERARCHICAL LEVELS THROUGH SMALL-ANGLE X-RAY SCATTERING (SAXS)

During the flame synthesis of industrial nanoparticles such as carbon black, silica and titania, etc., the first stable structures that form following nucleation are primary particles. The primary particles, which are on the order of 10 nm in size, serve as the building blocks for the hierarchical structure. These primary particles sinter downstream in the flame into ramified mass-fractal aggregates. The fractal aggregates that are on the order of 100 nm in size further cluster into agglomerates due to van der Waals forces. The clustered agglomerates which are on the order of microns in size are further compacted into powders that are visible to the naked eye. However, during shear mixing in a Banbury mixer or an extruder, these micron-scale agglomerates dissociate and disperse to aggregates. This inherent structural hierarchy impacts the percolation, network formation and dispersion while dictating the dynamic, dielectric and optical properties of the final nanocomposite (Donnet et al. 1993; Friedlander 2000; Brinker and Scherer 2013).

Structural characterization of these hierarchical structures is possible through advanced visualization techniques. As opposed to microscopy which is a direct imaging technique, the interpretation and analysis of Small-Angle X-ray Scattering (SAXS) occur in reciprocal space (q is the reciprocal space vector) which is a Fourier transform of real space. q is directly related to the sine of the scattering angle and inversely proportional to the incident X-ray wavelength, thus bearing units of inverse length. Analysis of SAXS data involves performing model fits to the reduced scattering intensity, $I(q)/\phi$, which is based on some assumption of the filler structure. However, despite its limitations, SAXS has numerous advantages. Firstly, the spatial resolution in combined small-angle scattering techniques can extend from Angstroms to microns, about four decades in size yielding information about the different structural levels in the filler hierarchy. To image this size range, one would have to use optical, scanning electron and transmission electron microscopes. Secondly, the sample preparation for SAXS measurements needs minimal effort whereas for electron microscopy one needs to either sputter coat the surface to prevent electron buildup in SEM or prepare extremely thin microtomes at cryogenic temperatures often with staining for transmission mode. AFM and SEM are limited to surface analysis. Thirdly, the data interpretation in microscopy is confined to 2D projections of 3D structures which limits a detailed understanding of the filler morphology. More importantly, for direct imaging on the nanoscale through transmission electron microscopy, the cryo-sectioned samples need to be about 100 nm thick, and any information thus gleaned is only a local measure of the filler structure. On the contrary, SAXS measurements are made over macroscopic dimensions resulting in an exploration of the nano to the macroscopic structure at 20 μm to 5 mm sample sizes. Although electron tomography to reconstruct a 3D picture of the aggregate is gaining traction (Liu et al. 2017; Song et al. 2017), it is limited to micron size structures. Finally, scattering measurements are free from operator bias.

The scattered intensity, $I(q) = \phi V \Delta\rho^2 P(q) S(q)$, depends on the form factor, $P(q)$, the interparticle structure factor, $S(q)$, the particle volume, V, the overall filler volume fraction, ϕ and the scattering contrast between the filler and the matrix, $\langle\Delta\rho\rangle^2$. When the filler concentration in the polymer is dilute (ϕ_0), $S(q) = 1$ and the above expression reduces to $\dfrac{I_0(q)}{\phi_0} = V\Delta\rho^2 P(q)$. The form factor for spherical particles, $P(q) = \dfrac{9\{\sin(qR) - (qR)\cos(qR)\}^2}{(qR)^6}$ is the square of the spherical amplitude function where R is the radius of the particles. This definition of the form factor is valid for perfectly spherical colloidal silica particles (Akcora et al. 2009). However, this choice would be insufficient to describe scattering in complex multi-level hierarchical, nano- to macro-scale structures. Industrial nanofillers typically display three structural levels comprised of the primary particle (level 1), mass-fractal aggregates of primary particles (level 2), and agglomerates of aggregates (level 3) due to their synthesis as discussed earlier (Jin et al. 2017; Hashimoto et al. 2019). To account for these multi-scale structural hierarchies, the concentration normalized dilute scattering can be interpreted using the Unified Scattering Function (Beaucage 1995; 2004; Beaucage et al. 2004) such that,

$$\frac{I_0(q)}{\phi_0} = \sum_{i=1}^{n} \left[G_i \exp\left(\frac{-q^2 R_{g,i}^2}{3}\right) + B_i (q_i*)^{-P_i} \exp\left(\frac{-q^2 R_{g,i=1}^2}{3}\right) \right] \tag{1}$$

where "i" is the structural level, G_i and B_i, normalized by the volume fraction, are the Guinier and Porod pre-factors that account for the particle volume, V and the scattering contrast, $\Delta\rho^2$; the radius of gyration, $R_{g,i}$, specifies the size of each structural level in the hierarchy; the power-law exponent, P_i, specifies the morphology of each structural level and is generally 4 for solid three-dimensional moieties with no surface roughness whereas it varies between 1 and <3 for mass-fractal objects. Additionally,

$q_i^* = q \left[\text{erf}\left(\dfrac{kqR_{g,i}}{\sqrt{6}} \right) \right]^{-3}$, wherein "erf" is the error function and k equals 1 for three-dimensional

structures and approximately 1.06 ± 0.005 for mass-fractal structures (Beaucage 1995). The advantage of using the Unified Scattering function is that it uses no additional variables other than the ones stipulated by the Guinier (Guinier and Fournet 1955) and Porod (Porod 1982) laws and yet accounts for the size-related dependence of adjacent structural levels in the nanofiller hierarchy.

Industrial nano-aggregates are convoluted mass-fractals (fractal dimension, for level 2 from the Unified fit (Beaucage 1995; 2004; Beaucage et al. 2004) in equation (1) with a tortuous topology as shown in Fig. 1.1. The topological parameters that can be derived from the Unified fit parameters in equation (1) include: the size of the primary particle or the equivalent Sauter mean diameter, $d_p = 6V/S$; the degree of aggregation or the average number of primary particles in an aggregate, $z = (G_2/G_1) + 1$; the aggregate end-to-end distance, $R_{eted} = d_p z^{1/d_f}$; the aggregate tortuosity, $d_{min} = B_2 R_{g,i}^{d_f}/(C_p \Gamma(d_f/2)G_2)$; where, C_p represents the aggregate polydispersity; the short circuit path, $p = (R_{eted}/d_p)^{d_{min}}$ the aggregate connectivity dimension, $c = d_f/d_{min}$; the connective path, $s = (R_{eted}/d_p)^c$; and the branch fraction, $\phi_{br} = (z - p)/z$ (Mulderig et al. 2017a; Rai et al. 2018; Beaucage 2004; Beaucage et al. 2004).

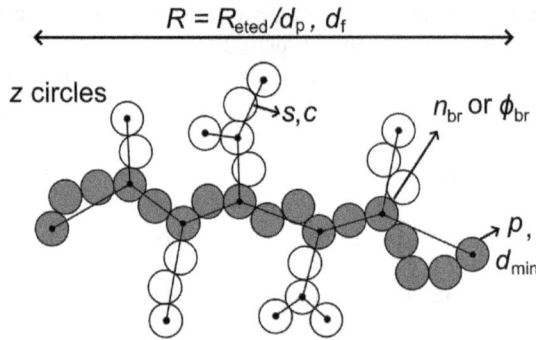

Figure 1.1 A fractal aggregate represented on a two-dimensional plane. The aggregate topology is described by the ratio (R) of the end-to-end distance (R_{eted}) to the primary particle size (d_p), its fractal dimension, (d_f), the number of primary particles (circles) that comprise the aggregate (z), the minimum path (p), the dimension of the short-circuit path (d_{min}), the connective path (s), the connectivity dimension (c), the number of branches (n_{br}) and branch fraction (ϕ_{br}). Reprinted from Journal of Aerosol Science, 109 (March), Mulderig et al. 2017a. Quantification of Branching in Fumed Silica, 28–37, Copyright 2017, with permission from Elsevier.

To image these filler nano-aggregates from scattering, a particle-cluster aggregation code was developed on a 3D grid (Mulderig et al. 2017a) that uses only the average number of primary particles in an aggregate as the input to the simulations. This measure of the degree of aggregation is used to truncate the growth of particles which is controlled only by a probability of sticking (ranging from 0 to 1). The simulation uses a path-finding procedure to determine the topological parameters of the aggregate wherein a neighbor list for each stuck particle in the aggregate is created and the distance between all combinations of end points is computed. The weight average distance between all possible end points equals R. R is a dimensionless number that equals the ratio of R_{eted} to d_p. To determine the minimum path, multiple percolation pathways along the aggregate are traced such that any given pathway commences and terminates at a random end point while being constrained to move backward. The weighted average of all such pathways equals the short circuit path, p. Since $R = z^{1/d_f} = p^{1/d_{min}}$, the other structural parameters d_{min}, d_f, c and ϕ_{br} from the simulation can be compared to the same parameters derived from scattering and the simulation iterated until a match between the simulated 3D structure and the measurement parameters is found.

Figure 1.2(a) compares the TEM images of single carbon black nanoaggregates in polybutadiene rubber at dilute loading to aggregates simulated from the fit to the reduced scattering intensity vs. q plot (inset). An exploded view of the two inset TEM micrographs show an R_{eted} of 120 nm that quantitatively agrees with the scattering result (Rishi et al. 2018). A qualitative comparison of the two micrographs in Fig. 1.2(a) to the simulated aggregates based on scattering indicates a good agreement although a quantitative comparison was not attempted since the micrograph is a 2D projection of the actual aggregate. Figure 1.2(b) shows a similar qualitative comparison for a porous precipitated silica aggregate with the aggregate simulated from scattering (McGlasson et al. 2020). The simulation procedure is available in the Irena software package provided by Jan Ilavsky at Argonne National Laboratory (Ilavsky and Jemian 2009; Ilavsky 2012; Ilavsky et al. 2013; 2018).

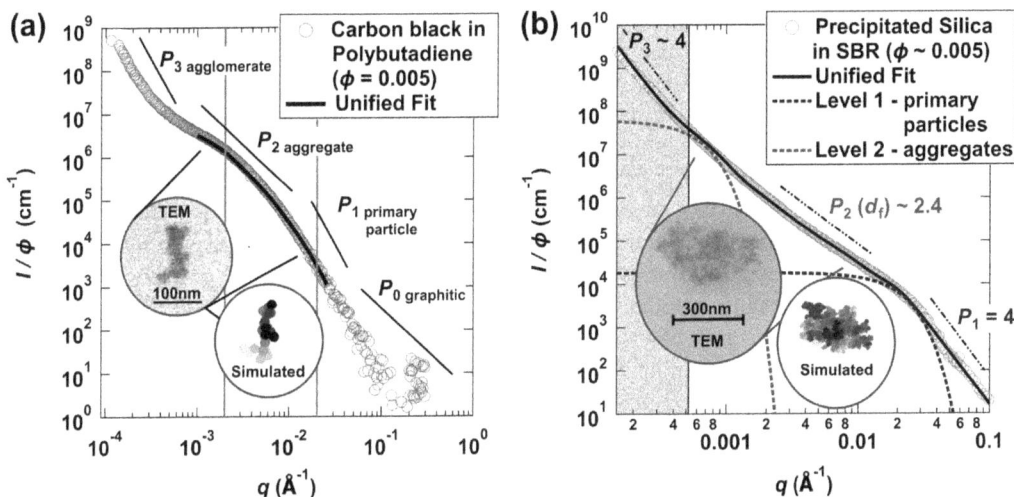

Figure 1.2 **(a)** Concentration (dilute) normalized scattering intensity, $I_0(q)/\phi_0$, as a function of reciprocal space vector, q fit using the Unified Scattering Function (Beaucage 1995; 2004; Beaucage et al. 2004) for carbon black filled polybutadiene. The inset images compare the micrographs of carbon black nanoaggregates to those simulated from the scattering result. Reprinted (adapted) with permission from Rishi et al. 2018. Impact of an Emergent Hierarchical Filler Network on Nanocomposite Dynamics. *Macromolecules* 2018, 51(20): 7893–7904. Copyright 2018 American Chemical Society **(b)** Concentration (dilute) normalized scattering intensity, $I_0(q)/\phi_0$, as a function of reciprocal space vector, q fit using the Unified Scattering Function (Beaucage 1995; 2004; Beaucage et al. 2004) for precipitated silica filled styrene-butadiene rubber. The inset image compares the micrograph of a silica nanoaggregate to the simulated aggregate based on scattering. Reprinted (adapted) with permission from McGlasson et al. 2020. Quantification of Dispersion for Weakly and Strongly Correlated Nanofillers in Polymer Nanocomposites. *Macromolecules* 2020, 53(6): 2235–2248. Copyright 2020, American Chemical Society.

3 SURFACE FUNCTIONALIZATION OF FRACTAL NANOPARTICLES

Filler-polymer incompatibility results from differences in the surface chemistry of nanofillers and the polymer. For example, the presence of surface hydroxyls on silica renders it polar due to which it is incompatible with non-polar polymers like styrene-butadiene rubber. Although, the surface of carbon black may contain many functional groups, the reactions involving oxygen complexes do not necessarily lead to improved rubber-carbon black interactions in SBR (Leblanc 2002). The density of functional groups on the surface plays a critical role in determining whether the nanoparticle surface is amenable to functionalization. The extent of polarity in nanofillers is characterized by differences in the polar and disperse components of surface energy (Wang et al. 1991a, b; Wolff and Wang 1992). Classical methods to improve filler-polymer interactions in incompatible systems were based on the addition of different coupling agents to the melt during processing. These coupling agents tether nanoparticles to the polymer matrix. Block co-polymers have also been used to nanoparticles compatible with polymeric matrices. For example, Raut et al. (2018) used poly(butadiene-graft-pentafluorostyrene) block copolymer as a compatibilizer in Styrene-Butadiene Rubber (SBR) carbon black nanocomposites to improve dispersion. The electron-rich aromatic rings of carbon black attract the electron-deficient pentafluorostyrene, whereas the butadiene backbone shows affinity towards SBR.

For chemically bonded coupling agents, estimating whether the coupling reaction occurred on the particle surface, limited by reaction kinetics and processing conditions, can only be determined by secondary gauging of improvements in the mechanical properties, tear strength, etc., (Hamed 1991). To overcome this challenge, chemical treatments to introduce different functionalities, and grafting polymers or short-chain oligomers to the surface of bare nanoparticles before melt compounding is performed. For example, lignosulfonates were modified with cyclohexylamine to enhance compatibility which led to improvements in nanocomposite strength (Bahl and Jana 2014). Similarly, the addition of carboxylic acid functionalities to the surface of zirconia nanoparticles via a silane coupling agent resulted in an improvement in Young's modulus of the polystyrene matrix composite as compared to neat polystyrene (Kockmann et al. 2018). Additionally, altering the surface of zinc oxide nanofillers led to an enhancement in dispersion in high-density polyethylene (Benabid et al. 2019). To improve the mixing of colloidal silica nanoparticles prepared by the Stöber process in a bismaleimide/diamine matrix surface modification was achieved by chemically introducing amine and epoxide functionalities (Sipaut et al. 2015). The impact of varying the graft chain length and particle size in PDMS-grafted silica particles on the interparticle interaction potential quantified via rheology has also been studied (McEwan and Green 2009). The different strategies for surface functionalization have been comprehensively detailed by Kango et al. (2013). For other *in situ* surface functionalization techniques, advanced characterization of the functionalized surfaces, and modeling the dynamics of these novel functionalized nanoparticles the reader is referred to a succinct review by Hore et al. (2020).

Compared to model colloidal silica nanoparticles for academic studies, industrial-grade nanofillers such as silica rely on simpler surface modification techniques that involve reacting the inherent silanol moieties. The surface of fumed silica particles is characterized by the presence of isolated silanol and hydrogen-bonded silanol groups, which are statistically distributed over the surface as shown in Fig. 1.3. This nature of the fumed silica surface results in strong hydrophilic characteristics with high surface energy which reduces its compatibility with highly non-polar polymer matrices. These monoenergetic surface hydroxyls can react with hexamethyldisilazane (HMDS) to cover the silica surface with trimethylsilyl groups (Hair and Hertl 1971) rendering it compatible with non-polar polymers, as shown in Fig. 1.3. A less common surface modification technique relies on the introduction of a carbon precursor during flame-spray pyrolysis (Mädler et al. 2002; Mueller et al. 2004) of silica, also shown in Fig. 1.3. Carbon-coated silica was synthesized using a commercial hydrogen-air burner, by oxidation of hexamethyldisiloxane (HMDSO) (Kammler et al. 2001). It was found that by increasing the production rate and lowering the hydrogen concentration in flame at a constant airflow, nano-aggregates of silica-carbon composite particles with varying surface carbon contents could be produced (Kammler et al. 2001). More recently, Okoli et al. (2021) have attempted to characterize the surface carbon content in terms of carbon monolayers and gauged the impact of varying the number of monolayers on the dynamic properties of the nanocomposite.

4 STRUCTURAL EMERGENCE IN INCOMPATIBLE NANOFILLER-POLYMER SYSTEMS

The Unified Scattering Function (Beaucage 1995; 2004; Beaucage et al. 2004), equation (1), can be used to model the structure of nanoaggregates at dilute concentrations as discussed earlier. Modeling the structure through SAXS is limited to dilute nanofillers in polymer since above the

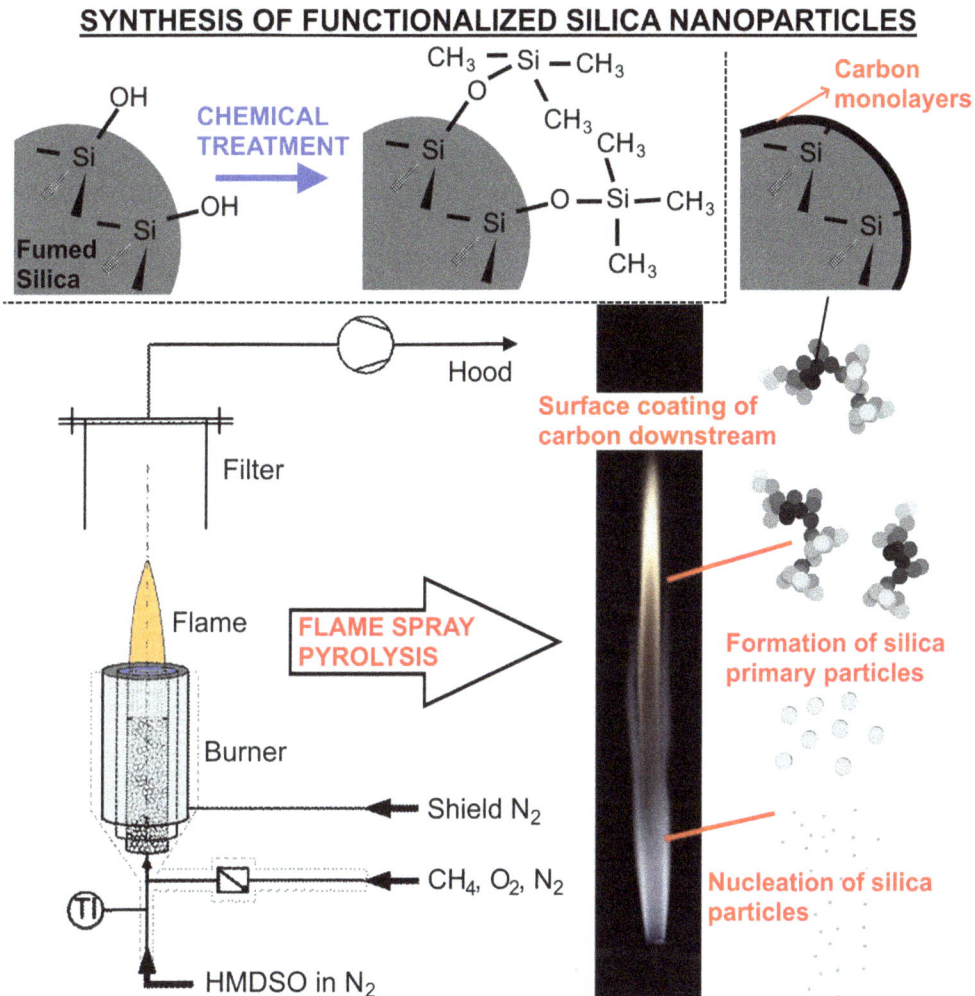

Figure 1.3 Routes of functionalizing the surface of industrial grade pyrogenic silica: chemical post-treatment with HMDS to generate trimethylsilyl functional groups, and introduction of carbon precursors during flame-spray pyrolysis. The carbon/soot is coated downstream in the flame. Reprinted (adapted) with permission from Okoli et al. 2021. Dispersion and Dynamic Response for In-Flame and Chemically Modified Fumed Silica Nanocomposites. *In*: Bulletin of the American Physical Society, 2021 (top figure). Reprinted with minor changes from Powder Technology, Vol 140/Issues 1-2, Mueller, R., Kammler, H.K., Pratsinis, S.E., Vital, A., Beaucage, G. and Burtscher, P. Non-agglomerated dry silica nanoparticles, Pages 40–48, Copyright 2004, with permission from Elsevier. (bottom schematic)

overlap concentration, ϕ^*, the aggregate structure can no longer be observed, by it being screened from view by the other, overlapping aggregates. Concentrations above ϕ^* but below concentrated conditions are referred to as semi-dilute solutions since substantial regions devoid of aggregates still exist. In the semi-dilute regime for immiscible fillers, aggregates form clusters. Typically, ϕ^* is about 3 to 5 weight percent filler (Jin et al. 2017). Percolation at the 100 nm scale does not lead to global percolation and the formation of a conductive pathway for immiscible systems. At a concentration of about 16 to 20 weight percent filler a second, global percolation of nanoclusters occurs which can lead to a bulk conductive pathway, ϕ^{cc}. This involves size scales of about 1 μm. At concentrations above ϕ^{cc} independent clusters can no longer be resolved. Following this motif, complex multi-hierarchical structures emerge when the nanofiller content in polymers increases.

Referring to the nanoscale semi-dilute regime above ϕ^* binary aggregate-aggregate interactions dominate,

$$S(q) = \frac{I(q)/\phi}{I_o(q)/\phi_0} \neq 1 \qquad (2)$$

over the entire q-range. Owing to the inherent structural hierarchy, $S(q)$ varies with the size scale of observation. Increasing concentration beyond a local percolation threshold, ϕ^*, is expected to have no impact on the structure of primary particles observed at high-q where $S(q) = 1$. However, $S(q)$ is expected to vary considerably form 1 in the intermediate-q and low-q regions reflecting the different configurations of nanoaggregates, agglomerates, clusters above ϕ^*; and networks above ϕ^{cc}.

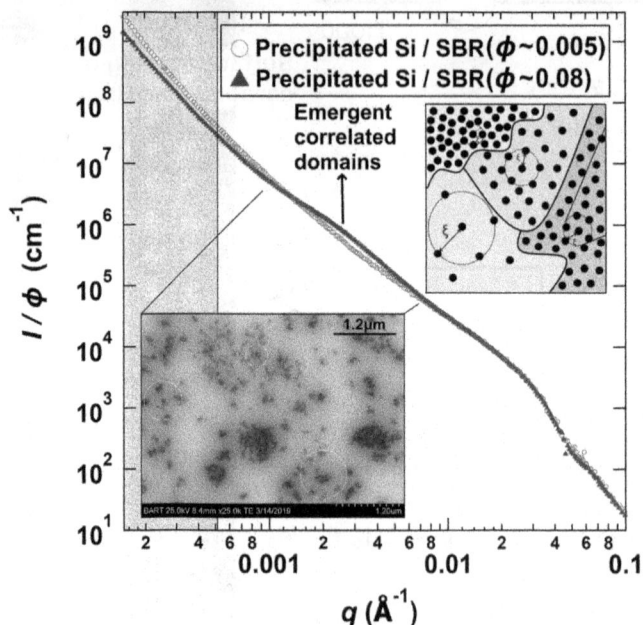

Figure 1.4   Reduced scattering intensities, $I_0(q)/\phi_0$ (dilute, light circles) and $I(q)/\phi$ (semi-dilute, dark triangles) as a function of reciprocal space vector, q for unmodified precipitated silica in SBR rubber. Notice the emergence of a broad peak in the dark curve at intermediate-q associated with aggregate correlations. The inset cartoon depicts these emergent structures in the same bulk with varying correlation distances as evidenced in the inset TEM micrograph (McGlasson et al. 2020). Reprinted (adapted) with permission from McGlasson et al. 2020. Quantification of Dispersion for Weakly and Strongly Correlated Nanofillers in Polymer Nanocomposites. *Macromolecules* 2020, 53(6): 2235–2248. Copyright 2020, American Chemical Society.

Although electrostatic effects leading to correlations in a low dielectric media such as in a polymer matrix are not obvious, several recent papers theoretically predict moderate electrostatic forces over about 4 microns between weakly charged spheres (Bichoutskaia et al. 2010; Stace and Bichoutskaia 2012). For industrial nanoparticles such as precipitated/fumed/colloidal silica, the electrostatic charge varies due to the density of silanol functional groups on the surface (Croissant and Brinker 2018). A larger surface charge leading to increased repulsive interactions between the nanoaggregates presents a weak correlation peak in the small-angle X-ray scattering pattern (Baeza et al. 2013; Bouty et al. 2014; 2016; Beaucage et al. 1995; Anderson and Zukoski 2007; 2010; Jethmalani et al. 1997; Jin et al. 2017). Figure 1.4 shows the emergence of this correlation peak in scattering for precipitated silica ($\phi \sim 0.08$) SBR nanocomposite. The surface hydroxyls on precipitated silica are polar which detracts from their compatibility with non-polar SBR. Whereas the concentration normalized scattered intensity for both dilute and semi-dilute concentrations match at high-q indicating no change in the primary particles, at intermediate-q the aggregates correlate. A peak in scattering indicates periodicity with a repeat on the order of $2\pi/q^*$ following Bragg's law. It is to be noted that this USAXS peak, unlike the sharp peak in diffraction/wide-angle scattering, is weak and broad as shown in Fig. 1.4.

Binary interactions for non-functionalized nanoparticles with charged surfaces like colloidal silica due to silanol moieties are described by the Ornstein and Zernike (O-Z) function, equation (3). The O-Z function describes the interactions of particles in the liquid state such that the total correlation function, (Egelstaff 1967)

$$h(r) = c(r) + \rho \int c\left(\left|\vec{r} - \vec{r'}\right|\right) h(r') d^3\vec{r'} \qquad (3)$$

is the sum of the direct binary correlations, $c(r)$, and the higher-order interactions expressed by a convolution $h(r)$ of and $c(r)$ over all distances r. Here, ρ indicates the number density of particles. While the O-Z expression describes the total interactions within a system, owing to the recursive nature of the function, obtaining a solution to the expression in equation (3) is difficult. To solve the expression various closure relationships are used. The Percus-Yevick (P-Y) approximation (Percus and Yevick 1958) is a commonly used closure relationship for the O-Z expression and has been used to describe the dispersion of hard spheres within a matrix such as monodisperse spherical colloidal silica. An analytic solution to the direct binary correlations, $c(r)$ in the O-Z expression in equation (3) was given by Wertheim (1963) as,

$$c(r) = -\alpha - \beta\left(\frac{r}{\xi}\right) - \delta\left(\frac{r}{\xi}\right)^3 \qquad (4)$$

where ξ is the mean distance between the centers of spherical particles and the coefficients,

$\alpha = \dfrac{(1+2\phi)^2}{(1-\phi)^4}$, $\beta = \dfrac{-6\phi(1+0.5\phi)^2}{(1-\phi)^4}$, and $\delta = \alpha\phi/2$ are related to the filler volume fraction. The Fourier

transform of the binary correlation function in equation (4) is, (Wertheim 1963)

$$\rho C(q) = -24\phi\left\{\left[\left(\frac{\alpha}{(q\xi)^3}\right)\left[\sin(q\xi) - (q\xi)\cos(qD)\right] - \right.\right.$$

$$-\left(\frac{\beta}{(q\xi)^4}\right)\left[(q\xi)^2\cos(q\xi) - 2(q\xi)\sin(q\xi) - 2\cos(q\xi) + 2\right] -$$

$$-\left(\frac{\delta}{(q\xi)^6}\right)\left[(q\xi)^4\cos(q\xi) - 4(q\xi)^3\sin(q\xi) - 12(q\xi)^2\cos(q\xi)\right.$$

$$\left.\left. + 24(q\xi)\sin(q\xi) + 24\cos(q\xi) - 24\right]\right\} \qquad (5)$$

From the convolution theorem, (Roe 2000) the Fourier transform of equation (3) given by $H(q) = C(q) + \rho H(q)C(q)$, is used to determine the inter-particle structure factor, $S(q)$ as

$$S(q) = \frac{H(q)}{C(q)} = \frac{1}{1 - \rho C(q)} \qquad (6)$$

McEwan et al. (2011) quantified the impact on particle interactions of polydimethylsiloxane (PDMS) brushes grafted on to model monodisperse colloidal silica nanospheres in PDMS matrices of varying molecular weights. Small-angle X-ray scattering was used to determine $S(q)$ and the radial distribution function, $g(r)$ through the P-Y approach described above. $S(q)$, determined by normalizing the measured scattered intensity by a simulated form factor (dilute scattering intensity vs. q), was fit with an arbitrary power-law function by Baeza et al. (2013) Instead of simulating the correlated aggregate structure and interactions within, Bouty et al. (2014) fit the structure factor to the analytical solution of the P-Y integral equation developed by Wertheim (1963) described in equations (5) and (6) for colloidal silica in SBR. In a later study, Bouty et. al. (2016) used a similar model system to understand the effects of chain conformation on the dispersion of colloidal silica nanoparticles without attempting a P-Y fit.

The semi-empirical Born-Green (B-G) approximation (Guinier 1955) of the O-Z equation was also proposed to describe the distribution of hard spheres. Beaucage et al. (1995) prepared aggregated silica nanoparticles in PDMS *in situ* via a sol-gel process in which silica was generated from the excess cross-linking agent, tetraethyl orthosilicate (TEOS). In these systems correlations are developed in a low viscosity oligomer during crosslinking so there is no residual stress or variable accumulated strain in the sample, resulting in close to monodisperse correlation distances between aggregates. Further, under these synthetic conditions, the interpenetration of aggregates is not possible since there is no mixing, aggregates form near crosslink sites or in regions where there is excess TEOS. To describe structural correlations of the B-G function,

$$S(q) = \frac{1}{1 + p\theta(q)} \tag{7}$$

was used. In equation (7), the dimensionless packing factor, p, is 8 times the ratio of the average "hard-core" occupied volume of the filler to the total volume. The packing factor determines the strength of the correlation peak in equation (7). $\theta(q)$ is the sphere amplitude function, $3\frac{(\sin q\xi - q\xi \cos q\xi)}{(q\xi)^3}$, reflecting spherical correlations about a central aggregate and ξ is the average correlation distance between particles as shown in the inset cartoon in Fig. 1.4. The packing factor reflects the degree of adherence to the organization in a spherical shell through the ratio of occupied to available volume. For spherical particles, p varies from 0, indicating no correlations, to 8(0.74) = 5.92, for closest packed structures where the packing factor for closely packed spheres is 0.74. Note that $\theta(q)$ parallels the first term in equation (5) of the P-Y approximation. At $q = 0$, $\theta = 1$ and the structure factor in equation (7) is $S(0) = 1/(1 + p)$ indicating that the strength of the correlation peak determines the $q \to 0$ intercept for the structure factor. The $q \to 0$ intercept is related to the second virial coefficient as is discussed below.

Although the P-Y and B-G closure relationships are able to sufficiently describe correlations for spherical colloidal silica particles (McEwan et al. 2011), and even aggregated silica structures with low dispersity in correlation distance and the absence of aggregate interpenetration using the Unified Function (Beaucage et al. 1995), manipulation of the two approximations are needed for melt blends of mass-fractal nanoaggregates. If the less cumbersome, B-G approximation is used for mass-fractal aggregates, one would expect the packing factor, p, to have higher values due to nanoaggregate asymmetry and the likelihood of aggregate overlap above ϕ^* in melt blends. For example, p, eight times the occupied to available volume, could be a larger number for highly asymmetric objects such as lamellae, up to a value of $p = 8$ for space-filling stacked lamellae. Asymmetric fractal nanoaggregates can align but can also interpenetrate resulting in a higher packing density than regular objects like lamellae. An estimate of the fraction interpenetration, I, is,

$$I = \frac{p}{8} - 1 \tag{8}$$

where the maximum value of p with interpenetration is 16, reflecting complete interpenetration of the aggregates. p can also be interpreted as a thermodynamic parameter since $S(0) = 1/(1 + p) = \dfrac{1}{d\Pi/d\phi} = \dfrac{1}{kT(1 + 2\phi B_2)}$. From the thermodynamic perspective, there is no upper limit to p.

When fractal nanoaggregates are kinetically mixed with highly viscous polymers, the correlations result from a balance between the accumulated strain and the repulsive charge. Note that colloidal silica (McEwan et al. 2011) and silica prepared *in-situ* (Beaucage et al. 1995) in low-viscosity PDMS represent systems with no accumulated strain and the mobile polymer phase would result in close to a regular separation of aggregates such that an average value for the correlation distance, ξ, in the P-Y and B-G equations can be used. On the contrary, milling of highly viscous nanocomposites results in non-uniform application of accumulated strain, as indicated by Baeza et al. (2013), such that the correlation distance could vary through the polymer bulk. This variation could be envisioned through the formation of domains with varying accumulated strains in the highly viscous nanocomposites. McGlasson et al. (2020) modified the B-G approximation to account for this variation in correlation distance over different domains in the polymer bulk due to the processing history of the nanocomposite by introducing a lognormal distribution for correlation distances, (Beaucage et al. 2004; Crow and Shimizu 1987) $P(\xi) = (1/\sqrt{2\pi}\xi\sigma)\exp(-\{\ln(\xi/m)\}^2/2\sigma^2)$ such that,

$$S(q, \xi) = \int_0^\infty P(\xi)\left[\frac{1}{1 + p\theta(q, \xi)}\right]d\xi \qquad (9)$$

Here, $m = \xi\exp(-\sigma^2/2)$ is the geometric mean of correlation lengths whereas, σ is the geometric standard deviation. The mean correlation length, ξ then represents an average distance between structural features across different domains of correlation. This modified B-G approximation by McGlasson et al. (2020) is compared to the P-Y and B-G fits for a precipitated silica SBR system in Fig. 1.5(a). Quite clearly, the modified B-G function fits the data at intermediate-q and high-q. The function is not expected to fit the low-q range due to agglomeration of aggregates above ϕ^{cc} as noted by the shaded region in the graph.

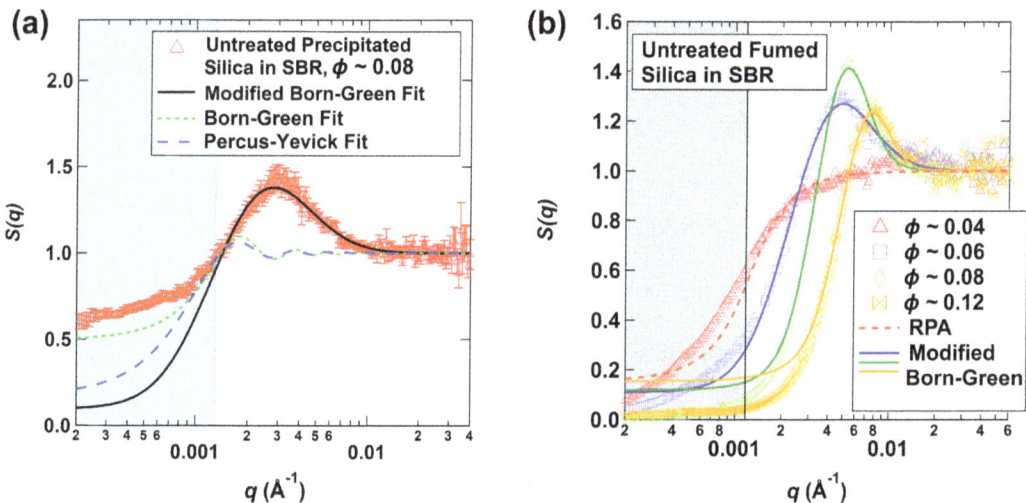

Figure 1.5 (a) $S(q)$ for precipitated silica in SBR at semi-dilute filler concentration fit using the P-Y function, B-G function, and the modified B-G function. Reprinted (adapted) with permission from McGlasson et al. 2020. Quantification of Dispersion for Weakly and Strongly Correlated Nanofillers in Polymer Nanocomposites. *Macromolecules* 2020, *53*(6): 2235–2248. Copyright 2020 American Chemical Society. (b) $S(q)$ for unmodified fumed silica in SBR at different semi-dilute filler loadings. At low concentrations, a correlation peak is absent, whereas correlated peaks appear above the critical ordering concentration that is dictated by the dielectric constant of the matrix. The low concentration curve (red triangles) was fit using an equation based on the Random Phase Approximation (RPA) model, described later. Reprinted from Rishi et al. 2020. Dispersion of Surface-Modified, Aggregated, Fumed Silica in Polymer Nanocomposites. *J. Appl. Phys.* 2020, *127*(17): 174702, with the permission of AIP Publishing.

The correlated aggregate structural emergence, described above, occurs above a local percolation threshold, ϕ^*. This threshold was defined at a filler volume fraction, ϕ, where $S(q) \neq 1$ in scattering for the aggregate and agglomerate regions in q. The question that arises then is whether these correlated structures emerge at all concentrations above the local percolation threshold, ϕ^*? The emergence of correlations in Fig. 1.4 was observed at $\phi \sim 0.08$ for precipitated silica in SBR. Rishi et al. (2020) compared the $S(q)$ for untreated fumed silica with a high surface hydroxyl content mixed with SBR at different loadings above ϕ^*, as shown in Fig. 1.5(b). Note that $S(q) \neq 1$ in the intermediate- and low-q regions for all the concentrations. A complete absence of a correlation peak in $S(q)$ at filler volume fractions up to $\phi \sim 0.04$ was observed. Since for low dielectric materials, the Debye screening length is small, $\lambda_D \sim \kappa^{1/2}$, it was hypothesized that the repulsive forces due to the charged aggregates were only felt at short distances (or at moderate concentrations). When the average aggregate separation distance approaches the Debye screening length of the polymer, λ_D, a critical ordering concentration is reached. The emergence of a correlation peak with increasing concentration for surface charged nanofillers such as precipitated and fumed silica is indicative of a free energy change on ordering, ΔG, analogous to the free energy change on micellization.

At low concentrations, the nanoaggregates are separated by distances larger than the Debye screening length, λ_D, and the effect of the charged surface hydroxyls on fumed silica is not felt by adjacent aggregates within a domain resulting in random distribution or lack of correlations. With increased concentration, the correlation length within the domains is reduced below λ_D such that the repulsive charges result in aggregate ordering within the domains. Although non-functionalized silica nanofillers mixed with SBR displayed a correlation peak with increasing concentration, this effect was not observed when fumed silica was mixed with polystyrene and PDMS at similar loading levels. It was argued that the low dielectric constant for polystyrene (Debye and Bueche 1951) ($\kappa_{PS} \sim 2.5$) and PDMS (Mark 1999) ($\kappa_{PDMS} \sim 2.56$) as opposed to SBR (Karásek et al. 1996; Gunasekaran et al. 2008) ($\kappa_{SBR} \sim 6.25$) meant that the Debye screening length is about 1.58 times larger for SBR (Rishi et al. 2020). Consequently, the critical ordering concentration for charged nanofillers, above which charges lead to repulsion and order would be about four times lower for SBR.

5   STRUCTURAL EMERGENCE IN COMPATIBLE SURFACE-FUNCTIONALIZED NANOFILLER-POLYMER SYSTEMS

For compatible nanofillers in non-polar polymers such as carbon black in SBR, $S(q) \neq 1$ above local percolation at ϕ^* (McGlasson et al. 2020). The same holds true for surface-functionalized nanoparticles mixed with compatible polymers (McGlasson et al. 2019; Rishi et al. 2020; Okoli et al. 2021). Nanoaggregates with chemically modified surfaces tend to not correlate, although the amount of surface treatment or carbon coating could impact the structural emergence (Okoli et al. 2021). The scattered intensity at intermediate-q shows a pronounced decrement instead of a correlation peak. Since no specific filler interactions dominate the scattering signal, the interactions of these randomly dispersed nanoaggregates can be approximated through a mean field. It should be noted that this random distribution of nanoaggregates could also show up with polar aggregates wherein the emergence is governed by the dielectric constant of the polymer as mentioned earlier. One way to describe mean-field interactions when the filler concentration is semi-dilute is through the Random Phase Approximation (RPA) as described by de Gennes (Anderson 1958; de Gennes 1979). This model was used to describe structural screening in polymer blends by Pedersen et al. (Zimm 1948; Benoit and Benmouna 1984; Pedersen and Schurtenberger 1999; Graessley 2002; Pedersen and Sommer 2005) and has been adapted for worm-like micelles by Vogtt et al. (2017) and highly-viscous polymer nanocomposites by Jin et al. (2017). Scattering from a homogeneous phase arises from concentration fluctuations that are dampened by the osmotic compressibility, $d\Pi/d\phi$. The $q \Rightarrow 0$ scattered intensity from a uniform phase is proportional to $k_B T/(d\Pi/d\phi)$. At large size-scales or low-q, screening is likely since a uniform phase is observed. However, at smaller scales, higher-q, the structure can be resolved, and the phases are no long uniform. This is like a cloth that is made of fibers. At high magnification, the individual fibers can be seen, but at large scales a uniform fabric is observed. The random phase approximation quantifies a single mean-field parameter, υ, that approximates the systemic interactions. This RPA is modeled as,

$$\left(\frac{I(q)}{\phi}\right)^{-1} = \left(\frac{I_0(q)}{\phi_0}\right)^{-1} + \phi\upsilon \tag{10}$$

At intermediate-q, the first term in equation (10), $\dfrac{I_0(q)}{\phi_0}$ reflects scattering from dilute nanoaggregates in the absence of structural screening. The second term, $\phi\upsilon$ is a measure of the structural screening at large sizes. The second term's contribution to the reduced inverse intensity, $\left(\dfrac{I(q)}{\phi}\right)^{-1}$, increases linearly with concentration, dampening the low-q reduced intensity as shown in Fig. 1.6(a) and 1.6(b). The mean-field parameter, υ, that bears the units of inverse intensity is used such that $1/\phi\upsilon$ approximates the plateau intensity at low-q for scattering from nanofillers in

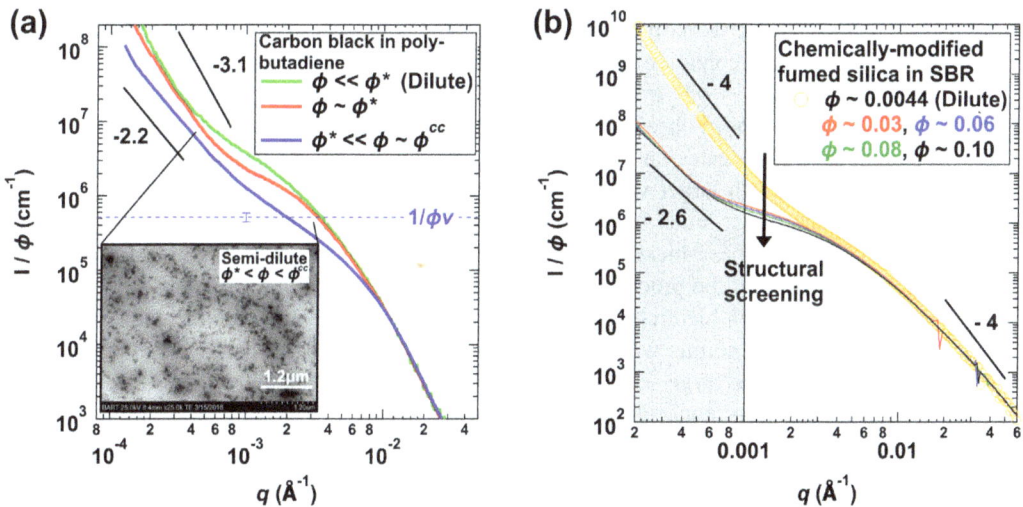

Figure 1.6 (a) Reduced scattering intensities, $I_0(q)/\phi_0$ (dilute, solid green curve), $I(q)/\phi$ (semi-dilute, solid red curve) and $I(q)/\phi$ (concentrated, solid blue curve) as a function of reciprocal space vector, q for carbon black in polybutadiene (PBD) rubber. At semi-dilute concentrations, the red curve shows screening at intermediate-q with surface fractal scaling at low-q. When the filler is concentrated, the screening at intermediate-q increases, and a mass-fractal scaling at low-q appears indicating the formation of a bulk network. Reprinted (adapted) with permission from Rishi et al. 2018. Impact of an Emergent Hierarchical Filler Network on Nanocomposite Dynamics. *Macromolecules* 2018, 51(20): 7893–7904. Copyright 2018, American Chemical Society. (b) Reduced scattering intensities, $I_0(q)/\phi_0$ (dilute, orange circles) and $I(q)/\phi$ (semi-dilute, all solid as a function of reciprocal space vector, q for chemically modified fumed silica in SBR. With increasing concentration, the screening increases at intermediate-q, although the large-scale network develops at a much lower concentration and does not change with increase concentration. Below the vertical line in q level 3 dominates the scattering so equation (10) is not appropriate. From Rishi et al. 2020. Dispersion of Surface-Modified, Aggregated, Fumed Silica in Polymer Nanocomposites. *J. Appl. Phys.* 2020, 127(17): 174702, with the permission of AIP Publishing.

the semi-dilute regime as shown in Fig. 1.6(a). Note that for surface compatibilized, disorganized particles, local aggregate overlap develops, and the aggregates are no longer isolated entities. Thus, under semi-dilute conditions above ϕ^*, the mean-field approach predicts no change in the concentration normalized scattering intensity, $I(q)/\phi$ from primary particles (level 1), but a reduction in $I(q)/\phi$ at larger size scales (level 2). The agglomerate and network levels (level 3) are not considered so the region of scattering at lowest-q is excluded from fits. The corresponding structure factor using the RPA approach is obtained by rearranging equation (10),

$$S(q) = 1 / \left(1 + \phi \upsilon \frac{I_0(q)}{\phi_0} \right) \tag{11}$$

υ is only valid for screening of nano-aggregates on the 100 nm size scale. For micron-scale agglomerates and networks of clusters above ϕ^{cc} the structural emergence can be visualized directly through scattering by monitoring the change in the power-law slope at low-q. In this region, an observed power-law slope lying between –3 and –4 is associated with surface scattering from these coarse agglomerates of nano-aggregates. Once the nano-aggregates screen at semi-dilute concentrations above ϕ^*, the agglomerates may or may not remain as well-separated entities as shown in Fig. 1.6(a), red curve. On further increase in a concentration above ϕ^{cc}, a mass-fractal slope appears (blue curve in Fig. 1.6(a) and all solid curves in Fig. 1.6(b)) indicating the formation of an agglomerate cluster network.

From the above observations, it is inferred that compatible nano-filled polymer systems present two hierarchically related filler networks, a micron-scale network of filler agglomerates composed of a nanoscale network of filler nanoaggregates. The nanoscale network forms once the filler aggregates locally percolate on the nanoscale with an increase in concentration from dilute ($\phi < \phi^*$) to semi-dilute ($\phi^* < \phi < \phi^{cc}$). Percolation of the local network at ϕ^* occurs well below global percolation and has been observed at about 3 volume percent for colloidal silica (Banc et al. 2014). The percolated aggregates on a nanoscale, agglomerate into clusters that further percolate on a micron-scale at a much larger volume fraction ($\phi > \phi^{cc}$) on the order of 20 volume percent. Although this could occur at much lower concentrations for surface-functionalized nanoparticles as shown in Fig. 1.6(b). These hierarchical levels of percolation are sensitive to the filler type, the specific surface area and the processing conditions. Rishi et al. (Rishi et al. 2018) likened this emergent dual-level network hierarchy to the arrangement of tables covered with tablecloths in a restaurant. If the restaurant ceiling were to be chosen as the point of perspective, the arrangement of tables would constitute a large-scale network. The arrangement of tables can be dilute when the tables do not touch (low concentration, bottom two images in Fig. 1.7, left) and semi-dilute otherwise (high concentration, bottom two images in Fig. 1.7, right). Converse to this large-scale network, if the perspective is from an individual table an observer on such a table would only be able to see the tablecloth with fixed fiber density on the weave which corresponds to the nano-scale aggregate network (top two images in Fig. 1.7, left). In fact, there is a fixed maximum semi-dilute concentration for the aggregate networks since they are bonded and cannot pack more densely than their internal, aggregate overlap concentration (top two images in Fig. 1.7, right). At higher concentrations than that the micron-scale network must cluster to compensate for the addition of more material.

Under semi-dilute conditions, above percolation, ϕ^*, individual aggregates cannot be observed due to overlap. A size scale emerges, the local mesh size, ξ, the length below the structure appears identical to the dilute condition and above which a uniform fabric network is observed. For systems that present a correlation peak, the aggregate correlation distance can be obtained by fitting $S(q)$ using equation (9). The mesh size of the emergent network for surface-modified compatibilized nanofillers is the average distance between the centers of mass of the aggregates. This mesh size is on the order of the nanoaggregate size at ϕ^* and reduces in size with increasing concentration following a fractal scaling law (Mulderig et al. 2017b). Attempts have been made to ascertain the

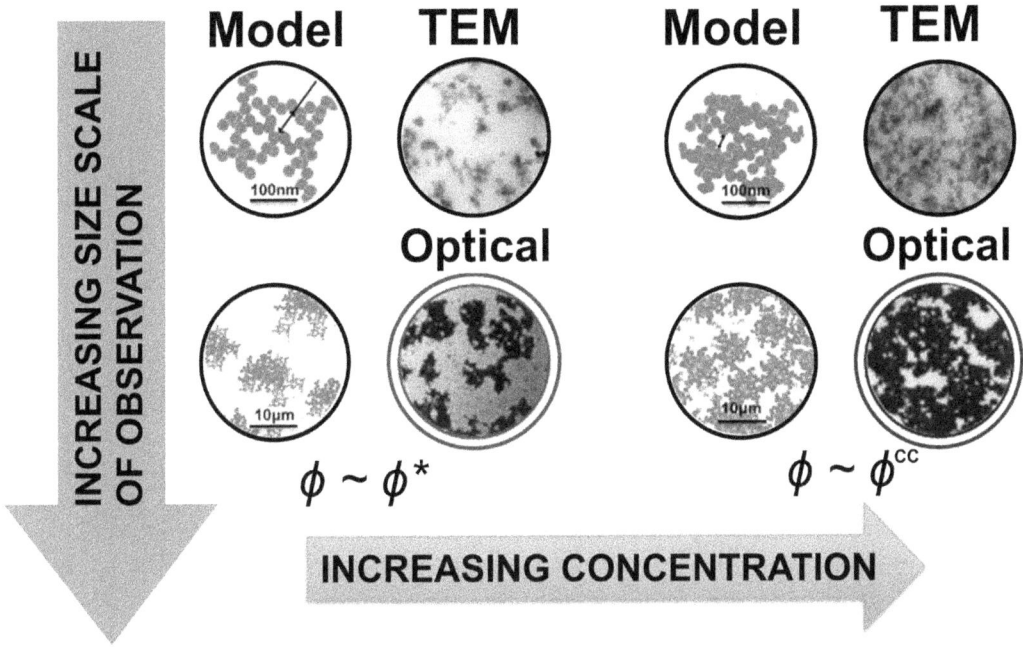

Figure 1.7 Sketch of structural hierarchies under semi-dilute ($\phi^* < \phi < \phi^{cc}$) and concentrated ($\phi > \phi^{cc}$) conditions. The local percolation is associated with a nano-scale network. All models and TEM micrographs in single circle: Reprinted (adapted) with permission from Rishi et al. 2018. Impact of an Emergent Hierarchical Filler Network on Nanocomposite Dynamics. *Macromolecules* 2018, 51(20): 7893–7904. Copyright 2018 American Chemical Society. Whereas, the global percolation is associated with a micron-scale network. Optical micrographs in double concentric circles: Reprinted (figure) with permission from Trappe and Weitz 2000. Physical Review Letters, *85*(2), 449–452, 2000. Copyright 2000 by the American Physical Society. https://doi.org/10.1103/PhysRevLett.85.449.

interaggregate distance or mesh size by likening the aggregates to spherical particles packed in a cubic lattice (Polley and Boonstra 1957; McDonald and Hess 1977). Based on this model Polley and Boonstra (Polley and Boonstra 1957) proposed the separation distance between aggregates,

$$d_{a-a} = d_{agg}\{[(200 + \phi_{PHR}/1.91\phi_{PHR})]^{1/3} - 1\} \qquad (12)$$

where, ϕ_{PHR} is the filler loading in parts (by weight) per hundred, and d_{agg} is the diameter of the aggregate. Here the filler to elastomer density ratio was assumed to be 2:1. Caruthers et al. (1976) defined a loading-interfacial area parameter, ψ, that represents the total filler surface area per unit volume, $\psi = \rho\phi S$ where, ρ is the filler density measured in g/cm^3 and S is the specific surface area of the filler measured in cm^2/g. The dimension of ψ is the inverse length and can be related to the inter-aggregate distance, though the authors themselves claimed the idea to be speculative. Tokita et al. (1994) modified the average separation distance between spherical entities by considering the distribution in aggregate size and aggregate arrangement,

$$d_{a-a} = f_{dist}d_{St}\{[(c/\phi_{eff})]^{1/3} - 1\} \qquad (13)$$

where, d_{St} is the Stokes diameter for the aggregate and f_{dist} is the distribution factor that can be computed from the median ($\Delta D50$) via a disk centrifuge photo sedimentometer. c, in equation (13), is the atomic packing factor for the aggregates which are considered as equivalent spheres. This parameter is generally considered to be an average of the simple cubic (0.52) and FCC arrangements (0.74). In equation (13), ϕ_{eff} is the effective volume fraction that accounts for rubber occlusion due to the bound rubber. Wang et al. (1993) argued that the apparent Stokes' diameter in equation (13) exhibits a less meaningful correlation compared with the geometrical diameter of the aggregates and proposed,

$$d_{a-a} = d_{agg}\{[(c/\phi_{eff})]^{1/3} - 1\} \tag{14}$$

where the aggregate diameter was normalized by the ratio of the effect to the actual filler volume fraction, β, $d_{agg} = (6000/\rho S)\beta^{1.43} = d_p z^{0.436}$. The diameter of the primary particle, d_p, or equivalently the Sauter mean diameter, was estimated from the filler density and the specific surface area assuming spherical primary particles (Kraus and Gruver 1965; Wang et al. 1993). Medalia determined the degree of aggregation, z, from the projected areas of the aggregate and the primary particles and found that the effective loading parameter was related to z, such that $\beta = z^{0.3}$ (Medalia 1970). This estimate of d_{agg} has limited use mainly due to the experimentally determined value of the exponent, 0.436 (Medalia 1972). The equation can be generalized using the end-to-end distance of the aggregate, $R_{eted} = d_p z^{1/d_f}$ as discussed earlier. Here, d_f is the mass fractal dimension of the aggregate. Zhang et al. (2001) proposed a model for the interparticle distance based on simple cubic packing by Wu (1985) that resembles equation (14). However, the radius of gyration of the aggregates, $R_{g,2}$, rather than R_{eted} was considered. Silica aggregates have been likened to prolate spheroids such that d_{a-a} depends on an empirical fit parameter, A, related to the major and minor radii of the spheroids (Staniewicz et al. 2014),

$$d_{a-a} = \frac{A}{\{\phi(1+8\phi)\}} \tag{15}$$

In mean-field systems, the local mesh size, ξ, of the aggregate filler network can be computed from the reciprocal space vector, $\xi = 2\pi/q^*$, corresponding to the point where the horizontal line associated with screening in the RPA equation, $1/\phi\upsilon$, intersects the reduced dilute curve, $I_0(q)/\phi_0$. For practical loading levels, the mesh size, representing the pore size of the primary nanoscale network, is expected to scale between the size of the agglomerated super-structure (much greater than the aggregate size) and the primary particle size. This local mesh size can be computed by equating the Unified Function in equation (1) (truncated to structural level 2, that is ignoring the micron-scale network and agglomerates or clusters) to $1/\phi\upsilon$ when q satisfies,

$$G_2 \exp\left(\frac{-q^2 R_{g,2}^2}{3}\right) + B_2 (q_2^*)^{-P_2} \exp\left(\frac{-q^2 R_{g,1}^2}{3}\right) + G_1 \exp\left(\frac{-q^2 R_{g,1}^2}{3}\right) + B_1 (q_1^*)^{-P_1} = \frac{1}{\phi\upsilon} \tag{16}$$

The above estimate of the mesh size was contrasted against the geometric models developed for the interaggregate distance by Rishi et al. (2018). It was observed that likening, fractal nanofiller aggregates to prolate spheroids (Staniewicz et al. 2014) approximates the mesh size at low concentrations but deviates as the concentration increases (Rishi et al. 2018).

6 QUANTIFYING DISPERSION IN FUNCTIONALIZED NANOPARTICLES

During mechanical mixing of hierarchical nanoparticles in viscous polymers, shear forces break up loosely bound agglomerates into aggregates which are dispersed. These dispersed aggregates re-cluster into agglomerates/clusters, which can be randomly arranged or correlated based on interfacial interactions as discussed earlier. Clustering of filler particles is opposed by the application of mixing energy or accumulated strain. One would expect the energy input or accumulated strain to have a smaller impact at smaller sizes since the lever arm is smaller.

On the nanoscale kinetic mixing has a limited impact. In mechanically-dispersed, immiscible nanocomposites some spontaneous diffusion of nanoparticles is known to occur (Ceccia et al. 2010). At large and intermediate length scales, to the contrary, thermal transport is insignificant and the lever arm for mixing is much larger.

This scale-dependent mixing process is generally observed by monitoring the variation of torque with mixing duration (Cotten 1984; Kondo 2014) or by process conductivity measurements for conductive fillers (Le et al. 2004; 2005). In the mixing process, the matrix is first thermally softened for suitable uptake of fillers followed by incorporation of the filler. A filler wetting process is evidenced by a prominent peak in both the mixing torque (Cotten 1987) and conductivity curves (Le et al. 2004; 2005) as a function of mixing duration. Following wetting and incorporation, the average powder size, initially on the order of several hundred microns, decreases until it reaches the size of nanoaggregates, a few hundred nanometers, and cannot be broken down further under the application of mixing shear forces. The reduction in agglomerate size as a function of mixing time was described by Shiga and Furuta (1982) through Scanning Electron Microscopy (SEM) micrographs. It was observed that the nano-aggregates peel off from the surface of agglomerates in a manner analogous to how an onion would peel. The later stages of the mixing process involve the distribution of these nanoaggregates as shown by Cotton (Cotten 1987). It is hypothesized that although suitable filler wetting is associated with filler-polymer compatibility, the structural emergence is associated with incompatibility which drives local clustering followed by the formation of cluster networks on the micron scale at later stages (Filippone et al. 2010; Filippone and Salzano de Luna 2012; Song et al. 2017; Rishi et al. 2018).

Dispersion encompasses at least four processes: (1) the reduction in agglomerate size into smaller nanoaggregates; (2) their subsequent distribution in the matrix; (3) formation of clusters of dispersed nanoaggregates, and (4) emergence of a micron-scale network of clusters/agglomerates. A size-based quantification of the first two stages was achieved by Leigh-Dugmore (1956) through a series of micrographs. By visual comparison, a dispersion rating based on the percentage of filler agglomerates below a certain size was proposed by Medalia (1961). Coran and Donnet (1992a, b) modeled the change in Medalia's dispersion rating as a function of mixing duration through the *DR* function,

$$DR(t) = DR(\infty) - \frac{DR(0)}{\exp(\beta t)} \qquad (17)$$

The most optimal dispersion, $DR(\infty)$, was based on the critical flaw size of the matrix polymer from Griffith's theory of crack propagation since an improvement in mechanical properties is observed below the flaw size. A higher percentage of agglomerates smaller than the critical flaw size in a micrograph resulted in a better dispersion rating. In equation (17), $DR(0)$ is related to the percentage of agglomerates larger than the critical flaw size prior to incorporation. β is an analogue to the specific rate in a first-order chemical reaction. One would expect β to vary with material and processing properties such as viscosity, mixing rate and temperature as well as mixing geometry. Since the dispersion process commences once all the filler particles have been incorporated into or wetted by the matrix, there exists a time delay known as the wetting time, before which a dispersion rating cannot be assigned (Coran et al. 1994; Leblanc 2002). This characteristic time delay or filler incorporation time can be computed at $DR(t_0) = 0$. Improvements to this model were made by assessing the area ratio of agglomerates of a specific size although the techniques were limited to optical micrographs of these nanofillers (Le et al. 2014; Faraguna et al. 2017) Bohin et al. (1996) proposed a simpler model to monitor the erosion of a filler agglomerate under simple shear flow,

$$DR(t) = \frac{DR(\infty)}{1 + (\gamma t)^{-1}} \qquad (18)$$

The rate of agglomerate erosion, γ, indicates a competition between the hydrodynamic shearing force and the cohesive force that holds the agglomerate together and is related to the matrix viscosity and the shear rate. However, Bohin's model assumed a negligibly small incorporation time. Yamada et al. (1998) showed that increased matrix viscosity slowed polymer matrix infiltration into the filler. A slower infiltration process at the same applied shear is associated with longer incorporation time indicating that the characteristic time delay would be longer if the matrix viscosity is higher.

To quantify the distribution of agglomerates and nanoaggregates, optical, SEM and TEM micrographs have been analyzed using numerical, statistical models and advanced computational geometry (Bakshi et al. 2009; Pegel et al. 2009; Khare and Burris 2010; Bray et al. 2012; Li et al. 2012; Lively et al. 2012; Glaskova et al. 2011; Glaskova et al. 2012; Haslam and Raeymaekers 2013; Fu et al. 2013). The strength and weaknesses of quantifying dispersion through these techniques have been reviewed (Krishnamoorti 2007). It has been argued that although the assessment of micrographs depends on a robust statistical method, proper identification of the filler by image thresholding to completely eliminate the inhomogeneous background is generally overlooked (Li et al. 2012). As an alternative, electrical conductivity (or inversely resistivity) could be used to quantify the extent of filler dispersion although the technique is limited to conductive fillers such as carbon black (Kondo 2014). For example, O'Farrell et al. (2000) noticed an increase in volume resistivity with increasing mixing times close to the filler percolation threshold indicating that the number of direct contacts between the filler particles reduced with increasing mixing time. At longer mixing times, the appearance of a plateau in resistivity indicated a state of terminal dispersion such that the number of direct contacts remained unchanged. However, for filler concentrations above percolation, the change in volume resistivity was insignificant indicating that this measure of dispersion is only suitable in the proximity of the global percolation limit.

The extent of dispersion depends on the size scale of observation. Macro-dispersion is related to size scales that are large enough so that some key aspects such as aggregate structure and specific surface area are averaged out. At large size scales, the reduction in agglomerate size is a suitable way to quantify dispersion. On the contrary, nano-dispersion involves observation on size scales comparable to the filler aggregate and primary particle sizes. Jin et al. (2017) used an analogy with thermally dispersed colloids to quantify dispersion using the second virial coefficient. Dispersion in colloidal mixtures, such as polymer solutions, printing inks, milk, red blood cells, etc., is dictated by Brownian motion (kT). Generally, such thermally-dispersed colloidal particles have favorable surface interactions with the matrix, i.e., they may be either stabilized by a surfactant or have a natural interfacial compatibility, (Rishi et al. 2019b; Vogtt et al. 2019) although phase separation can be achieved by sedimentation in a centrifuge to overcome the thermally driven Brownian motion. On the contrary, mechanically-dispersed polymer nanocomposites exist in a kinetically locked-in state. The nanoparticles in such systems would normally flocculate, settle or cluster except for the action of shear fields coupled with a viscous, glassy or semi-crystalline matrix phase. Both functionalized and non-functionalized nanofillers in viscous polymers are examples of kinetically dispersed colloids. Nanoparticle dispersion and the resulting properties are tied to the interfacial compatibility between the dispersed phase and the matrix. It is well known that clustering of colloidal particles in thermal equilibrium is opposed by the osmotic compressibility, $d\pi/d\phi$, or the build-up of osmotic pressure, π, with concentration, ϕ. Under dilute conditions, the osmotic pressure is estimated by the van't Hoff equation, $\pi = \rho_{num}k_BT$. A virial expansion of osmotic pressure can describe deviations from ideality through the second virial coefficient, B_2 as,

$$\frac{\pi}{k_BT} = \rho_{num} + B_2\rho_{num}^2 + B_3\rho_{num}^3 + \cdots \qquad (19)$$

Here B_2, expressed in cm^3 per particle, reflects deviations in osmotic pressure due to binary particle-particle interactions. The second virial coefficient, B_2, has been used to determine the stability of colloidal thermal dispersions (Mulderig et al. 2017b). For a stable colloid, with $B_2 > 0$, the dispersed phase remains evenly distributed throughout the solution, whereas particle clustering, sedimentation or flocculation occurs for unstable colloids with $B_2 < 0$. Vogtt et al. (2017) used the mean-field model in equation (10) for a two-level hierarchical worm-like micelle system to obtain B_2 from υ,

$$B_2 = \frac{1}{2}(zV_{pp})^2 \upsilon \Delta\rho^2 \tag{20}$$

Here $\Delta\rho^2$ is the squared difference in scattering length density between the nanofiller and the nanocomposite matrix or the scattering contrast, V_{pp} is the volume of the primary subunit and z is the degree of aggregation. The use of equation (20) was extended to hierarchical nanofiller particle systems by Jin et al. (2017). Similar methods to quantify dispersion using interaction parameters that measure filler-polymer compatibility have been explored. For example, Stöckelhuber et al. (2010) proposed the use of the free energy of immersion, ΔG_i, that can be represented as a function of the surface energies of the filler and the polymer-filler interface to quantify the wettability of the filler particles by the polymer. Hassinger et al. (2016) developed a quantitative tool to incorporate mechanical processing conditions to predict interfacial thermodynamics. A set of descriptors to quantify the interfacial energy based on the ratio of the work of adhesion between polymer-filler and filler-filler following Natarajan et al. (2013), the total power consumption during the mixing process, and the volume fraction normalized filler surface area based on TEM micrographs were proposed. Using correlations between the three descriptors resulted in adequate quantification of dispersion under a given set of conditions.

The first derivative of osmotic pressure, in the virial expansion (the osmotic compressibility), is proportional to the second virial coefficient. The temperature dependence of the second virial coefficient, B_2, for thermally-dispersed, colloidal solutions can be obtained by describing the second virial coefficient of osmotic pressure, with the van der Waals equation of state, $B_2(T) = b - \dfrac{a}{RT}$, where b is the excluded volume of the colloidal particles and a is the pressure correction term associated with inter-particle attraction. Rishi et al. (2020) considered the overall accumulated strain, γ, to be an analogue to thermal energy for mechanically-mixed, kinetically dispersed filler-elastomer nanocomposites and expressed B_2 as,

$$B_2(t, N, \psi) = b^* - \frac{a^*}{\gamma} = b^* - \frac{a^*}{N\Psi t} \tag{21}$$

Here b^* represents the excluded volume of the nano-aggregates and consequently depends on the degree of aggregation, z, and the primary particle size, d_p. Interestingly, the difference of this experimental estimate of and the actual hard-sphere excluded volume of an aggregate could provide an estimate of the bound rubber layer on the aggregate (Rishi et al. 2019a). N represents the mixing speed whereas, t indicates the mixing duration for batch type geometries and the residence time for continuous mixers such as extruders. Ψ reflects the mixing geometry. For a traditional twin-rotor internal mixer such as a Brabender or Banbury mixer, Ψ can be approximated using simple Couette flow following Bousmina et al. (1999) Similarly, for a single-screw extruder this geometric constant can be approximated following Hassinger et al. (2016). For vortex mixing of nanoparticles in low viscous matrices, this constant can be approximated by an equivalent bob and cup geometry (Mezger 2006). The functions are listed in Table 1.1.

Table 1.1 Instrument geometry-related constants for estimating the accumulated stain

Mixing equipment	Approximate for mixing geometry	Details
Brabender mixer	$\Psi = \dfrac{4\pi(\beta)^{2/n}}{n\{(\beta)^{2/n}-1\}}$	β : ratio of wall to rotor diameter; n: power-law index under shear flow (Sadhu and Bhowmick 2005)
Single-screw extruder	$\Psi = \dfrac{\pi\{d-2H(L)\}}{H(L)}$	d is the screw diameter and $H(L)$ is the channel depth that depends on the screw length, L (Hassinger et al. 2016)
Vortex mixer	$\Psi = \dfrac{2\pi(2R^2)}{(R^2-r^2)}$	R is the cup radius and r is the radius of the bob

These simple estimates for accumulated strain have shown to serve as good approximations for laboratory-scale equipment which are bulk fed. On the contrary, industrial-grade twin-screw extruders are starve-fed and an estimate of the accumulated strain is far more involved due to variations in the fill factor with each rotation. Veigel et al. (2021) estimated $A_2 = (N_A/M^2)B_2$ for carbon black polystyrene nanocomposites by varying the mixing speeds, residence times and the mixing geometries. Note that M is the aggregate mass and N_A is the Avogadro's constant. The functional dependence of nanoparticle dispersion on accumulated strain in equation (21) was manageable as shown in Fig. 1.8.

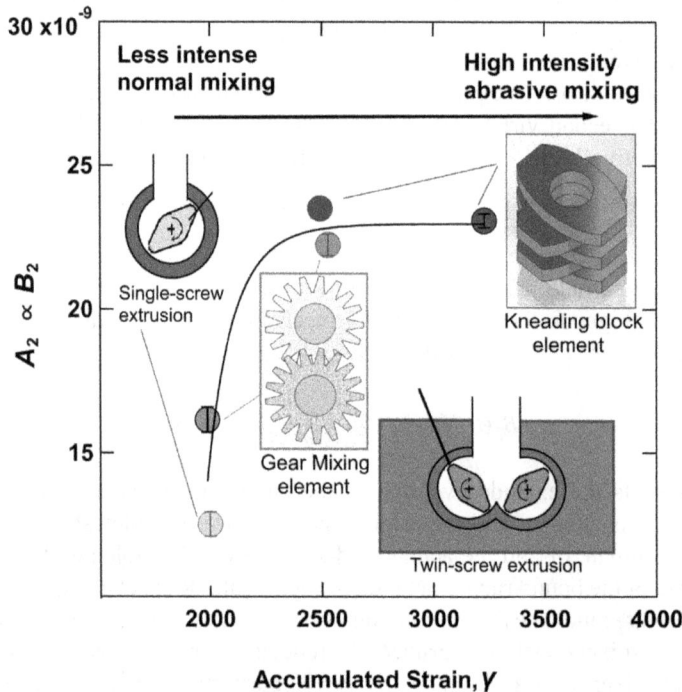

Figure 1.8 Dependence of nanoscale dispersion, quantified via A_2 on the accumulated strain in carbon black-polystyrene nanocomposites for different mixing geometries (single screw vs twin-screw extruder), screw element (gear mixer vs kneading blocks), and mixing speeds (points to the left for a specific screw element represent lower mixing speed). Reprinted (adapted) with permission from Veigel et al. 2021. "Nanocomposite Dispersion in Melt Mixers." *Bulletin of the American Physical Society*, 2021.

a^* is a measure of the interaction potential and is sensitive to the surface chemistry of functionalized nanoparticles as shown by Rishi et al. (2020). Figure 1.9 shows the dependence of

a^* on the natural log of the surface concentration of silanols and methyl groups for neat, fumed silica (solid symbols) and silane-coupled fumed silica (open symbols) in polystyrene, styrene-butadiene rubber and polydimethyl siloxane compounded using different mixers and processing conditions. A straight-line fit was obtained by varying the constants k' and k'' which represent the relative influence of silanol and methyl functional groups on the interaction potential, whereas C_2 represented a scaling factor. Additionally, the fit parameter, C_1, was found to be proportional to the total accumulated strain based on the mixing geometry, and the zero-shear rate viscosity of the polymers, η, at the processing temperature. The functional dependence of the interaction energy, ΔG, normalized by the accumulated strain which is an analogue to kT in these viscous nanocomposite systems on the surface chemistry of both neat and modified silica bears semblance to free energy of micellization in surfactant systems.

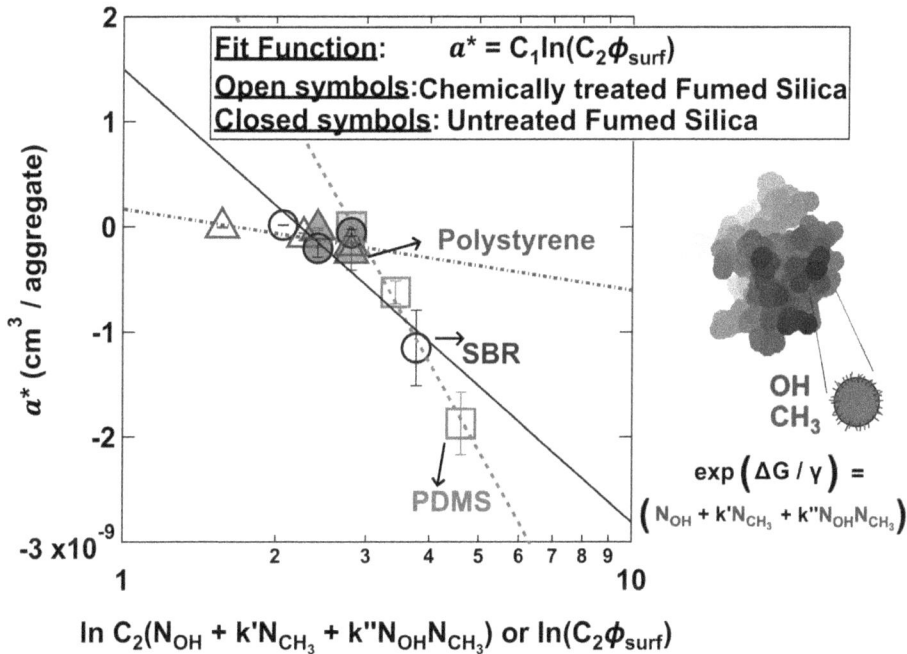

Figure 1.9 The dependence of the binary interaction potential, a^*, on the surface concentration of chemical species (N_{OH} and N_{CH_3}) for surface-functionalized and bare silica nanofiller grades in SBR (solid line), polystyrene (dot-dash line), and PDMS (dashed line) matrices. The constants, k' and k'' describe the relative effects of different surface functionalities whereas, C_2 is a scaling constant. C_1 is directly related to the accumulated strain sensitive to the mixing conditions and the zero-shear rate viscosity of the matrix. A logarithmic dependence of a^* normalized by the accumulated strain ($\gamma \propto C_1$) on ϕ_{surf} is analogous to the free energy of micellization. Reprinted from Rishi et al. 2020. Dispersion of Surface-Modified, Aggregated, Fumed Silica in Polymer Nanocomposites. *J. Appl. Phys.* 2020, 127(17): 174702, with the permission of AIP Publishing.

Quantification of dispersion through the pseudo-second-order virial coefficient, B_2 (or A_2), accounts for the material-specific variables viz. particle size, the degree of aggregation, the interaction potential which is sensitive to the surface chemistry of the nanofillers and the matrix viscosity. The dependence of B_2 on the accumulated strain, which accounts for the mixing duration, mixing speed, viscosity and the instrument geometry presents a large phase-space within which the effects of processing on polymer nanocomposites can be specified. B_2 is a direct, quantitative measure of the dispersion with larger values indicating better dispersion. Negative B_2 indicates phase segregation and $B_2 = 0$ is a critical value. Understanding the dependencies of B_2 indifferent polymer-filler systems are important to developing predictive techniques for control of structural emergence in polymer-filler systems. B_2 can be used to calculate binary filler interaction potentials for coarse-grained computer simulations of complex multi-level hierarchical filler mixtures. Control over this complex multi-hierarchical structure can be achieved through manipulation of filler-polymer interactions such as by varying the silanol surface density, by chemically-tailoring the surface and by grafting low molecular weight polymers. It is expected that these modifications can control dispersion and the associated emergent multi-hierarchy.

7 THE IMPACT OF EMERGENT STRUCTURES ON THE DYNAMIC RESPONSE OF FUNCTIONALIZED NANOPARTICLE POLYMER COMPOSITES

In commercial products such as tires, specific properties such as traction, wear-resistance and rolling resistance depend on the filler morphology, the filler-elastomer surface chemistry as well as the degree of filler dispersion. With the increase in filler concentration, well-dispersed nanoparticle aggregates in polymers form a continuous network (Warring 1950; Fletcher and Gent 1954). The presence of this filler network has been evidenced in the dynamic mechanical response with varying strain amplitude (Payne 1962; Payne and Watson 1963; Payne 1963; Payne and Whittaker 1971). This network is associated with the Payne effect and bulk conductivity is on the micron-scale (Karásek et al. 1996; Yurekli et al. 2001) and commences at the global percolation limit, ϕ^{cc} in the structural hierarchy discussed earlier. For many types of nanofillers, ϕ^{cc} can be estimated from the dibutyl phthalate (DBP) absorption number (Janzen 1975) although, only conductive fillers like carbon black with lower DBP numbers agree experimentally (Li et al. 2016).

The complex dynamic modulus for a nanocomposite is estimated by applying an oscillatory perturbation, $\gamma(t) = \gamma_0 \sin(\omega t)$, with a fixed strain amplitude, γ_0 and monitoring the subsequent stress response. In the linear viscoelastic regime at low strain, the stress varies linearly as

$$\sigma(t) = \sigma_0 \sin(\omega t + \delta) = \gamma_0 \{G' \sin(\omega t) + iG'' \cos(\omega t)\} \qquad (22)$$

In equation (22), $G'(\omega) = G \sin \delta$ and $G''(\omega) = G \cos \delta$ represent the storage and loss shear moduli, respectively. The phase difference between the applied strain and the resulting stress, δ, varies between $0°$ (Hookean elastic response) and $90°$ (Newtonian viscous response) (Ferry 1980). The ratio of the two moduli is the damping factor, $\tan \delta = G''(\omega)/G'(\omega)$.

Most commercially available instruments are limited in the accessible frequency range. Accessing the response over a wider frequency range necessitates the use of the time-temperature superposition principle. The William-Landel-Ferry (WLF) equation (Williams et al. 1955), is one such example that allows for estimation of this inaccessible frequency range through horizontal shifting (a_T) of isothermal frequency curves at different temperatures to a reference temperature, T_{ref}. The logarithm of the horizontal shift factor, $\log a_T = -C_1(T - T_{ref})/(C_2 + T - T_{ref})$ such that C_1 and C_2 are constants for a particular polymer. In practice, least-square fitting methods are used to arbitrarily shift the isothermal response curves and the estimated a_T is fit to the WLF equation to determine C_1 and C_2. The construction of the moduli master curves for polymer nanocomposites requires both frequencies, a_T, and moduli, $b_T = \rho_{ref}T_{ref}/\rho T$, shift factors.

The addition of nanofillers to a polymer matrix increases the shear modulus, although a_T, is expected to be independent of the filler loading. The Einstein-Smallwood equation (Smallwood 1944), written in analogy to the Einstein equation for the viscosity of colloidal suspensions, is used to describe the enhancement of modulus under dilute conditions such that, $G = G_0(1 + K\phi)$. Here, G and G_0 are the moduli of the filled elastomer and neat elastomer, respectively whereas, ϕ is the volume fraction of filler, and K is a factor related to filler structure (analogous to the intrinsic viscosity [η]). For mono-disperse rigid spheres under dilute conditions, K is 2.5 (using the Einstein equation for [η] of spheres). This linear dependence on volume fraction arises from amplification of the observed strain rate due to volumetric displacement of elastomer by rigid filler. For mass-fractal nanofiller aggregates, a larger effective volume is displaced as compared to a sphere for a given volume fraction, so values of K greater than 2.5 can be used. Additionally, under semi-dilute filler concentrations, the modulus enhancement can be generalized by a Taylor series expansion,

$$G = G_0(1 + K\phi + K'\phi^2) \qquad (23)$$

where, the coefficients, K and K' are related to the filler structure and interactions. For low structure carbon blacks, that are more compact and allow closer packing, K equals 2.5 and K' equals 14.1, as described by Guth (1945). For mass-fractal nanoparticles, the ratio of excess modulus to the bare polymer modulus, $(G/G_0) - 1$, was found to vary linearly with filler concentration until aggregate overlap, ϕ^*, based on linear elastic theory (Huber and Vilgis 1999; Huber and Vilgis 2002). It was predicted that beyond percolation, ϕ^*, this dependence on volume fraction would change to a power-law which is not necessarily 2 (equation (23)).

Increasing filler-elastomer interaction by surface functionalization could result in the formation of a bound polymer layer. The amount of bound polymer is related to the amount of occluded polymer owing to the fractal nature of the fillers (Robertson et al. 2007). Medalia (1970) postulated that the amount of occluded polymer within the interstices of the filler aggregates would increase the effective filler volume fraction. The ratio of effective to actual filler volume fraction, $\beta = (1 + e)/(1 + \varepsilon)$, was directly related to the effective loading parameter, e, that represented the void space in a single aggregate divided by the volume of the solid aggregate. The void space is related to the volume of DBP (dibutyl phthalate) that fills the aggregates during a crushed DBP absorption, (Kraus 1971; Medalia 1972) such that $e = \rho DBPA$. ρ is the filler density whereas, ε is the aggregate packing factor ranging from 0.35 to 0.92. It was experimentally determined that a part of the occluded polymer volume, $\phi_{\mathrm{eff}} - \phi$ deformed under the application of stress, (Medalia 1972) necessitating the need to introduce an occlusion effectiveness parameter, F, to adequately account for the effective filler volume fraction in the nanocomposite as,

$$\phi_{\mathrm{eff}} = \phi\{F(\beta - 1) + 1\} \qquad (24)$$

F is approximately 0.5 under low-strain conditions meaning that only 50% of the occluded rubber remains immobilized and contributes to modulus enhancement and is independent of the carbon black structure (Medalia 1972).

The enhancement in modulus based on equation (23) considers the hydrodynamic effect and the binary filler interactions whereas, equation (24) accounts for the immobilization layer and the effective volume fraction. However, the contribution due to filler network formation is not considered in equation (24), and thus, a deviation in the predicted relative modulus, G/G_0, is expected when the filler network emerges above ϕ^*. Rishi et al. (2018) attempted to relate the dynamic response in the plateau and terminal flow regions at very low-strain amplitudes in the linear viscoelastic regime to features of the hierarchical network obtained from static X-ray scattering. The nanoscale mesh size discussed previously was found to relate to a characteristic transition frequency in the dynamic spectrum. Figure 1.10(a) shows a plot of storage modulus for a semi-dilute carbon black polybutadiene nanocomposite and bare polymer after correction for the hydrodynamic reinforcement and effective filler volume fraction of the filler. Since the semi-dilute filler concentration is on the order of 20 volume percent, a nano-scale network based on the hierarchical model discussed previously has developed. In the high-frequency plateau region, the response follows the Einstein-Smallwood equation (Smallwood 1944) with modifications based on the effective filler content (Medalia 1972). A clear deviation in the two master curves is observed at low frequency and the transition frequency between the two opposing regimes is related to the mesh size from scattering as shown in Fig. 1.10(b). This simple frequency-length relationship was found to depend on the spectral dimension (Vilgis and Winter 1988) which is associated with the network connectivity (Rishi et al. 2018). The static aggregate connectivity dimension has been described as an intrinsic measure of the aggregate structure with a value of 1 for linear aggregates and greater than 1 for branched aggregates (Meakin et al. 1984). For homogenous percolation clusters, the spectral dimension is predicted to be 4/3 (Alexander and Orbach 1982). In Fig. 1.10(b), this spectral dimension, c ~ 1/0.8 ~ 1.25 was in agreement with the aggregate connectivity dimension from scattering.

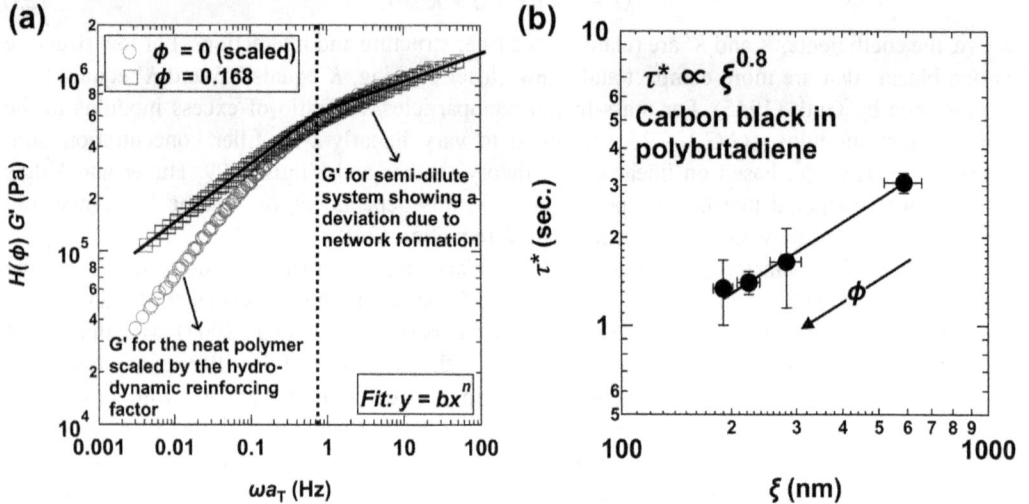

Figure 1.10 (a) Storage modulus master curve for a semi-dilute (squares) compatible nanocomposite (carbon black in polybutadiene) compared to a master curve for the same nanocomposite based on the modulus of the neat polymer, the hydrodynamic reinforcement and effective filler content (circles). The dashed line indicates the transition frequency (time) associated with the filler network. (b) Dependence of this characteristic time scale, τ^* and the network mesh size, ξ that are related by the spectral dimension. Reprinted (adapted) with permission from Rishi et al. 2018. Impact of an Emergent Hierarchical Filler Network on Nanocomposite Dynamics. *Macromolecules* **2018**, *51*(20), 7893–7904. Copyright 2018 American Chemical Society.

Wet grip, described as the handling efficiency of an automobile tire on the road, and rolling resistance, defined as the energy loss induced by the deformation over the contact area of an automobile tire with the road, are two essential properties that dictate tire performance (Sae-oui et al. 2017). The dynamic response parameter, tan δ is generally employed in the evaluation of wet grip and rolling resistance. The hysteresis of the polymer compound during continuous deformation at various temperatures is the underlying principle behind the functionalities of these parameters, which are directly related to driving safety and fuel economy (Araujo-Morera et al. 2019). In the industry, the dynamic response at 60 to 70°C is a measure of the tire rolling resistance whereas, the response at about 0°C is associated with tire braking/grip. tan δ values at these temperatures are used to parameterize the performance (Lee et al. 2013). Elastomeric nanocomposites with relatively lower tan δ values at 60°C have reduced rolling resistance while higher tan δ values at 0°C improve wet grip. Consequently, reaching a compromise between reducing the tan δ values at 60°C and increasing the tan δ values at 0°C becomes an elusive goal in tire research (Lei et al. 2016). tan δ measures the combined effect of energy lost and stored owing to the viscoelastic nature of the silica SBR nanocomposites. A lower rolling resistance is characterized by a reduction in tan δ or a relative increase in the storage modulus whereas, a better-wet grip is achieved at larger tan δ values or a relative increase in the energy lost. To achieve this synergistic effect, tire compounders use both carbon black and silica (Sattayanurak et al. 2019; Gabriel et al. 2019). Figure 1.11(a) shows how the wet grip changes with the surface silanol content in carbon-coated and surface silanized pyrogenic silica in SBR rubber. A lower surface silanol content resulted in an improved wet grip. Similarly, a lower surface carbon content resulted in lower rolling resistance in Fig. 1.11(b). It was argued that although the reduction in the surface hydroxyl groups by silanization could improve the wet grip of the resulting nanocomposite, it would concomitantly result in an increase in the surface carbon or surface methyl content (Okoli et al. 2021). An optimal solution could result if the surface of silica were coated with a critical number of carbon monolayers that would also result in improved dispersion.

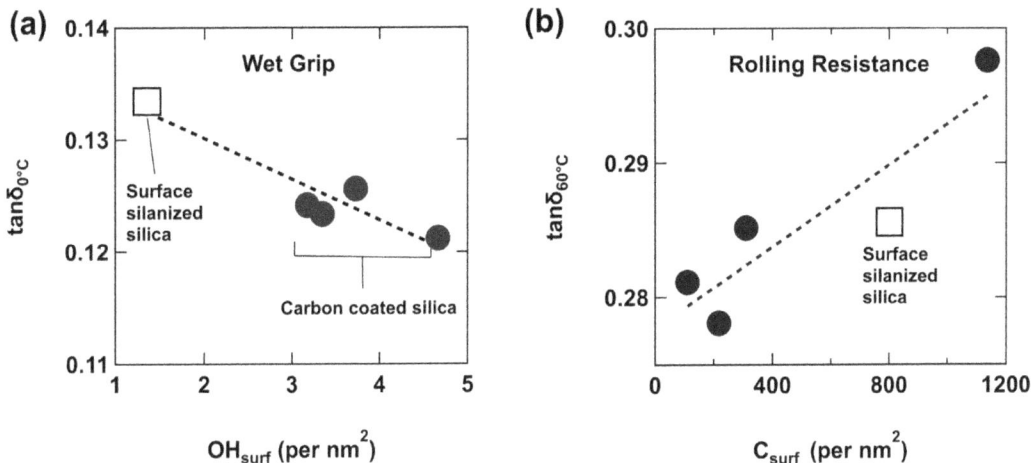

Figure 1.11 (a) Plot of $\tan \delta$ at 0°C as a function of surface silanol content and (b) $\tan \delta$ at 60°C as a function of surface carbon content for functionalized fumed silica in SBR rubber. Reprinted (adapted) with permission from Okoli et al. 2021. "Dispersion and Dynamic Response for In-Flame and Chemically Modified Fumed Silica Nanocomposites." In *Bulletin of the American Physical Society*, 2021.

8 CONCLUSION

Surface functionalization of nanoparticles can be used to manipulate the complex structural emergence that leads to control over mechanical, electrical and viscometric properties of polymer nanocomposites. For industrial filled materials containing immiscible aggregated nanoparticles, a balance between compatibility and immiscibility is needed to influence the emergence of a robust macroscopic network, while retaining accessible nanoscale surface area at low volume fractions of the filler. The emergent structure with a multi-hierarchy can influence both high-frequency response (rolling resistance) and moderate frequency (wet grip) as well as low frequency (bulk modulus and tear resistance). Dramatic advances have been made in the past 5 years in this area that have allowed the direct correlation for the first time of the mechanical frequency spectrum with the structural spectrum as measured with X-ray scattering and X-ray tomography. This multi-hierarchical approach shows promise for the design of new materials for solid electrolytes, paints, inks, flame retardants and filled elastomers among other areas where features on the macroscopic scale are desired from nanoscale polymeric additives. Some general observations have been found associated with the influence of the Debye screening length coupled with concentration for polar additives such as silica. A critical concentration was found for aggregated nanoparticles to display correlated organization. Dispersion on the nanoscale is improved for larger particles if the mixing is dominated by kinetics, while it is improved for smaller particles if the mixing is dominated by thermal diffusion. A quantitative function has been reviewed that relates specific surface functionality to nanoscale dispersion. This function predicts how surface organic and polar content will influence mixing at a specified accumulated strain which is related to polymer viscosity and mixing time and geometry. The impact of accumulated strain on nano-dispersion for polar and non-polar fillers was also described . It is generally understood that compatibilization can be important to nanocomposites. In simple model systems composed of spherical, monodisperse particles mixed in monodisperse polymers by solvent blending from single phase colloidal suspensions, idealized mappings of phase behavior have been reported. These predictive tools are of little use in the commercial setting where incompatible, aggregated, polydisperse multi-component melt blended materials are encountered. This chapter has indicated some of the fundamental tools that have been made available in the past few years to predict and design these commercial materials which are of increasing importance to the advancement of a wide range of fields.

ACKNOWLEDGMENTS

The research work detailed in this chapter was supported by the National Science Foundation (NSF) through Grant No. CMMI-1635865. Small-angle X-ray Scattering measurements were conducted at the Advanced Photon Source (APS), Argonne National Laboratory, an Office of Science User Facility operated for the U.S. Department of Energy (DOE) under Contract No. DE-AC02-06CH11357. The SAXS/USAXS data were collected at the APS on beamline 9-ID-C operated by Jan Ilavsky and Ivan Kuzmenko in the X-ray Science Division. We gratefully acknowledge their vital assistance.

The authors declare no competing financial interest. Mention of company names and products does not constitute endorsement by NIOSH, CDC. The findings and conclusions in this report are those of the authors and do not necessarily represent the views of NIOSH, CDC.

■ References

Akcora, P., H. Liu, S.K. Kumar, J. Moll, Y. Li, B.C. Benicewicz, et al. 2009. Anisotropic self-assembly of spherical polymer-grafted nanoparticles. Nature Materials 8(4): 354–359. doi:10.1038/nmat2404.

Alexander, S. and R. Orbach. 1982. Density of states on fractals : « fractons ». Journal de Physique Lettres 43(17): 625–631. doi:10.1051/jphyslet:019820043017062500.

Anderson, P.W. 1958. Random-phase approximation in the theory of superconductivity. Physical Review 112(6): 1900–1916. doi:10.1103/PhysRev.112.1900.

Anderson, B.J. and C.F. Zukoski. 2007. Nanoparticle stability in polymer melts as determined by particle second virial measurement. Macromolecules 40(14): 5133–5140. doi:10.1021/ma0624346.

Anderson, B.J. and C.F. Zukoski. 2010. Rheology and microstructure of polymer nanocomposite melts: Variation of polymer segment–surface interaction. Langmuir 26(11): 8709–8720. doi:10.1021/la9044573.

Araujo-Morera, J., M.H. Santana, R. Verdejo and M.A. López-Manchado. 2019. Giving a second opportunity to tire waste: An alternative path for the development of sustainable self-healing styrene–butadiene rubber compounds overcoming the magic triangle of tires. Polymers 11(12): 2122. doi:10.3390/polym11122122.

Baeza, G.P., A.-C. Genix, C. Degrandcourt, L. Petitjean, J. Gummel, M. Couty, et al. 2013. Multiscale filler structure in simplified industrial nanocomposite silica/SBR systems studied by SAXS and TEM. Macromolecules 46(1): 317–329. doi:10.1021/ma302248p.

Bahl, K. and S.C. Jana. 2014. Surface modification of lignosulfonates for reinforcement of styrene-butadiene rubber compounds. Journal of Applied Polymer Science 131(7): 1–9. doi:10.1002/app.40123.

Bakshi, S.R., R.G. Batista and A. Agarwal. 2009. Quantification of carbon nanotube distribution and property correlation in nanocomposites. Composites Part A: Applied Science and Manufacturing 40(8): 1311–1318. doi:10.1016/j.compositesa.2009.06.004.

Banc, A., A.C. Genix, M. Chirat, C. Dupas, S. Caillol, M. Sztucki, et al. 2014. Tuning structure and rheology of silica-latex nanocomposites with the molecular weight of matrix chains: A coupled SAXS-TEM-simulation approach. Macromolecules 47(9): 3219–3230. doi:10.1021/ma500465n.

Beaucage, G. 1995. Approximations leading to a unified exponential/power-law approach to small-angle scattering. Journal of Applied Crystallography, International Union of Crystallography. 28(6): 717–728. doi:10.1107/S0021889895005292.

Beaucage, G., T.A. Ulibarri, E.P. Black and D.W. Schaefer. 1995. Multiple size scale structures in silica—siloxane composites studied by small-angle scattering. pp. 97–111. *In*: J.E. Mark C.Y-C Lee and P.A. Bianconi (eds). Hybrid Organic-Inorganic Composites. American Chemical Society. doi:10.1021/bk-1995-0585.ch009.

Beaucage, G. 2004. Determination of branch fraction and minimum dimension of mass-fractal aggregates. Physical Review E 70(3): 031401. doi:10.1103/PhysRevE.70.031401.

Beaucage, G., H.K. Kammler and S.E. Pratsinis. 2004. Particle size distributions from small-angle scattering using global scattering functions. Journal of Applied Crystallography, International Union of Crystallography. 37(4): 523–535. doi:10.1107/S0021889804008969.

Benabid, F.Z., N. Kharchi, F. Zouai, A.-H.I. Mourad and D. Benachour. 2019. Impact of co-mixing technique and surface modification of ZnO nanoparticles using stearic acid on their dispersion into HDPE to produce HDPE/ZnO nanocomposites. Polymers and Polymer Composites 27(7): 389–399. doi:10.1177/0967391119847353.

Benoit, H. and M. Benmouna. 1984. Scattering from a polymer solution at an arbitrary concentration. Polymer 25(8): 1059–1067. doi:10.1016/0032-3861(84)90339-2.

Bichoutskaia, E., A.L. Boatwright, A. Khachatourian and A.J. Stace. 2010. Electrostatic analysis of the interactions between charged particles of dielectric materials. Journal of Chemical Physics 133(2): 1–10. doi:10.1063/1.3457157.

Bohin, F., D.L. Feke and I. Manas-Zloczower. 1996. Analysis of power requirements and dispersion quality in batch compounding using a dispersion model for single agglomerates. Rubber Chemistry and Technology 69(1): 1–7. doi:10.5254/1.3538355.

Bousmina, M., A. Ait-Kadi and J.B. Faisant. 1999. Determination of shear rate and viscosity from batch mixer data. Journal of Rheology 43(2): 415–433. doi:10.1122/1.551044.

Bouty, A., L. Petitjean, C. Degrandcourt, J. Gummel, P. Kwaśniewski, F. Meneau, et al. 2014. Nanofiller structure and reinforcement in model silica/rubber composites: A quantitative correlation driven by interfacial agents. Macromolecules 47(15): 5365–5378. doi:10.1021/ma500582p.

Bouty, A., L. Petitjean, J. Chatard, R. Matmour, C. Degrandcourt, R. Schweins, et al. 2016. Interplay between polymer chain conformation and nanoparticle assembly in model industrial silica/rubber nanocomposites. Faraday Discussions 186: 325–343. doi:10.1039/c5fd00130g.

Bray, D.J., S.G. Gilmour, F.J. Guild and A.C. Taylor. 2012. Quantifying nanoparticle dispersion by using the area disorder of delaunay triangulation. Journal of the Royal Statistical Society. Series C: Applied Statistics 61(2): 253–275. doi:10.1111/j.1467-9876.2011.01009.x.

Brinker, C.J. and G.W. Scherer. 2013. Sol-Gel Science: The Physics and Chemistry of Sol-Gel Processing. San Diego, CA: Academic Press Inc.

Caruthers, J.M., R.E. Cohen and A.I. Medalia. 1976. Effect of carbon black on hysteresis of rubber vulcanizates: equivalence of surface area and loading. Rubber Chemistry and Technology 49(4): 1076–1094. doi:10.5254/1.3534990.

Ceccia, S., F. Bellucci, O. Monticelli, A. Frache, G. Traverso and A. Casale. 2010. The effect of annealing conditions on the intercalation and exfoliation of layered silicates in polymer nanocomposites. Journal of Polymer Science Part B: Polymer Physics 48(23): 2476–2483. doi:10.1002/polb.22146.

Coran, A.Y. and J.-B. Donnet. 1992a. The dispersion of carbon black in rubber part I. Rapid method for assessing quality of dispersion. Rubber Chemistry and Technology 65(5): 973–997. doi:10.5254/1.3538655.

Coran, A.Y. and J.-B Donnet. 1992b. The dispersion of carbon black in rubber Part II. The kinetics of dispersion in natural rubber. Rubber Chemistry and Technology 65(5): 998–1015. doi:10.5254/1.3538656.

Coran, A.Y., F. Ignatz-Hoover and P.C. Smakula. 1994. The dispersion of carbon black in rubber Part IV. The kinetics of carbon black dispersion in various polymers. Rubber Chemistry and Technology 67(2): 237–251. doi:10.5254/1.3538671.

Cotten, G.R. 1984. Mixing of carbon black with rubber I. Measurement of dispersion rate by changes in mixing torque. Rubber Chemistry and Technology 57(1): 118–133. doi:10.5254/1.3535988.

Cotten, G.R. 1987. Mixing of carbon black with rubber IV. Effect of carbon black characteristics. Plastics and Rubber Processing and Applications 7(3): 173–178.

Croissant, J.G. and C.J. Brinker. 2018. Biodegradable silica-based nanoparticles: Dissolution kinetics and selective bond cleavage. pp. 181–214. *In*: F. Tamanoi (ed.). Mesoporous Silica-based Nanomaterials and Biomedical Applications, Part A, Vol. 43. The Enzymes Series. Elsevier: Amsterdam, The Netherlands. doi:10.1016/bs.enz.2018.07.008.

Crow, E.L. and K. Shimizu (eds). 1987. Lognormal Distributions: Theory and Applications. Boca Raton, Florida, US: CRC Press.

de Gennes, P.G. 1979. Scaling Concepts in Polymer Physics. Ithaca, New York: Cornell University Press.

Debye, P. and F. Bueche. 1951. The dielectric constant of polystyrene solutions. Journal of Physical and Colloid Chemistry 55(2): 235–238. doi:10.1021/j150485a011.

Donnet, J.-B., R.C. Bansal and M.-J. Wang (eds). 1993. Carbon Black: Science and Technology, 2nd Ed. New York, US: Marcel Dekker, Inc. https://books.google.com/books?id=SPpx6MkRYwMC&pgis=1.

Egelstaff, P.A. 1967. An Introduction to The Liquid State. London, New York, US: Academic Press. https://books.google.com/books?id=hkdRAAAAMAAJ&source=gbs_ViewAPI.

Faraguna, F., P. Pötschke and J. Pionteck. 2017. Preparation of polystyrene nanocomposites with functionalized carbon nanotubes by melt and solution mixing: Investigation of dispersion, melt rheology, electrical and thermal properties. Polymer (United Kingdom) 132: 325–341. doi:10.1016/j.polymer.2017.11.014.

Ferry, J.D. 1980. Viscoelastic Properties of Polymers, 3rd Ed. New York: John Wiley & Sons.

Filippone, G., G. Romeo and D. Acierno. 2010. Viscoelasticity and structure of polystyrene/fumed silica nanocomposites: Filler network and hydrodynamic contributions. Langmuir 26(4): 2714–2720. doi:10.1021/la902755r.

Filippone, G. and M.S. de Luna. 2012. A unifying approach for the linear viscoelasticity of polymer nanocomposites. Macromolecules 45(21): 8853–8860. doi:10.1021/ma301594g.

Fletcher, W.P. and A.N. Gent. 1954. Nonlinearity in the dynamic properties of vulcanized rubber compounds. Rubber Chemistry and Technology 27(1): 209–222. doi:10.5254/1.3543472.

Friedlander, S.K. 2000. Smoke, Dust and Haze: Fundamentals of Aerosol Behavior. New York, US: Oxford University Press.

Fu, X., J. Wang, J. Ding, H. Wu, Y. Dong and Y. Fu. 2013. Quantitative evaluation of carbon nanotube dispersion through scanning electron microscopy images. Composites Science and Technology 87: 170–173. doi:10.1016/j.compscitech.2013.08.014.

Gabriel, C.F.S., A. de A.P. Gabino, A.M.F. de Sousa, C.R.G. Furtado and R.C.R. Nunes. 2019. Tire tread rubber compounds with ternary system filler based on carbon black, silica and metakaolin: Contribution of silica/metakaolin content on the final properties. Journal of Elastomers & Plastics 51(7–8): 712–726. doi:10.1177/0095244318819196.

Glaskova, T., M. Zarrelli, A. Borisova, K. Timchenko, A. Aniskevich and M. Giordano. 2011. Method of quantitative analysis of filler dispersion in composite systems with spherical inclusions. Composites Science and Technology 71(13): 1543–1549. doi:10.1016/j.compscitech.2011.06.009.

Glaskova, T., M. Zarrelli, A. Aniskevich, M. Giordano, L. Trinkler and B. Berzina. 2012. Quantitative optical analysis of filler dispersion degree in MWCNT-epoxy nanocomposite. Composites Science and Technology 72(4): 477–481. doi:10.1016/j.compscitech.2011.11.029.

Gogia, A., K. Rishi, A. McGlasson, G. Beaucage and V.K. Kuppa. 2021. Dissipative particle dynamics (DPD) simulation to understand the nanoparticle dispersion and aggregation behavior in polymer nanocomposites. *In*: Bulletin of the American Physical Society. American Physical Society. https://meetings.aps.org/Meeting/MAR21/Session/M71.90.

Graessley, W.W. 2002. Scattering by modestly concentrated polymer solutions. Macromolecules 35(8): 3184–3188. doi:10.1021/ma0121518.

Guinier, A. and Gerard Fournet (ed.). 1955. Small Angle Scattering of X-rays. Translation by C.B. Walker. New York, US: John Wiley & Sons.

Gunasekaran, S., R.K. Natarajan, A. Kala and R. Jagannathan. 2008. Dielectric studies of some rubber materials at microwave frequencies. Indian Journal of Pure and Applied Physics 46(10): 733–737.

Guth, E. 1945. Theory of filler reinforcement. Journal of Applied Physics 16(1): 20–25. doi:10.1063/1.1707495.

Hair, M.L. and W. Hertl. 1971. Reaction of hexamethyldisilazane with silica. The Journal of Physical Chemistry 75(14): 2181–2185. doi:10.1021/j100683a020.

Hamed, G.R. 1991. Energy dissipation and the fracture of rubber vulcanizates. Rubber Chemistry and Technology 64(3): 493–500. doi:10.5254/1.3538566.

Hashimoto, T., N. Amino, S. Nishitsuji and M. Takenaka. 2019. Hierarchically self-organized filler particles in polymers: Cascade evolution of dissipative structures to ordered structures. Polymer Journal 51(2): 109–130. doi:10.1038/s41428-018-0147-2.

Haslam, M.D. and B. Raeymaekers. 2013. A composite index to quantify dispersion of carbon nanotubes in polymer-based composite materials. Composites Part B: Engineering 55(1): 16–21. doi:10.1016/j.compositesb.2013.05.038.

Hassinger, I., X. Li, H. Zhao, H. Xu, Y. Huang, A. Prasad, et al. 2016. Toward the development of a quantitative tool for predicting dispersion of nanocomposites under non-equilibrium processing conditions. Journal of Materials Science 51(9): 4238–4249. doi:10.1007/s10853-015-9698-1.

Hore, M.J.A., La Shanda T.J. Korley and S.K. Kumar. 2020. Polymer-grafted nanoparticles. Journal of Applied Physics 128(3): 030401. doi:10.1063/5.0019326.

Huber, G. and T.A. Vilgis. 1999. Universal properties of filled rubbers: Mechanisms for reinforcement on different length scales. Kgk Kautschuk Gummi Kunststoffe 52(2): 102–107.

Huber, G. and T.A. Vilgis. 2002. On the mechanism of hydrodynamic reinforcement in elastic composites. Macromolecules 35(24): 9204–9210. doi:10.1021/ma0208887.

Ilavsky, J. and P.R. Jemian. 2009. Irena: Tool suite for modeling and analysis of small-angle scattering. Journal of Applied Crystallography 42(2): 347–353. doi:10.1107/S0021889809002222.

Ilavsky, J. 2012. Nika: Software for two-dimensional data reduction. Journal of Applied Crystallography, International Union of Crystallography 45(2): 324–328. doi:10.1107/S0021889812004037.

Ilavsky, J., F. Zhang, A.J. Allen, L.E. Levine, P.R. Jemian and G.G. Long. 2013. Ultra-small-angle X-ray scattering instrument at the advanced photon source: History, recent development and current status. Metallurgical and Materials Transactions A 44(1): 68–76. doi:10.1007/s11661-012-1431-y.

Ilavsky, J., F. Zhang, R.N. Andrews, I. Kuzmenko, P.R. Jemian, L.E. Levine, et al. 2018. Development of combined microstructure and structure characterization facility for in situ and operando studies at the advanced photon source. Journal of Applied Crystallography, International Union of Crystallography 51: 867–882. doi:10.1107/S160057671800643X.

Janzen, J. 1975. On the critical conductive filler loading in antistatic composites. Journal of Applied Physics 46(2): 966–969. doi:10.1063/1.321629.

Jethmalani, J.M., W.T. Ford and G. Beaucage. 1997. Crystal structures of monodisperse colloidal silica in Poly(Methyl Acrylate) films. Langmuir 13(13): 3338–3344. doi:10.1021/la9708795.

Jin, Y., G. Beaucage, K. Vogtt, H. Jiang, V. Kuppa, J. Kim, et al. 2017. A Pseudo-thermodynamic description of dispersion for nanocomposites. Polymer 129(October): 32–43. doi:10.1016/j.polymer.2017.09.040.

Kammler, H.K., R. Mueller, O. Senn and S.E. Pratsinis. 2001. Synthesis of silica-carbon particles in a turbulent H2-Air flame aerosol reactor. AIChE Journal 47(7): 1533–1543. doi:10.1002/aic.690470707.

Kango, S., S. Kalia, A. Celli, J. Njuguna, Y. Habibi and R. Kumar. 2013. Surface modification of inorganic nanoparticles for development of organic-inorganic nanocomposites – A review. Progress in Polymer Science 38(8): 1232–1261. doi:10.1016/j.progpolymsci.2013.02.003.

Karásek, L., B. Meissner, S. Asai and M. Sumita. 1996. Percolation concept: Polymer-filler gel formation, electrical conductivity and dynamic electrical properties of carbon-black-filled rubbers. Polymer Journal 28(2): 121–126. doi:10.1295/polymj.28.121.

Khare, H.S. and D.L. Burris. 2010. A quantitative method for measuring nanocomposite dispersion. Polymer 51(3): 719–729. doi:10.1016/j.polymer.2009.12.031.

Kockmann, A., J.C. Porsiel, R. Saadat and G. Garnweitner. 2018. Impact of nanoparticle surface modification on the mechanical properties of polystyrene-based nanocomposites. RSC Advances 8(20): 11109–11118. doi:10.1039/C8RA00052B.

Kondo, H. 2014. Evaluation of rubber processing for unvulcanized rubber. Nippon Gomu Kyokaishi 87(1): 16–21. doi:10.2324/gomu.87.16.

Kraus, G. and J.T. Gruver. 1965. Steady-state melt viscosity of plasticized hydrocarbon elastomers. Transactions of the Society of Rheology 9(2): 17–34. doi:10.1122/1.548994.

Kraus, G. 1971. A carbon black structure-concentration equivalence principle. Application to stress-strain relationships of filled rubbers. Rubber Chemistry and Technology 44(1): 199–213. doi:10.5254/1.3547354.

Krishnamoorti, R. 2007. Strategies for dispersing nanoparticles in polymers. MRS Bulletin 32(4): 341–347. doi:10.1557/mrs2007.233.

Le, H.H., S. Ilisch, B. Jakob and H-J Radusch. 2004. Online characterization of the effect of mixing parameters on carbon black dispersion in rubber compounds using electrical conductivity. Rubber Chemistry and Technology 77(1): 147–160. doi:10.5254/1.3547808.

Le, H.H., M. Tiwari, S. Ilisch and H.J. Radusch. 2005. Effect of molecular structure on carbon black dispersion in rubber compounds. Kautschuk Gummi Kunststoffe, no. November: 575–580.

Le, H.H., E. Hamann, S. Ilisch, G. Heinrich and H.J. Radusch. 2014. Selective wetting and dispersion of filler in rubber composites under influence of processing and curing additives. Polymer 55(6): 1560–1569. doi:10.1016/j.polymer.2014.02.002.

Leblanc, J. 2002. Rubber–filler interactions and rheological properties in filled compounds. Progress in Polymer Science 27(4): 627–687. doi:10.1016/S0079-6700(01)00040-5.

Lee, H.-G., H.-S. Kim, S.-T. Cho, I.-T. Jung and C.-T. Cho. 2013. Characterization of solution styrene butadiene rubber (SBR) through the evaluation of static and dynamic mechanical properties and fatigue in silica-filled compound. Asian Journal of Chemistry 25(8): 5251–5256. doi:10.14233/ajchem.2013.F27.

Lei, W., X. Zhou, T.P. Russell, K.-C. Hua, X. Yang, H. Qiao, et al. 2016. High performance bio-based elastomers: Energy efficient and sustainable materials for tires. Journal of Materials Chemistry A 4(34): 13058–13062. doi:10.1039/C6TA05001H.

Leigh-Dugmore, C.H. 1956. Measurement of dispersion in black-loaded rubber. Rubber Chemistry and Technology 29(4): 1303–1308. doi:10.5254/1.3542632.

Li, Z., Y. Gao, K.-S. Moon, Y. Yao, A. Tannenbaum and C.P. Wong. 2012. Automatic Quantification of Filler Dispersion in Polymer Composites. Polymer 53(7). Elsevier Ltd: 1571–1580. doi:10.1016/j.polymer.2012.01.048.

Li, X., H. Deng, Q. Zhang, F. Chen and Q. Fu. 2016. The effect of DBP of carbon black on the dynamic self-assembly in a polymer melt. RSC Advances, Royal Society of Chemistry 6(30): 24843–24852. doi:10.1039/c5ra28118k.

Liu, M., M. Kang, Y. Mou, K. Chen and R. Sun. 2017. Visualization of filler network in silicone rubber with confocal laser-scanning microscopy. RSC Advances, Royal Society of Chemistry 7(84): 53578–53586. doi:10.1039/c7ra09773e.

Lively, B., P. Smith, W. Wood, R. Maguire and W.H. Zhong. 2012. Quantified stereological macrodispersion analysis of polymer nanocomposites. Composites Part A: Applied Science and Manufacturing 43(6): 847–855. doi:10.1016/j.compositesa.2012.01.012.

Mädler, L., H.K. Kammler, R. Mueller and S.E. Pratsinis. 2002. Controlled synthesis of nanostructured particles by flame spray pyrolysis. Journal of Aerosol Science 33(2): 369–389. doi:10.1016/S0021-8502(01)00159-8.

Mark, J.E. 1999. Polymer Data Handbook. New York, US: Oxford University Press.

McDonald, G.C. and W.M. Hess. 1977. Carbon black morphology in rubber. Rubber Chemistry and Technology 50(4): 842–862. doi:10.5254/1.3535180.

McEwan, M. and D. Green. 2009. Rheological impacts of particle softness on wetted polymer-grafted silica nanoparticles in polymer melts. Soft Matter 5(8): 1705–1716. doi:10.1039/b816975f.

McEwan, M.E., S.A. Egorov, J. Ilavsky, D.L. Green and Y. Yang. 2011. Mechanical reinforcement of polymer nanocomposites: Theory and ultra-small angle X-Ray scattering (USAXS) Studies. Soft Matter 7(6): 2725. doi:10.1039/c0sm00393j.

McGlasson, A., K. Rishi, G. Beaucage, V. Narayanan, M. Chauby, A. Mulderig, et al. 2019. The effects of staged mixing on the dispersion of reinforcing fillers in elastomer compounds. Polymer 181(May): 121765. doi:10.1016/j.polymer.2019.121765.

McGlasson, A., K. Rishi, G. Beaucage, M. Chauby, V. Kuppa, J. Ilavsky, et al. 2020. Quantification of dispersion for weakly and strongly correlated nanofillers in polymer nanocomposites. Macromolecules 53(6): 2235–2248. doi:10.1021/acs.macromol.9b02429.

Meakin, P., I. Majid, S. Havlin and H.E. Stanley. 1984. Topological properties of diffusion limited aggregation and cluster cluster aggregation. Journal of Physics A: Mathematical and General 17(18): L975–L981. doi:10.1088/0305-4470/17/18/008.

Medalia, A.I. 1961. Dispersion of carbon black in rubber: Revised calculation procedure. Rubber Chemistry and Technology 34(4): 1134–1140. doi:10.5254/1.3540272.

Medalia, A.I. 1970. Morphology of aggregates: VI. Effective volume of aggregates of carbon black from electron microscopy; Application to vehicle absorption and to die swell of filled rubber. Journal of Colloid and Interface Science 32(1): 115–131. doi:10.1016/0021-9797(70)90108-6.

Medalia, A.I. 1972. Effective degree of immobilization of rubber occluded within carbon black aggregates. Rubber Chemistry and Technology 45(5): 1171–1194. doi:10.5254/1.3544731.

Mezger, T.G. 2006. The Rheology Handbook: For Users of Rotational and Oscillatory Rheometers, 2nd Ed. Hannover, Germany: Vincentz Network GmbH & Co KG. doi:10.1515/9783748600367-009.

Mueller, R., H.K. Kammler, S.E. Pratsinis, A. Vital, G. Beaucage and P. Burtscher. 2004. Non-agglomerated dry silica nanoparticles. Powder Technology 140(1): 40–48. doi: 10.1016/j.powtec.2004.01.004.

Mulderig, A., G. Beaucage, K. Vogtt, H. Jiang and V. Kuppa. 2017a. Quantification of branching in fumed silica. Journal of Aerosol Science 109(March): 28–37. doi:10.1016/j.jaerosci.2017.04.001.

Mulderig, A., G. Beaucage, K. Vogtt, H. Jiang, Y. Jin, L. Clapp, et al. 2017b. Structural emergence in particle dispersions. Langmuir 33(49): 14029–14037. doi:10.1021/acs.langmuir.7b03033.

Natarajan, B., Y. Li, H. Deng, L.C. Brinson and L.S. Schadler. 2013. Effect of interfacial energetics on dispersion and glass transition temperature in polymer nanocomposites. Macromolecules 46(7): 2833–2841. doi:10.1021/ma302281b.

O'Farrell, C.P., M. Gerspacher and L. Nikiel. 2000. Carbon black dispersion by electrical measurements. KGK Kautschuk Gummi Kunststoffe 53(12): 701–710.

Okoli, U., K. Rishi, G. Beaucage, H.K. Kammler, A. McGlasson, C. Michael, et al. 2021. Dispersion and dynamic response for in-flame and chemically modified fumed silica nanocomposites. In: Bulletin of the American Physical Society. American Physical Society. https://meetings.aps.org/Meeting/MAR21/Session/M71.95.

Payne, A.R. 1962. The dynamic properties of carbon black-loaded natural rubber vulcanizates. Part I. Journal of Applied Polymer Science 6(19): 57–63. doi:10.1002/app.1962.070061906.

Payne, A.R. 1963. Dynamic properties of heat-treated butyl vulcanizates. Journal of Applied Polymer Science 7(3): 873–885. doi:10.1002/app.1963.070070307.

Payne, A.R. and W.F. Watson. 1963. Carbon black structure in rubber. Rubber Chemistry and Technology 36(1): 147–155. doi:10.5254/1.3539533.

Payne, A.R. and R.E. Whittaker. 1971. Low strain dynamic properties of filled rubbers. Rubber Chemistry and Technology 44(2): 440–478. doi:10.5254/1.3547375.

Pedersen, J.S. and P. Schurtenberger. 1999. Static properties of polystyrene in semidilute solutions: A comparison of monte carlo simulation and small-angle neutron scattering results. Europhysics Letters 45(6): 666–672. doi:10.1209/epl/i1999-00219-7.

Pedersen, J.S. and C. Sommer. 2005. Temperature dependence of the virial coefficients and the chi parameter in semi-dilute solutions of PEG. pp. 70–78. In: Scattering Methods and the Properties of Polymer Materials. Berlin, Heidelberg: Springer Berlin Heidelberg. doi:10.1007/b107350.

Pegel, S., P. Pötschke, T. Villmow, D. Stoyan and G. Heinrich. 2009. Spatial statistics of carbon nanotube polymer composites. Polymer 50(9): 2123–2132. doi:10.1016/j.polymer.2009.02.030.

Percus, J.K. and G.J. Yevick. 1958. Analysis of classical statistical mechanics by means of collective coordinates. Physical Review 110(1): 1–13. doi:10.1103/PhysRev.110.1.

Polley, M.H. and B.B.S.T. Boonstra. 1957. Carbon blacks for highly conductive rubber. Rubber Chemistry and Technology 30(1): 170–179. doi:10.5254/1.3542660.

Porod, G. 1982. General theory. pp. 17–51. In: O. Glatter and O. Kratky (eds). Small Angle X-Ray Scattering. New York, US: Academic Press Inc.

Rai, D.K., G. Beaucage, K. Vogtt, J. Ilavsky and H.K. Kammler. 2018. In situ study of aggregate topology during growth of pyrolytic silica. Journal of Aerosol Science 118(September 2017): 34–44. doi:10.1016/j.jaerosci.2018.01.006.

Raut, P., N. Swanson, A. Kulkarni, C. Pugh and S.C. Jana. 2018. Exploiting arene-perfluoroarene interactions for dispersion of carbon black in rubber compounds. Polymer 148: 247–258. doi:10.1016/j. polymer.2018.06.025.

Rishi, K., G. Beaucage, V. Kuppa, A. Mulderig, V. Narayanan, A. McGlasson, et al. 2018. Impact of an emergent hierarchical filler network on nanocomposite dynamics. Macromolecules 51(20): 7893–7904. doi:10.1021/acs.macromol.8b01510.

Rishi, K., V. Narayanan, G. Beaucage, A. McGlasson, V. Kuppa, J. Ilavsky, et al. 2019a. A thermal model to describe kinetic dispersion in rubber nanocomposites: The effect of mixing time on dispersion. Polymer 175(June): 272–282. doi:10.1016/j.polymer.2019.03.044.

Rishi, K., A. Mulderig, G. Beaucage, K. Vogtt and H. Jiang. 2019b. Thermodynamics of hierarchical aggregation in pigment dispersions. Langmuir 35(40): 13100–13109. doi:10.1021/acs.langmuir.9b02192.

Rishi, K., L. Pallerla, G. Beaucage and A. Tang. 2020. Dispersion of surface-modified, aggregated, fumed silica in polymer nanocomposites. Journal of Applied Physics 127(17): 174702. doi:10.1063/1.5144252.

Robertson, C.G., R. Bogoslovov and C.M. Roland. 2007. Effect of structural arrest on Poisson's ratio in nanoreinforced elastomers. Physical Review E - Statistical, Nonlinear, and Soft Matter Physics 75(5): 1–7. doi:10.1103/PhysRevE.75.051403.

Roe, R.-J. 2000. Methods of X-Ray and Neutron Scattering in Polymer Science. New York, US: Oxford University Press.

Sadhu, S. and A.K. Bhowmick. 2005. Unique rheological behavior of rubber based nanocomposites. Journal of Polymer Science, Part B: Polymer Physics 43(14): 1854–1864. doi:10.1002/polb.20469.

Sae-oui, P., K. Suchiva, C. Sirisinha, W. Intiya, P. Yodjun and U. Thepsuwan. 2017. Effects of blend ratio and SBR type on properties of carbon black-filled and silica-filled SBR/BR tire tread compounds. Advances in Materials Science and Engineering 2017: 1–8. doi:10.1155/2017/2476101.

Sattayanurak, S., J.W.M. Noordermeer, K. Sahakaro, W. Kaewsakul, W.K. Dierkes and A. Blume. 2019. Silica-reinforced natural rubber: Synergistic effects by addition of small amounts of secondary fillers to silica-reinforced natural rubber tire tread compounds. Advances in Materials Science and Engineering 2019(February): 1–8. doi:10.1155/2019/5891051.

Shiga, S. and M. Furuta. 1982. Processability of EPR in an internal mixer (II) morphological changes of carbon black agglomerates during mixing. Nippon Gomu Kyokaishi 55(8): 491–503. doi:10.2324/gomu.55.491.

Sipaut, C.S., R.F. Mansa, V. Padavettan, I. Ab. Rahman, J. Dayou and M. Jafarzadeh. 2015. The effect of surface modification of silica nanoparticles on the morphological and mechanical properties of bismaleimide/diamine matrices. Advances in Polymer Technology 34(2): 10p. doi:10.1002/adv.21492.

Smallwood, H.M. 1944. Limiting law of the reinforcement of rubber. Journal of Applied Physics 15(11): 758–766. doi:10.1063/1.1707385.

Song, L., Z. Wang, X. Tang, L. Chen, P. Chen, Q. Yuan, et al. 2017. Visualizing the toughening mechanism of nanofiller with 3D X-ray nano-CT: Stress-induced phase separation of silica nanofiller and silicone polymer double networks. Macromolecules 50(18): 7249–7257. doi:10.1021/acs.macromol.7b00539.

Stace, A.J. and E. Bichoutskaia. 2012. Absolute electrostatic force between two charged particles in a low dielectric solvent. Soft Matter 8(23): 6210–6213. doi:10.1039/c2sm25602a.

Staniewicz, L., T. Vaudey, C. Degrandcourt, M. Couty, F. Gaboriaud and P. Midgley. 2014. Electron tomography provides a direct link between the payne effect and the inter-particle spacing of rubber composites. Scientific Reports 4: 1–7. doi:10.1038/srep07389.

Stöckelhuber, K.W., A. Das, R. Jurk and G. Heinrich. 2010. Contribution of physico-chemical properties of interfaces on dispersibility, adhesion and flocculation of filler particles in rubber. Polymer 51(9): 1954–1963. doi:10.1016/j.polymer.2010.03.013.

Tokita, N., C.H. Shieh, G.B. Ouyang and W.J. Patterson. 1994. Carbon black-elastomer interaction modelling. KGK-Kautschuk Und Gummi Kunststoffe 47(6): 416–420.

Trappe, V. and D.A. Weitz. 2000. Scaling of the viscoelasticity of weakly attractive particles. Physical Review Letters 85(2): 449–452. doi:10.1103/PhysRevLett.85.449.

Veigel, D., K. Rishi, G. Beaucage, U. Okoli, J. Galloway, J. Ilavsky, et al. 2021. Nanocomposite dispersion in melt mixers. In: Bulletin of the American Physical Society. American Physical Society. https://meetings.aps.org/Meeting/MAR21/Session/M71.89.

Vilgis, T.A. and H.H. Winter. 1988. Mechanical selfsimilarity of polymers during chemical gelation. Colloid & Polymer Science 266(6): 494–500. doi:10.1007/BF01420759.

Vogtt, K., G. Beaucage, M. Weaver and H. Jiang. 2017. Thermodynamic stability of worm-like micelle solutions. Soft Matter 13(36). Royal Society of Chemistry: 6068–6078. doi:10.1039/C7SM01132F.

Vogtt, K., G. Beaucage, K. Rishi, H. Jiang and A. Mulderig. 2019. Hierarchical approach to aggregate equilibria. Physical Review Research 1(3): 033081. doi:10.1103/PhysRevResearch.1.033081.

Wang, M.-J., S. Wolff and J.-B. Donnet. 1991a. Filler-elastomer interactions. Part I: Silica surface energies and interactions with model compounds. Rubber Chemistry and Technology 64(4): 559–576. doi:10.5254/1.3538573.

Wang, M.-J., S. Wolff and J.-B. Donnet. 1991b. Filler—elastomer interactions. Part III. Carbon-black-surface energies and interactions with elastomer analogs. Rubber Chemistry and Technology 64(5): 714–736. doi:10.5254/1.3538585.

Wang, M.-J., S. Wolff and E.-H. Tan. 1993. Filler-elastomer interactions. Part VIII. The role of the distance between filler aggregates in the dynamic properties of filled vulcanizates. Rubber Chemistry and Technology 66(2): 178–195. doi:10.5254/1.3538305.

Warring, J.R.S. 1950. Dynamic testing in compression: Comparison of the I.C.I. electrical compression vibrator and the I.G. mechanical vibrator in dynamic testing of rubber. Transactions of the Institution of the Rubber Industry 26: 4–26.

Wertheim, M.S. 1963. Exact solution of the percus-yevick integral equation for hard spheres. Physical Review Letters 10(8): 321–323. doi:10.1103/PhysRevLett.10.321.

Williams, M.L., R.F. Landel and J.D. Ferry. 1955. The temperature dependence of relaxation mechanisms in amorphous polymers and other glass-forming liquids. Journal of the American Chemical Society 77(14): 3701–3707. doi:10.1021/ja01619a008.

Wolff, S. and M.-J. Wang. 1992. Filler—elastomer interactions. Part IV. The effect of the surface energies of fillers on elastomer reinforcement. Rubber Chemistry and Technology 65(2): 329–342. doi:10.5254/1.3538615.

Wu, S. 1985. Phase structure and adhesion in polymer blends: A criterion for rubber toughening. Polymer 26(12): 1855–1863. doi:10.1016/0032-3861(85)90015-1.

Yamada, H., I. Manas-Zloczower and D.L. Feke. 1998. The influence of matrix viscosity and interfacial properties on the dispersion kinetics of carbon black agglomerates. Rubber Chemistry and Technology 71(1): 1–16. doi:10.5254/1.3538468.

Yurekli, K., R. Krishnamoorti, M.F. Tse, K.O. McElrath, A.H. Tsou and H.C. Wang. 2001. Structure and dynamics of carbon black-filled elastomers. Journal of Polymer Science, Part B: Polymer Physics 39(2): 256–275. doi:10.1002/1099-0488(20010115)39:2<256::AID-POLB80>3.0.CO;2-Z.

Zhang, Y., S. Ge, B. Tang, T. Koga, M.H. Rafailovich, J.C. Sokolov, et al. 2001. Effect of carbon black and silica fillers in elastomer blends. Macromolecules 34(20): 7056–7065. doi:10.1021/ma010183p.

Zimm, B.H. 1948. The scattering of light and the radial distribution function of high polymer solutions. The Journal of Chemical Physics 16(12): 1093–1099. doi:10.1063/1.1746738.

Application of Nanocomposites in the Automotive Industry

Surendra Maharjan and Ahalapitiya H. Jayatissa*

The University of Toledo, 2801 W. Bancroft St, Toledo, OH – 43606, USA

smaharj7@rockets.utoledo.edu; ahalapitiya.jayatissa@utoledo.edu

1 INTRODUCTION

Nanocomposites can be defined as multiphase solid materials where at least one of the phases has a dimension of fewer than 100 nanometers (nm), as manifested in Fig. 2.1. The concept behind a nanocomposite is to create novel material with unprecedented flexibility and improvement in properties by integrating nanoparticles in a bulk matrix, both having dissimilarities in chemistry and structure. The result can be observed in the enhancement of mechanical, electrical, thermal, optical, electrochemical, catalytic properties than the constituent materials. These properties are affected by the size limit of nanoparticles and are proposed as (Kamigaito 1991):

1. <5 nm for catalytic activity
2. <20 nm for making a hard-magnetic material soft
3. <50 nm for refractive index changes
4. <100 nm for achieving superparamagnetic, mechanical strengthening or restricting matrix dislocation movement

Nanocomposites are different from conventional composite material in the sense that the nanoscale reinforcement or filler material has an exceptionally high surface-to-volume ratio and/or high aspect ratio and transparency ($d < 30$ nm). This means, even a small amount of reinforcement can have a large interaction area with the matrix and have an observable effect on the macro scale properties. The nano-size materials may be of particles (e.g., minerals), sheets (e.g., exfoliated clay stacks), or fibers (e.g., carbon nanotubes or electrospun fibers) (Ben Hargreaves 2020) and are dispersed into the bulk matrix during processing. The number of nanoparticles introduced in the matrix remains very low on the order of 0.5 to 5% by weight. The advantages of designing novel nanocomposites can be seen in the improvement of properties:

---

*Corresponding Author

- Mechanical properties (strength, bulk modulus, hardness, toughness)
- Electrical conductivity
- Decreased gas, water and hydrocarbon permeability
- Decreased flammability and smoke generations
- Thermal stability or heat resistance
- Chemical resistance
- Surface appearance
- Optical clarity

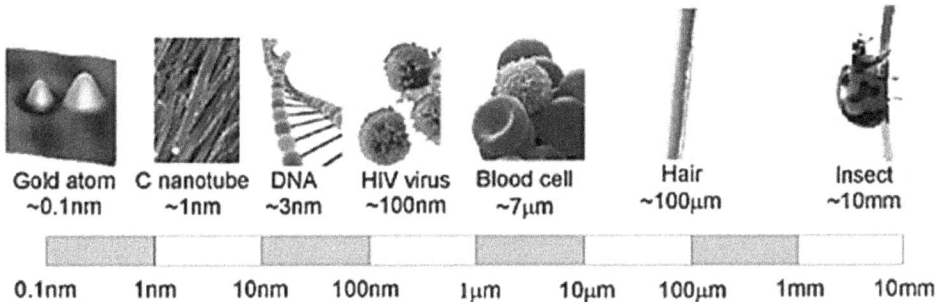

| Gold atom ~0.1nm | C nanotube ~1nm | DNA ~3nm | HIV virus ~100nm | Blood cell ~7μm | Hair ~100μm | Insect ~10mm |

| 0.1nm | 1nm | 10nm | 100nm | 1μm | 10μm | 100μm | 1mm | 10mm |

Figure 2.1 The representation of nanomaterials' length is measured in nano size (Serrano et al. 2009).

1.1 Background

The automotive industry consists of many organizations and companies involved in the design, prototyping, development, manufacturing, marketing and selling of vehicles. It is one of the largest economic sectors by revenue in the world.

A single car consists of roughly 30,000 parts, counting down to every smallest screw. Figure 2.2 shows the parts of a car. In a supply chain, there are Original Equipment Manufacturers (OEMs) and suppliers, that are categorized into a tier system (Tier 1, Tier 2, Tier 3). OEMs focus on designing, in-house part manufacturing, ordering from suppliers, assembling, quality inspection, promoting and selling vehicles. Tier 1 suppliers are companies that supply parts directly to OEMs and have a close relationship with them. Tier 2 suppliers are often experts in their domain. Generally, they supply their parts to Tier 1 companies but sometimes to OEMs too. The supplier of raw materials or close-to-raw, materials like metal, composites or plastic to OEMs, Tier 1, and Tier 2 companies are called Tier 3 suppliers. Car models show independent gradual changes within each line and internal components (e.g., electronics) (Fisher et al. 1999) have had only moderate effects on the overall external shape of cars.

Nanotechnology has been one of the most interesting and challenging fields for researchers from the last century and numerous developments have been achieved since then. Toyota Motor Company is the first automotive industry to use nanocomposite material. They used Nylon-6-clay nanocomposite to make timing belt cover for Toyota Camry in collaboration with Ube industries in 1991. Later Unitika Company of Japan made engine covers for Mitsubishi with Nylon-6-clay nanocomposites.

However, General Motors revealed the first commercial use of PP-based nanocomposite in automotive part production in 2002, with the Chevrolet Astro and GMC Safari vans. The material is composed of thermoplastic olefin filled with 3% nanoclays. Although the material was much lighter, and less brittle at cold temperature and drew much attention by the media, the application was terminated. GM applied nanoclays to manufacture body side trim for Chevrolet Impala in 2004–2005. Even achieving weight savings of 3 to 25%, their performance remained poor and was stopped. In 2005, GM used 3 kg of nanoclays per car for Hummer H2 SUT cargo bed trim.

Figure 2.2   Parts of a car (Source: *https://www.amatechinc.com/resources/blog/tier-1-2-3-automotive-industry-supply-chain-explained*)

Nanocomposites are expanding in the automotive market with increased interest from other car manufacturers. Maserati, Daimler Chrysler and Audi are the other automakers that are making nanoclay filled parts commercially. Maserati engine bay covers comprise Ube nylon-6 material and 2% of nanoclays by weight. Daimler Chrysler used polypropylene with nanofillers and conventional fillers for the inner door handle of the Smart for four whereas Audi and Ptsch GmbH made heater vent of the A3 with PP/PS reinforced with nanoclays.

2   NANOCOMPOSITE MATERIALS AND THEIR TYPES

Nanocomposites are of different types depending upon the kinds of matrix material and nanoparticles. Due to the ability to withstand high temperature, pressure and better mechanical properties nanocomposites are largely used in the automotive industry. Non-homogenous dispersion of the nanoparticles and weak links between matrix and the particles cause problems to its application. Though, broad research and development are going to overcome these challenges. In general, nanocomposites are classified into three groups based on the reinforcement on the matrix. Metal matrix nanocomposites, polymer matrix nanocomposites and ceramics matrix nanocomposites.

2.1   Metal Matrix Nanocomposites (MMNC's)

Metal Matrix Nanocomposites are a mixture of ductile metallic matrix and ceramic reinforcement nanoparticles. The major reinforcement divisions are oxides, nitrides, carbides, hydrides and borides. Among them, the popular nanoparticles are alumina (Al_2O_3) and yttria (Y_2O_3) in oxides, Si_3N_3 and AlN in nitrides, B_4C, SiC, and TiC in carbides. These nanocomposites have improved properties in density, hardness, abrasion, deformation and corrosion resistance.

Metals such as aluminum, magnesium and titanium (Ravi et al. 2017) are very popular in the automotive industry for their lightweight and high strength which give better performance. Among the matrices; aluminum, and aluminum alloys (such as Al-Si alloy) are the most investigated materials with the reinforcement of carbon and silicon carbide, and they have been found in many applications in the automotive industry. A typical application in the automotive industry can be found in car parts like cylinder head, pistons, crankcase, cylinder head cover, intake manifold, etc.

In general, the processing methods of MMNC's are classified into three groups: solid-state processing, liquid-state processing and semi-solid-state processing. In solid-state processing, *in situ* nanostructure material is formed by either simple blending of matrix and reinforcement material or milling of powders. Liquid-state processing is the large-scale processing technique in which *ex situ* fillers are mixed with molten metal using the stirring method. The semi-solid-state processing method adopts the advantages of solid and liquid state techniques (Rajan et al. 1998; Han et al. 2006; Mutale et al. 2010; Zhou et al. 2014). The selection of these manufacturing techniques for the automotive industry depends upon the industrial scalability and material properties.

2.2   Polymer Matrix Nanocomposites (PMNC's)

Polymer matrix nanocomposites are those materials in which nanoparticles are used as reinforcement in a polymer matrix. Polymers have some of the excellent properties such as low density, ease of machining, good surface finish and corrosion resistance, but they are weak in strength so they are generally not used in structural applications. When nanoparticles are reinforced in the polymer matrix in a proper way such as uniform dispersion and good interfacial adhesion between matrix and filler, they will impart the mechanical properties like strength and toughness, reduce the coefficient of thermal expansion, and increase resistance to cracking, propagation, abrasion and corrosion. Table 2.1 summarizes different types of polymer nanocomposites with their properties.

Table 2.1   Mechanical properties of a few representative polymer resins and their nanocomposites (Naskar et al. 2016).

Resin type	Neat resin strength (MPa)	Neat resin modulus (GPa)	Nano-composite strength (MPa)	Nano-composite modulus (GPa)	Remarks
Epoxy (thermoset)	102–110	3.0–3.4	120–140	3.5–4.2	Flexural properties of solvent-based functionalized graphene (0.1 wt.%) nanocomposite (Naebe et al. 2015)
	53–58	2.7–3.0	75–80	3.6–3.9	Tensile properties of solvent-based graphene (0.1 wt.%) nanocomposite (Rafiee et al. 2009)
Polyetherimide (amorphous thermoplastic)	79	2.6	126	4.7	*In situ* polymerization in 1.2 vol.% single-walled carbon nanotube suspension insolvent. Although the matrix is amorphous, nanocomposites exhibit semi-crystallinity (Hegde et al. 2013)
Polyether ether ketone (PEEK)	88–90	2.0–2.2	85–95	3.3–3.4	Tensile properties of melt-extruded and injection-molded components after melt mixing with 7.5 wt.% hydroxyapatite (Wang et al. 2010b)
Polypropylene (PP)	36–42	1.1–1.3	50–52	3.1–3.4	Flexural strength of solvent (isopropanol) based premixing of 10 vol.% graphene platelets with powder PP followed by melt-mixing and extrusion (Kalaitzidou et al. 2007)
Nylon	69	1.1	107	2.1	Nylon 6 per clay hybrid obtained by melt-mixing at 4.2 wt.% clay loading (Okada and Usuki 1995)

Resin type	Neat resin strength (MPa)	Neat resin modulus (GPa)	Nano-composite strength (MPa)	Nano-composite modulus (GPa)	Remarks
Polycarbonate (PC)	120	2.0	160	2.7	Rolled fibers of multi-layered (320 aligned layers) graphene per PC (0.08 vol.%) nanocomposite film (Liu et al. 2016)
Polymethyl methacrylate (PMMA)	70	2.1	77–86	3.6–4.0	Methanol-coagulated films from THF solution containing 1 wt.% functionalized (partially oxygenated) graphene sheets (Ramanathan et al. 2008)

End-user applications and the property requirement will determine the choice of polymer matrix and nanofiller. Nanosized clay is the most widely used nanofiller due to significant improvement in the mechanical property of polymer matrices. Besides this, silicate nanoparticles, nanofibers, nanotubes, graphite, graphene, polyhedral oligomeric silsesquioxane (POSS) and organosilanes are also used as nanofillers in elastomers, thermoplastics (polyolefin, polyamide, polypropylene sulfide, polyetheretherketone, polyethylene terephthalate, and polycarbonates) and thermosets (epoxy). Researchers have also developed metallic nanoparticles such as silver, gold, copper and nanoceramic oxides such as TiO_2, ZnO, and SiO_2 that can be used as reinforcement material in the polymer matrix (Ramanathan et al. 2008; Park et al. 2008; Vivek and Yashwant 2011). It has been reported that vehicle weight can be reduced by 20–40% and the fuel efficiency can be increased by 20% with the significant use of PMNC's (Tjong 2006; Ramanathan et al. 2008). The potential applications of PMNC's are shown in Fig. 2.3.

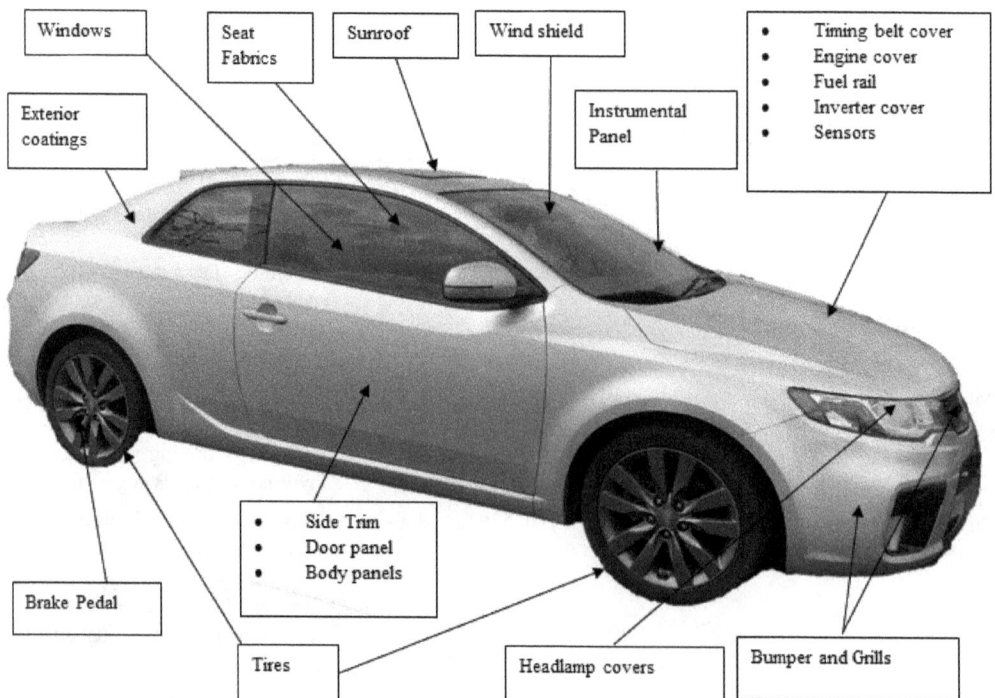

Figure 2.3   Potential automobile applications of PMNCs.

Polymer matrix nanocomposites can be manufactured in different ways depending upon the matrix and filler material and the most common types are solvent mixing, melt mixing and *in situ* polymerization (Garcés et al. 2000; Park et al. 2008). Researchers have investigated various modifications in the manufacturing process to improve the dispersion of nanofillers in the polymer matrix such as microwave-induced synthesis, one-pot synthesis, template-directed synthesis, electrochemical synthesis, etc. (Giannelis 1998; Privalko et al. 2005; Pandey et al. 2014).

2.3   Ceramics Matrix Nanocomposites (CMNC's)

Ceramics represent a class of promising material for a harsh environment such as high temperature and high corrosion. Ceramics matrix nanocomposites are the combination of ceramics matrix and nanoparticles as reinforcement which offer improved properties than conventional ceramic composites such as high hardness, strength, and toughness, creep resistance, flame retardancy, thermal shock resistance, magnetic and optical properties, wear resistance, chemical inertness and low density. Ceramic nanocomposites are grouped into four categories: intragranular, intergranular, hybrid and nano/nanocomposites as shown in Fig. 2.4. In intragranular, intergranular and hybrid types, the matrix is not in the nanoscale whereas, in nano/nanotype, both filler and matrix are in nanoscales.

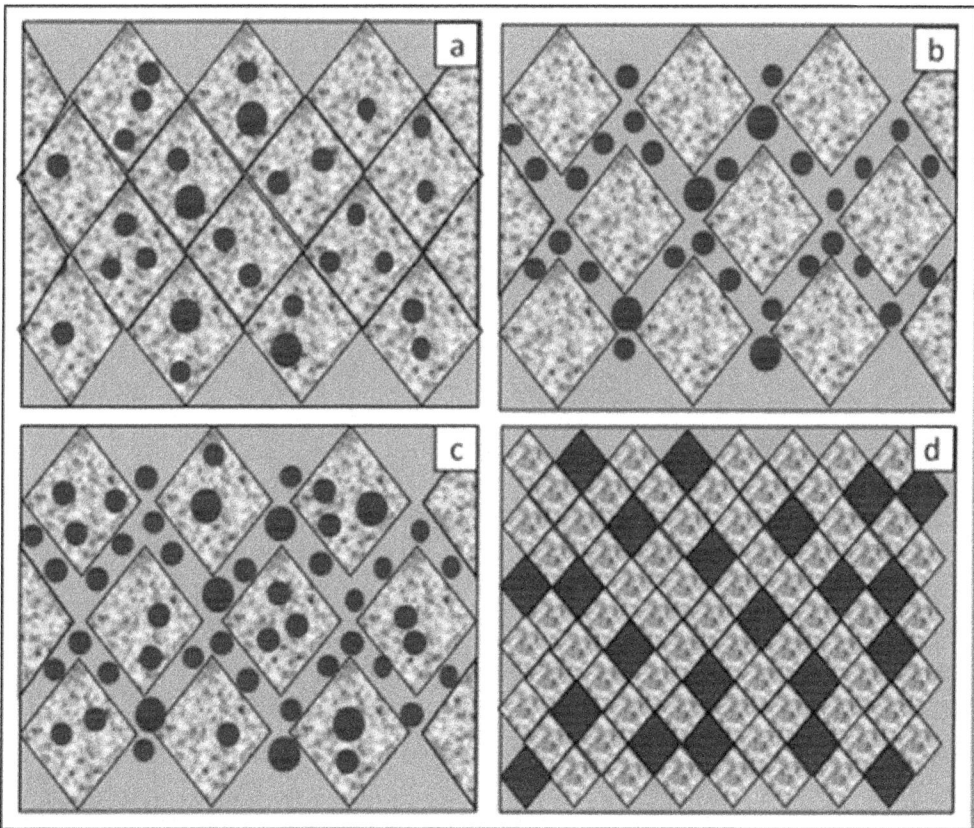

Figure 2.4 Classification of ceramic nanocomposites: (a) intratype, (b) intertype, (c) hybrid type, and (d) nano/nanotype (After Bhaduri and Bhaduri 1998).

The manufacturing of CMNC's is complicated and should be carefully modified according to the adaptability of nanometric particles. The major method for generating powder composites include plasma synthesis, Chemical Vapor Deposition (CVD), sputtering, sol-gel technique, combustion synthesis, intercalation and mechanical milling. The consolidation of powder composites is one of the important processes involved in the manufacturing of near-net-shape CMNC components and the techniques used in it are pressure-less sintering, reaction sintering, spark plasma sintering and hot pressing (Fukushima and Inagaki 1987; Bhaduri and Bhaduri 1998).

CMNC's possess the ability to create very complex shapes, the possibility of connecting the parts during the manufacturing process, reduce the after-treatment cost of parts, dimensional stability in extreme working conditions and can be fabricated as strong metals which make them applicable to different fields including automotive industry. Zirconium-based nanocomposites are used in thermal protection of turbo engines. Nanocarbon-carbon composites are used in brake disk material and brake lining. Sensors, transducers, capacitors and electric contacts are the common parts made up of ceramic nanocomposites for automobiles.

3 NANOCOMPOSITES IN AUTOMOTIVE PARTS

Most of the research and development activities based on nanotechnology are in the automobile sector as we depend on it more frequently compared to air or water transportation. Nanotechnology is applied to body parts, emissions, chassis and tires, automobile interiors, electrics and electronics, engines and drive trains. The main advantages of using nanocomposites in automobiles include lighter and stronger body parts, improving fuel efficiency and ultimately achieving better performance over a long period. The important parts of automobiles that are shaped by nanotechnology are depicted in Fig. 2.5.

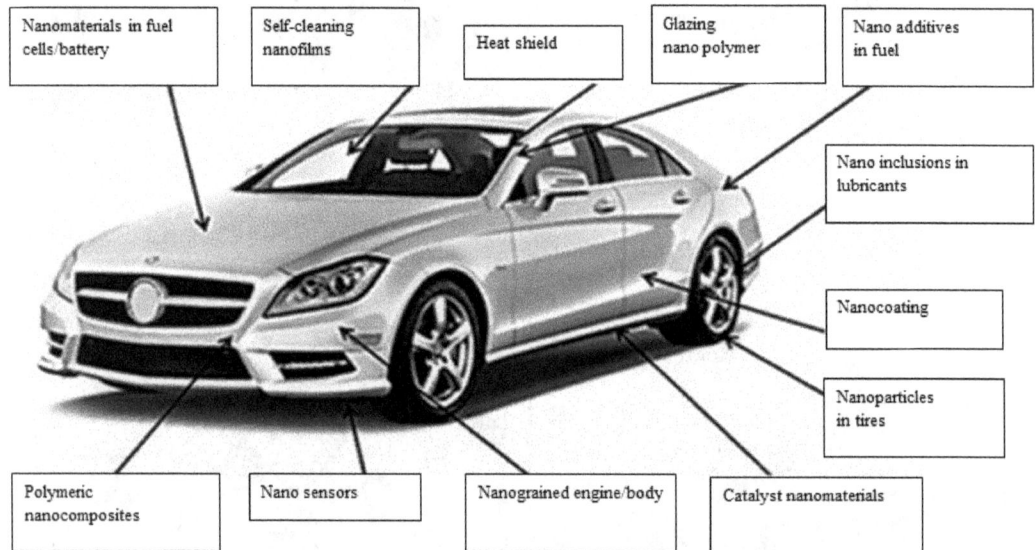

Figure 2.5   Various parts of the automobile in which nanotechnology is applied (Not representing the actual vehicle) (Asmatulu et al. 2013).

3.1 Body Parts

The application of nanocomposites for body parts includes paint coatings, self-cleaning, scratch-resistant and making the parts lighter as compared to the existing parts.

3.1.1   Paint Coatings

Automotive body parts are painted for decorative and protective purposes with three general coats: primer, the basecoat and a clear coat. However, it may vary with carmakers. Paint coatings are determined by various factors including corrosion resistance, appearance, aesthetic characteristics, cost and environmental requirements, durability and ease of mass production. The coating with nanoparticles is found to be an effective strategy to enhance the protection and scratch resistance on the body surfaces over a long period even when exposed to extreme weather conditions (Mohseni et al. 2012; Seubert et al. 2012; Asmatulu et al. 2013). Researchers have found that the addition of nano-SiO_2 in the polymer coating improves hardness, which protects the layer against cracking, wear and abrasion (Zhou et al. 2002; Song et al. 2007; Akafuah et al. 2016). In automotive headlights, the scratch in transparent glass is reduced by polysiloxane or acrylate paints which embed nanoparticles with polycarbonate material. Moreover, other types of nanoparticles include SiC, ZrO_2, ZnO, Al_2O_3, and TiO_2 (Hessen Agentur 2008; Gornicka et al. 2010; Srinivasan and Kumar 2016). Kotnarowska et al. (2014) analyzed the performance of the unmodified epoxy-polyurethane coating and epoxy-polyurethane coatings modified with silica or alumina nanoparticles over 3 years and found that the modified coatings indicated higher erosive wear resistance as shown in Fig. 2.6.

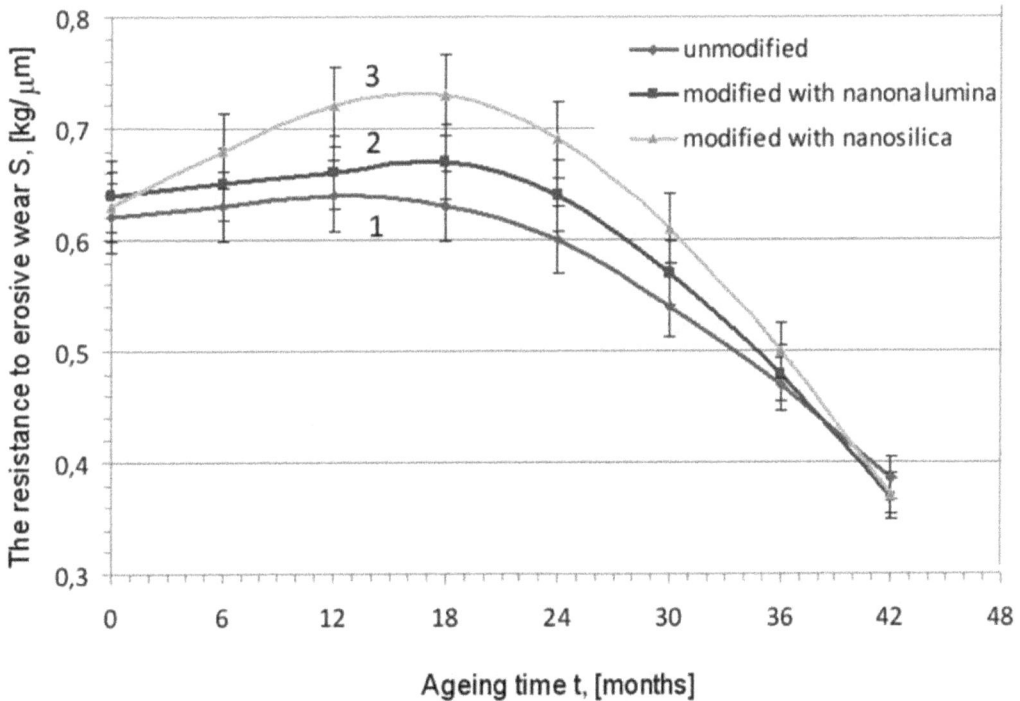

Figure 2.6   Representation of resistance to erosive wear of various coatings against aging time (Kotnarowska et al. 2014).

The performance of traditional coatings which is used to protect the surface against corrosion, degrades over a period. To overcome this problem, coatings with inhibitors such as cathodic inhibitors, anodic inhibitors and mixed inhibitors are used which could release the agent in the coating matrix. Such an active agent performs the self-repairing activity by nano passivation and is shown in Fig. 2.7.

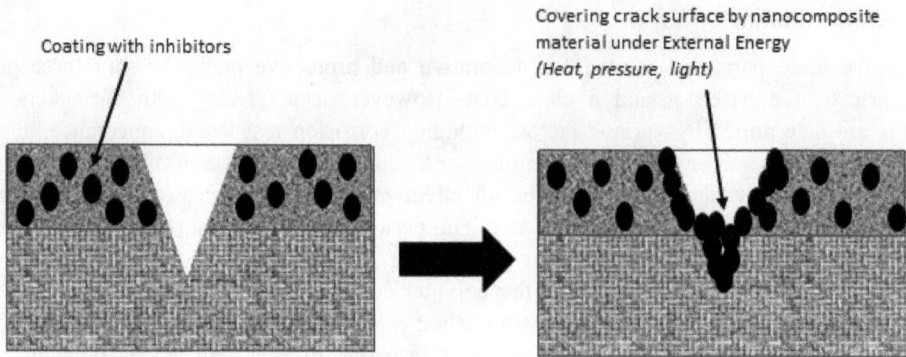

Coating with inhibitors

Covering crack surface by nanocomposite
material under External Energy
(Heat, pressure, light)

Figure 2.7 Self-repairing paint coating (After Shchukin and Möhwald 2007).

3.1.2 Lightweight Body Parts

One of the much-concerned factors in the automotive industry is the weight of the vehicle which plays an important role in fuel efficiency, CO_2 emission, stability, production cost, crash resistance and performance. The research by Coelho et al. (2012) and Goyal et al. (2014) showed that the reduction of weight by 10% increases the fuel efficiency by 7%. On the one hand, weight reduction achieves better fuel efficiency, and reduction in emission and cost while on the other hand, stability and crash resistance become poor.

The structural parts should possess good mechanical strength while the parts around the engine should have a high thermal resistance. CNTs are found to be a good replacement for steel in structural parts as they have very lightweight but are 150 times stronger than steel (Sequeira 2015). Clay nanocomposites with Polypropylene (PP), Polyamide (PA), Polybutylene terephthalate (PBT), and Polycarbonate (PC) have good flame retardance and thermal resistance and are used to make parts around the engine (Asmatulu et al. 2013; Sequeira 2015). Luo et al. (2014) investigated that the nanoparticles such as Y_2O_3 or Al_2O_3 when dispersed to magnesium and its alloy improved mechanical properties. Cu nanoparticles with magnesium enhanced 104% in 0.2% yield strength with a slight loss of ductility. Besides this, magnesium nanocomposites offer other properties such as high-temperature resistance, hardness, fatigue and wear resistance and are used in engine blocks and gear housings. Nanocomposites have reduced 900 kg of steel and other metal in a vehicle to 300 kg, offering 25% weight saving in polymers and 80% over steel (Coelho et al. 2012).

3.2 Interior Parts

3.2.1 Seat Protection

Car seats often come in contact with rainwater, snow and wet or dirty clothing causing water or dirt stains on the seats which cause customer dissatisfaction. Such unwanted items can be avoided or minimized by diffusing certain types of fluids including inorganic-organic hybrid materials based on aqueous or alcoholic solution into fabric and leather coverings. The impregnated layers have a hydrophobic and fat repellent effect which reduces the penetration rate of humidity by the entry of water and pollutants.

For leather seats, dispersion of aqueous microcapsules made up of polycarbonate which is only a few nanometers thick, penetrates the leather reaching different depths and produces diverse fragrances. The scent is placed inside the capsule which will burst out under mechanically stressed conditions, releasing the fragrance. Such capsules need to be small enough to penetrate the leather on one hand whereas, on the other hand, they have to be big enough to stick between the fibers. For the unused seats, the capsules may be intact, and the scent remains inside the capsule. Such a technique can also be applied to textiles.

3.2.2   Lighting and Displays

The interior electronic devices of a car must provide compatible interfaces to the cutting-edge technology as it is the techno hub of individual passengers. Interior lighting design will influence the psychology of the people inside the car. So, lighting and displays are considered key elements for interior design.

Organic Light-Emitting Diode (OLED) electroluminescent foils and transparent electrodes provide freedom on display geometries and ambient light design in modern cars. The report suggested that a 3D effect instrumental panel with transparent displays has been introduced in a car. Quantum dot displays made up of flexible materials offer higher efficiency and color variations than LCD which has been using in commercial televisions since 2015. Carbon nanotube (CNT) inks can be used to print electronics to make cost-efficient displays.

With the use of nanocomposites, the screen of the rear-view camera will be able to integrate into the rear mirror. The driver's health condition can also be monitored and improved with the application of nano-elements such as antiallergic or smell filtering, local flexible heating elements and sensors.

3.3   Combustion Engine

The efficiency of the engine plays a vital role in the better performance of the vehicle and the engine efficiency is determined by the functions of the engine oil, radiator and coolant. Ten to 15% of the engine energy is consumed by the friction of the moving parts such as piston, cylinder wall, connecting rod, crankshaft and bearing, camshaft and valves. The cooling of the engine is another problem as the cooling system consumes energy from the engine. The application of nanocomposites to the engine oil to reduce friction and radiator and coolant to reduce the heat generation, and offer better functions (Srinivasan and Kumar 2016).

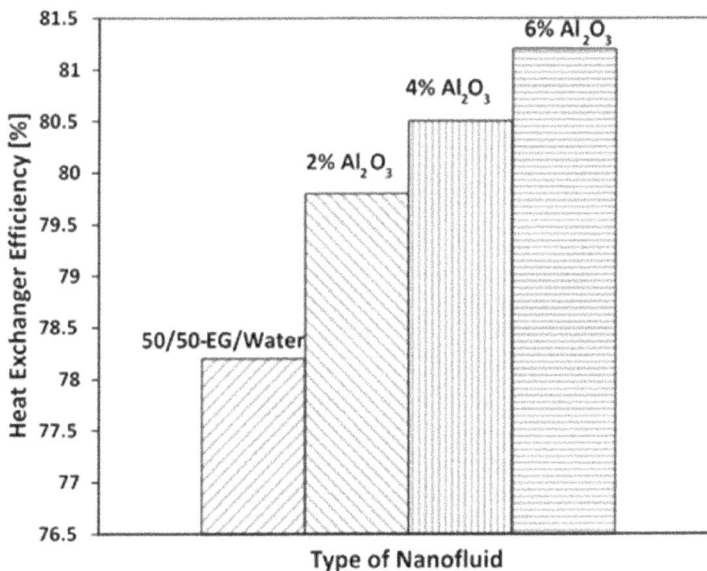

Figure 2.8   Illustration of the effect of various concentrations of Al_2O_3 nanofluid on the heat exchange the efficiency of heat recovery of an engine (Kulkarni et al. 2008).

Coating of the cylinder wall and aluminum crankshaft with nanocrystalline materials such as boride and iron carbide with the size of 50 nm to 120 nm result in an extremely hard surface with very low friction and abrasion (Hessen Agentur 2008). Lubricants that reduce friction between moving parts are made up of base oils such as petroleum oil, mineral oil, silicones, esters, etc. and some additives are added to enhance characteristic properties. Gornickaa et al. (2010), Shahnazar et al. (2016), and Rajendhran et al. (2018) studied the effect of nanoparticle additives to the lubricants such as CuO_2, TiO_2, MoS_2 nanosheets, and nano diamond and it was found that the Coefficient Of Friction (COF) of lubricant with TiO_2 nanoparticles was reduced by 52% for a load of 4 kg due to the rolling action of sphere-shaped nanoparticles.

Coolants such as water, ethylene glycol and mineral oil are the heat transfer fluids that run through the radiator and engine to remove heat generated in the engine. When nanoparticles like CuO, alumina, carbon nanotube, silica and titanium oxide are dispersed into the carrier liquid, the heat transfer rate increases significantly as compared to the carrier liquid alone (Satyamkumar et al. 2015). The experimental studies (Kole and Dey 2010; Satyamkumar et al. 2015) showed that the increase in the concentration of Al_2O_3 nanoparticles in water coolant will increase the heat exchange efficiency of the engine, shown in Fig. 2.8.

Furthermore, another study revealed that 0.4 vol.% of SiO_2 water-based coolant increased the heat transfer rate by 9.3% than without nanoparticles (Peyghambarzadeh et al. 2011). Table 2.2. displays the comparison of cooling performance of various nanofluids.

The addition of nanofluids not only improves the heat transfer rate but also reduces pollutant emissions. The experimental study found that the emission rates of NO_x and CO were reduced by 13 and 20.5% respectively by adding silver nanoparticles to pure diesel fuel (Soukht Saraee et al. 2015). Another study revealed that the smoke concentrations can be significantly reduced when aluminum nanoparticles are mixed with diesel fuel (Mitchell et al. 2008).

Table 2.2 Experimental studies of nanofluids for vehicle system cooling (Shafique and Luo 2019)

Nano fluid types	Advantages	References
Al_2O_3-EG	Enhances thermal conductivity by about 0.5% with the addition of AL_2O_3 nanoparticles (1.5 vol.%).	(Kole and Dey 2010)
Nanodiamond-engine oil	Enhances the engine performance by increasing the engine power by about 1.15% and reducing the fuel consumption by about 1.27% compared to simple engine oil.	(Liu et al. 2009)
Al_2O_3-water, Al_2O_3-EG, Al_2O_3-EG/water (5–20 vol.% of EG)	Heat transfer performance was enhanced by about 40% with the addition of 1.0 vol.% of nanoparticles of Al_2O_3 compared to the pure fluid.	(Peyghambarzadeh et al. 2011)
Al_2O_3-water	The maximum improvements of coolant heat transfer coefficient, heat transfer rate, and Nusselt number were 14.7, 14.8, and 9.5%, respectively.	(Ali et al. 2014)
CuO-water, Fe_2O_3-water	0.65 vol.% CuO-water nanoparticles enhanced the heat transfer coefficient by up to 9%.	(Peyghambarzadeh et al. 2013)
CuO-water	CuO-water is beneficial to improve the overall heat transfer coefficient. With a 0.4 vol.% CuO concentration of nanofluid the heat transfer coefficient was enhanced about 8% as compared to pure water.	(Naraki et al. 2013)
SiO_2-water, TiO_2-water	The maximum Nusselt number improvements for SiO_2 and TiO_2 nanofluids were 22.5 and 11% respectively.	(Hussein et al. 2014)
SiO_2-water	With 0.4 vol.% of SiO_2 nanoparticles at $60°C$ the heat transfer enhancement was about 9.3% as compared to the pure fluid.	(Ebrahimi et al. 2014)

3.4 Chassis and Tires

Rubber materials play an important role in the performance of tire cover that hits the road surface. For tire optimization, it should fulfill some contradictory requirements. Tires should have a good grip on the road while at the same time, rolling resistance has to be low. Furthermore, it needs to possess abrasion resistance, resistance against tear and tears propagation and slip-proof. Such unique properties can be achieved by highly complex chemical and physical interactions between reinforcing filler and rubber materials where filler materials comprise 30% of the cover. The size and form of the fillers along with chemical bonding to the rubber material determines the properties of the tire.

Usually, there are three types of filler materials: soot, silica and organosilane that improve the properties of rubber material significantly. Soot and silica, having a dimension in the nanometer range, are bonded chemically by organosilica with the rubber. The nanostructured soot particles have a coarser surface which provides excellent abrasion resistance in tires along with higher fuel efficiency and prolonged durability due to which it prevails in utility sports vehicles whereas silica is preferred more in passenger cars. The high surface energy possessed by the nanoparticles will provide good bonding with natural rubber that helps to reduce inner friction and consequently to better rolling resistance. At the same time, it reduces strain vibrations within the material during high speeds and ultimately provides superior traction on wet roads.

Driving comfort is one of the key factors to convince customers to use the car. It can be assisted by providing advanced damping systems that can be adjusted based on road conditions and driving situations. They use magneto or electrorheological fluids, whose viscosity can be controlled by an electric or magnetic field with time constants just in a few milliseconds. Such a system is compact and simple to be built and offers weight reduction as well.

3.5 Brakes

The disk brake and drum brake surfaces need to be hard and wear-resistant for their prolonged use. The application of CNT nanoparticles dispersed in the aluminum A356 matrix enhances the properties of brake material. The experiment performed by Sundaram M. and Mahamane U. has revealed that the increase in mass fraction of CNT nanoparticles in aluminum A356 leads to the improvement in hardness and wear resistance reducing the friction coefficient of the material (Meenakshi and Mahamani 2015).

3.6 Ultra-reflecting Mirrors and Glasses

The mirrors and headlights of the automobiles are made up of glass and polymer materials. When the sunlight falls on our eyes through the glasses or the glare of the lights from the other vehicle while driving, it causes difficulty or discomfort in driving and can lead to an accident. In the last few years, superior coating with nanomaterials in the mirror and glasses has been developed to cope with this problem.

The coating of the ultra-thin reflective layer of aluminum oxide, having a thickness less than 100 nm to the surface of mirrors, glasses, and headlamps of a vehicle (Goyal et al. 2014; Suresh et al. 2016) provides fat, dirt and water repellent features. These nanoparticle layers so-called hydrophobic and oleophobic are deposited by the chemical vapor deposition (CVD) method (Suresh et al. 2016). Figure 2.9 shows the mirror with and without nanoparticle coating.

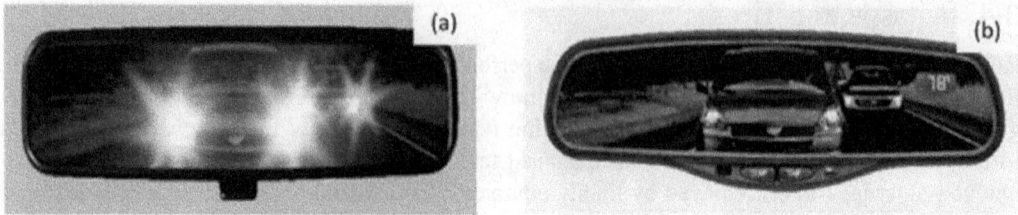

Figure 2.9   Representing the surface properties on glass plates in (a) conventional mirror (untreated surfaces) and (b) modern antiglare mirror (hydrophobic) (Mohseni et al. 2012).

3.7   Vehicle Indoor Environment Safety

The indoor environment of the vehicle needs to be clean and safe for the comfort and healthy life of the passenger and driver because various bacteria and microbes may be present inside the vehicle. Such a safe and clean indoor environment can be achieved using nano-agents such as gold, titanium oxide, silver, liposomes loaded with nanoparticles, titania nanotubes and copper (Charpentier et al. 2012; Dakal et al. 2016; Wang et al. 2016). For instance, gold and silver nanoparticle-based antimicrobial agents are positively charged biocidal which interact with the negatively charged cell membrane of the microorganism and are destroyed. Also, silver nanoparticles are biocompatible and function as antibacterial agents (Seubert et al. 2012, Wang et al. 2016). Similarly, air filters with nanofibers are an effective way to supply high-quality air inside the vehicle (Suresh et al. 2016).

Several studies have shown that nanomaterials such as CNTs and silver nanoparticles as filler in automotive fabrics, act as flame retardant agents for safety during accidents (Wang et al. 2006; Liu et al. 2009; Charpentier et al. 2012; Dakal et al. 2016; Wang et al. 2016). Thus, the application of proper nanomaterials inside the automobile enhances safety and cleanliness.

4   OUTLOOK ON AUTONOMOUS DRIVING

Today, car manufacturers are competing in the world for autonomous driving which can increase road safety and drivers' convenience. For safety and comfort, sensor systems and signal processing units need to be integrated into the vehicle which can identify the vehicle condition and surrounding situation such as obstacle detection, faster decision-making, self-learning capabilities, cruise control. The data from the sensors such as cameras, ultrasound, radar and lidar are intelligently combined and analyzed. The situation is then responded by activating actuators.

With the increasing number of driving conditions to be included, the data to be processed will increase exponentially, whereas the reaction time to the situation should be the minimum possible. To manage such huge data with reliability, new components, design, materials and communication systems have been developed. To address these concerns, automakers have come up with nanotechnology using nanocomposite materials in sensors, and processing units.

A higher level of autonomous driving will require faster and precise sensors, sophisticated systems for obstacle detection, lane assist, cruise control, faster decision-making, autonomously braking the vehicle if required, self-learning capabilities, and much more even with high reliability and fail-safe functionality. For the real-time solutions of these things to be performed, computing performance on board should be increased which calls for data security or deep learning IC chips. And computing power will affect the vehicle driving range due to which energy-efficient systems need to be developed. Till now, most of the automakers have been able to develop semiautonomous vehicles and for fully autonomous, research and development work is going on rapidly.

5 BENEFITS OFFERED BY NANOCOMPOSITE IN THE AUTOMOTIVE INDUSTRY

Nanocomposites have exhibited many impressive characteristics in terms of mechanical, electrical and thermal which will inevitably be a pacemaker for the automotive industry. Besides progressive designs, higher comfort and safety and innovative communication systems in an affordable manner, nanocomposites are expected to contribute significantly to environmental and climate protection by saving raw materials, and energy and water as well as by reducing greenhouse gases and hazardous wastes to sustain the environment. The main benefits are:

(a) Lightweight materials will reduce the overall vehicle weight improving vehicle performance.
(b) Saving fuel consumption by improving the combustion process and a higher level of controls with more mechatronics solutions.
(c) Reducing vehicle emissions by improving combustion engines, exhaust filters, catalysts, energy harvesting and energy recuperation.
(d) Development of electric vehicle and energy storage devices which are considered zero-emission. Polymer nanocomposites such as nanoclay incorporated and polyamide-based nanocomposites reduce water absorption in a polymer from the water-laden atmosphere, thus preserving properties.
(e) Improvement of traffic surveillance, driver assistance systems and durable nanocomposites will reduce traffic accidents or the accidents caused by parts failure.
(f) Sustainable products that will reduce cost and time for repair, maintenance or replacement.

6 ENVIRONMENTAL HEALTH AND SAFETY CONCERNS

Nanocomposites have spread widely in the world due to their various advantages in the science and engineering field, including the automotive industry for the improvement of vehicle performance, safety and comfort. However, there are big chances of nanomaterial exposure to the manufacturing workers and the environment (Valsami-Jones and Lynch 2015) which harm humans, animals and the environment. According to a study (Roco 2011), around 6 million people are anticipated to be exposed to nanoparticles in 2020. Some of the nanocomposites have been identified as hazardous to health and the environment including TiO_2, carbon-containing nanomaterials (Roco 2011; Valsami-Jones and Lynch 2015) Cu and ZnO nanoparticles, etc. (Bondarenko et al. 2013). However, the effect of most of the nanocomposites is still unknown and hence, more attention should be given to the careful selection of nanomaterials regarding the safety of nanocomposites applications and their long-term consequences.

6.1 Environmental Health

Nanoparticles are very small and more active so that they can easily be inhaled or enter the cells through other mediums such as skin and eyes. They can cause toxic effects such as inflammation, DNA damage, immune toxicity which can further be extended to immunosuppression, damage to the respiratory system. Some of such toxic nanoparticles include TiO_2, CuO nanoparticles, carbon nanotubes, SiO_2, etc. Table 2.3 shows the nanoparticles and their toxic effects.

Hansen et al. (2008) studied the potential risk of exposure of several materials. The study showed that unclassified products are exposed more than any others which means a lack of knowledge about nanomaterial exposure can be health hazardous for consumers. It is still a challenge to control the limits of nanoparticle exposure to human health and other living organisms due to the difficulty to understand the physio-chemical parameters of those nanoparticles.

Table 2.3   The toxicity of the various nanomaterials (Shafique and Luo 2019)

Nanomaterials	Toxic Effects	References
Carbon nanotubes	Antibacterial, Damage of cell membrane, necrosis/ apoptosis, Hinder the respiratory functions, DNA damage, Induce granulomas and atherosclerotic lesion, Lung damage	(Kato et al. 2009; Wang et al. 2010a; Bondarenko et al. 2013)
SiO_2	The slightly toxic effect, Toxic to marine algae, Apoptosis, Up-regulation of tumor necrosis factor-alpha genes, Inflammatory and immune responses	(Lu et al. 2008; Sergent et al. 2012; Bondarenko et al. 2013)
C_{60} derivatives	Bactericidal for Gram-positive bacteria, Oxidative cytotoxicity, Accumulation in the liver, Induces gliomas, sarcomas in mice as well as in human cells	(Wang et al. 2010a; Yu et al. 2012)
TiO_2	Growth inhibition and acute lethality, Bactericidal for gram-positive bacteria, Elimination of photosynthetic activity, Oxidative damage due to ROS, Liver damage	(Yang et al. 2002; Rossi et al. 2010; Jin et al. 2011)
CuO nanoparticles	Freshwater algae toxicity, Yeast toxicity, Damaging DNA, Acute toxicity to kidney, spleen, and liver	(Yu et al. 2007; Midander et al. 2009; Ivask et al. 2014; Hong and Zhang 2016; Zheng et al. 2016)

On the other hand, polymer-based nanocomposites do not last for a very long period. While replacing metals and metal alloys with polymers reinforced with nanoparticles can be beneficial to the environment.

6.2   Safety Concerns

Nanocomposites are used in the automotive industry to enhance the materials' functions and their durability. However, when toxic nanocomposites are exposed to human health and the environment, they will adversely affect them. Thus, there are significant safety concerns to the use of nanocomposites. Many researchers have considered the issue that the potential threat by nanoparticles may change during their Life Cycle Assessments (LCA) (Hansen et al. 2008; Nel et al. 2011; Hendren et al. 2015; Erdely et al. 2016) due to their physicochemical properties. For instance, nanoforms can exhibit different hazard behavior leading to different health concerns. Other methods have also been implemented for risk assessments such as standardized testing, benchmarking of materials and in silico approaches (Oomen et al. 2018).

To characterize and track nanoparticles in the environment, there needs to be the formulation of nanoparticles database to investigate LCA of nanoproducts; development of standardized protocols for handling nanomaterials in the workplace, and the risk assessment methods to enhance their overall safety (Sellers et al. 2015; Oomen et al. 2018). At the national and international levels, a systematic strategy framework should be developed to prevent risks associated with each nanomaterial for both workers and consumers (Türk et al. 2008; Cinelli et al. 2016; Glisovic et al. 2017).

7   FUTURE TREND OF NANOCOMPOSITES IN THE AUTOMOTIVE INDUSTRY

The automotive industry is investing a large portion of its revenue in research and development for the improvement of vehicle performance by reducing weight, fuel consumption and greenhouse gas emission. One way to achieve their target can be reached by replacing conventional materials with nanocomposites. The properties of nanocomposites, processes and products possess a multitude of the governments from many countries such as China, Japan, Vietnam and Indonesia have implemented various policies regarding emission standards due to the growing number of vehicles. Car manufacturers believe that replacing existing material with nanocomposites is one of the most effective solutions to reduce vehicle weight for ensuring compliance and reducing emissions.

According to the studies of the global vehicle market, automobiles are projected to grow by 5.3% from 2018 to 2023. The increase in vehicle production also demands improvement in ride quality, advanced systems of safety and comfort.

The European Commission has set a new standard for fuel economy, which require that passenger cars must meet the CO_2 emission target of 95 g/km (equivalent to 57.9 mpg) by 2021, and to produce light vehicles, is 147 g/km (equivalent to 43.3 mpg) by 2020 years. There is a possibility of growth of automotive composites for a variety of applications, parts replacement or improvement for the body, exterior, interior, chassis and others. Automotive composites of the European market are anticipated to reach US$ 4.1 billion by 2021 and will grow by 5.8% from 2016 to 2021 (Growth Opportunities for the European Automotive Composites Markets 2017).

Nanocomposites in the automotive industry manifest extremely attractive characteristics for safer and sustainable use, and good quality to ride, however, there are still research gaps and opportunities for future work in some areas including multifunctional nanomaterials, cost-effective nanomaterials, appropriate regulatory framework and environmentally friendly nanocomposites. Addressing those research gaps will help for the sustainable use of nanocomposites in the automotive industry, producing vehicles of the next level.

8 SUMMARY

The applications of nanocomposites are expanding in broad fields due to their unique properties such as huge surface area, large surface energy and very low weight to volume ratio that are different than those of bulk materials. There are different categories of nanocomposites such as PMNCs, MMNCs and CMNCs which have their own properties and application areas. Nanocomposites are used as nanofilms, nanoflakes, nanotubes, nanofibers, nanoparticles and offer improved mechanical, electrical and thermal properties, wear and corrosion resistance, flame retardance, self-cleaning and sensing abilities.

One of the major fields, taking advantage of nanocomposite materials is the automotive industry. The main catalysts for the high demand of nanocomposites in auto parts production are their lightweight and strength. When they are dispersed in matrix components with the proper amount, they exhibit excellent structural and operational functions and can replace metal parts, reducing the weight of the parts and the overall vehicle. The weight reduction improves fuel efficiency and performance along with the reduction of emission of hydrocarbons, CO and nitrogen oxides. Starting from polymer nanocomposites for the limited automotive parts production a few decades ago, now many automakers are competitively using variety of nanocomposites for a wide range of automotive parts manufacturing such as brakes, engine oil, tires, glasses and mirrors, seat and door trims, engine covers, coatings, etc. In the coming years, it is expected that the to increase the demand of variety of nanocomposites in the automotive industry.

Nanocomposites offer numerous benefits over a wide range of applications; however, attention should be paid to their impact on health and the environment for sustainable use. The toxic nanocomposites can cause various health hazards to humans as well as other living creatures. Thus, proper recycling methods and handling techniques of applicable nanocomposites should be developed to mitigate their effect on health and the environment.

References

Akafuah, N., S. Poozesh, A. Salaimeh, G. Patrick, K. Lawler and K. Saito. 2016. Evolution of the automotive body coating process—A review. Coatings 6: 1–24.

Ali, M., A.M. El-Leathy and Z. Al-Sofyany. 2014. The effect of nanofluid concentration on the cooling system of vehicles radiator. Advances in Mechanical Engineering 6: 1–13.

Asmatulu, R., P. Nguyen and E. Asmatulu. 2013. Nanotechnology safety in the automotive industry. pp. 57–72. In: R. Asmatulu (ed.). Nanotechnology Safety. Elsevier Publications, CA, USA.

Ben Hargreaves. 2020. What are Polymer Nanocomposites? https://coventivecomposites.com/explainers/what-are-polymer-nanocomposites/

Bhaduri, S. and S.B. Bhaduri. 1998. Recent developments in ceramic nanocomposites. Journal of Materials (Minerals, Metals & Materials Society) 50: 44–51.

Bondarenko, O., K. Juganson, A. Ivask, K. Kasemets, M. Mortimer and A. Kahru. 2013. Toxicity of Ag, CuO and ZnO nanoparticles to selected environmentally relevant test organisms and mammalian cells in vitro: A critical review. Archives of Toxicology 87: 1181–1200.

Charpentier, P.A., K. Burgess, L. Wang, R.R. Chowdhury, A.F. Lotus and G. Moula. 2012. Nano-TiO_2/Polyurethane composites for antibacterial and self-cleaning coatings. Nanotechnology 23.

Cinelli, M., S.R. Coles, O. Sadik, B. Karn and K. Kirwan. 2016. A framework of criteria for the sustainability assessment of nanoproducts. Journal of Cleaner Production 126: 277–287.

Coelho, M.C., G. Torrão, N. Emami and J. Gŕcio. 2012. Nanotechnology in automotive industry: Research strategy and trends for the future—Small objects, big impacts. Journal of Nanoscience and Nanotechnology 12: 6621–6630.

Dakal, T.C., A. Kumar, R.S. Majumdar and V. Yadav. 2016. Mechanistic basis of antimicrobial actions of silver nanoparticles. Frontiers in Microbiology 7.

Ebrahimi, M., M. Farhadi, K. Sedighi and S. Akbarzade. 2014. Experimental investigation of force convection heat transfer in a car radiator filled with SiO_2-water nanofluid. International Journal of Engineering, Transactions B: Applications 27: 333–340.

Erdely, A., M.M. Dahm, M.K. Schubauer-Berigan, B.T. Chen, J.M. Antonini and M.D. Hoover. 2016. Bridging the gap between exposure assessment and inhalation toxicology: Some insights from the carbon nanotube experience. Journal of Aerosol Science 99: 157–162.

Fisher, M., K. Ramdas and K. Ulrich. 1999. Component sharing in the management of product variety: A study of automotive braking systems. Management Science 45: 297–315.

Fukushima, Y. and S. Inagaki. 1987. Synthesis of an intercalated compound of montmorillonite and 6-polyamide. pp. 365–374. In: J.L. Atwood and J.E. Davies (eds). Inclusion Phenomena in Inorganic, Organic, and Organometallic Hosts. Netherlands: Springer.

Garcés, J.M., D.J. Moll, J. Bicerano, R. Fibiger and D.G. McLeod. 2000. Polymeric nanocomposites for automotive applications. Advanced Materials 12: 1835–1839.

Giannelis, E.P. 1998. Polymer-layered silicate nanocomposites: Synthesis, properties, and applications. Applied Organometallic Chemistry 12: 675–680.

Glisovic, S., D. Pesic, E. Stojiljkovic, T. Golubovic, D. Krstic, M. Prascevic, et al. 2017. Emerging technologies and safety concerns: A condensed review of environmental life cycle risks in the nano-world. International Journal of Environmental Science and Technology 14: 2301–2320.

Gornicka, B., M. Mazur, K. Sieradzka, E. Prociow and M. Lapinski. 2010. Antistatic properties of nanofilled coatings. Acta Physica Polonica A 117: 869–872.

Goyal, R., M. Sharma and U.K. Amberiya. 2014. Innovative nano composite materials and applications in automobiles. International Journal of Engineering Research & Technology 3: 3001–3009.

Growth Opportunities for the European Automotive Composites Markets. 2017. Reportlinker. https://www.prnewswire.com/news-releases/growth-opportunities-for-the-european-automotive-composites-market-300390444.html

Han, Y., Y. Dai, D. Shu, J. Wang and B. Sun. 2006. First-principles calculations on the stability of $AlTiB_2$ interface. Applied Physics Letters 89: 144107.

Hansen, S.F., E.S. Michelson, A. Kamper, P. Borling, F. Stuer-Lauridsen and A. Baun. 2008. Categorization framework to aid exposure assessment of nanomaterials in consumer products. Ecotoxicology, 17: 438–447.

Hegde, M., U. Lafont, B. Norder, S.J. Picken, E.T. Samulski, M. Rubinstein and T. Dingemans. 2013. SWCNT induced crystallization in an amorphous all-aromatic poly(ether imide). Macromolecules 46: 1492–1503.

Hendren, C.O., G.V. Lowry, J.M. Unrine and M.R. Wiesner. 2015. A functional assay-based strategy for nanomaterial risk forecasting. Science of The Total Environment 536: 1029–1037.

Hessen Agentur 2008. Nanotechnologies in Automobiles – Innovation Potentials in Hesse for the Automotive Industry and its Subcontractors. pp. 1–56.

Hong, J. and Y.-Q. Zhang. 2016. Murine liver damage caused by exposure to nano-titanium dioxide. Nanotechnology 27: 112001.

Hussein, A.M., R.A Bakar, K. Kadirgama and K.V. Sharma. 2014. Heat transfer enhancement using nanofluids in an automotive cooling system. International Communications in Heat and Mass Transfer 53: 195–202.

Ivask, A., I. Kurvet, K. Kasemets, I. Blinova, V. Aruoja, S. Suppi, et al. 2014. Size-dependent toxicity of silver nanoparticles to bacteria, yeast, algae, crustaceans and mammalian cells *in vitro*. PLoS ONE 9: e102108.

Jin, C., Y. Tang, F.G. Yang, X.L. Li, S. Xu, X.Y. Fan, et al. 2011. Cellular toxicity of TiO_2 nanoparticles in anatase and rutile crystal phase. Biological Trace Element Research 141: 3–15.

Kalaitzidou, K., H. Fukushima and L.T. Drzal. 2007. A new compounding method for exfoliated graphite–polypropylene nanocomposites with enhanced flexural properties and lower percolation threshold. Composites Science and Technology 67: 2045–2051.

Kamigaito, O. 1991. What can be improved by nanometer composites? Journal of the Japan Society of Powder and Powder Metallurgy 38: 315–321.

Kato, S., H. Aoshima, Y. Saitoh and N. Miwa. 2009. Biological safety of lipofullerene composed of squalane and fullerene-C60 upon mutagenesis, photocytotoxicity, and permeability into the human skin tissue. Basic & Clinical Pharmacology & Toxicology 104: 483–487.

Kole, M. and T.K. Dey. 2010. Viscosity of alumina nanoparticles dispersed in car engine coolant. Experimental Thermal and Fluid Science 34: 677–683.

Kotnarowska, D., M. Przerwa and T. Szumiata. 2014. Resistance to erosive wear of epoxy-polyurethane coating modified with nanofillers. Journal of Materials Science Research 3: 52–58.

Kulkarni, D.P., R.S. Vajjha, D.K. Das and D. Oliva. 2008. Application of aluminum oxide nanofluids in diesel electric generator as jacket water coolant. Applied Thermal Engineering 28: 1774–1781.

Liu, Z., G. Ren, T. Zhang and Z. Yang. 2009. Action potential changes associated with the inhibitory effects on voltage-gated sodium current of hippocampal CA1 neurons by silver nanoparticles. Toxicology 264: 179–184.

Liu, P., Z. Jin, G. Katsukis, L.W. Drahushuk, S. Shimizu, C.-J. Shih, et al. 2016. Layered and scrolled nanocomposites with aligned semi-infinite graphene inclusions at the platelet limit. Science 353: 364–367.

Lu, Z., C.M. Li, H. Bao, Y. Qiao, Y. Toh and X. Yang. 2008. Mechanism of antimicrobial activity of CdTe quantum dots. Langmuir 24: 5445–5452.

Luo, T., X. Wei, X. Huang, L. Huang and F. Yang. 2014. Tribological properties of Al_2O_3 nanoparticles as lubricating oil additives. Ceramics International 40: 7143–7149.

Meenakshi S.U. and A. Mahamani. 2015. Development of carbon nanotube reinforced aluminum matrix composite brake drum for automotive applications. pp. 67–83. *In:* Dr. Brahim Attaf (ed.). Research and Innovation in Carbon Nanotube-Based Composites. The World Academic Publishing Co. Ltd., Admiralty, Hong Kong.

Midander, K., P. Cronholm, H.L. Karlsson, K. Elihn, L. Möller, C. Leygraf, et al. 2009. Surface characteristics, copper release, and toxicity of nano- and micrometer-sized copper and copper(II) oxide particles: A cross-disciplinary study. Small 5: 389–399.

Mitchell, M.R., R.E. Link, M.-J Kao, C.-C. Ting, B.-F. Lin and T.-T. Tsung. 2008. Aqueous aluminum nanofluid combustion in diesel fuel. Journal of Testing and Evaluation, 36: 100579.

Mohseni, M., B. Ramezanzadeh, H. Yari and M. Moazzami. 2012. The role of nanotechnology in automotive industries. pp. 3–54. *In:* J. Carmo (ed.). New Advances in Vehicular Technology and Automotive Engineering. InTech, Janeza Trdine 9, 51000 Rijeka, Croatia.

Mutale, C.T., W.J. Krafick and D.A. Weirauch. 2010. Direct observation of wetting and spreading of molten aluminum on TiB_2 in the presence of a molten flux from the aluminum melting point up to 1033 K (760°C). Metallurgical and Materials Transactions B 41: 1368–1374.

Naebe, M., J. Wang, A. Amini, H. Khayyam, N. Hameed, L.H. Li, 2015. Mechanical property and structure of covalent functionalised graphene/epoxy nanocomposites. Scientific Reports 4: 4375.

Naraki, M., S.M. Peyghambarzadeh, S.H. Hashemabadi and Y. Vermahmoudi. 2013. Parametric study of overall heat transfer coefficient of CuO/water nanofluids in a car radiator. International Journal of Thermal Sciences 66: 82–90.

Naskar, A.K., J.K. Keum and R.G. Boeman. 2016. Polymer matrix nanocomposites for automotive structural components. Nature Nanotechnology 11: 1026–1030.

Nel, A., D. Grainger, P.J. Alvarez, S. Badesha, V. Castranova, M. Ferrari, et al. 2011. Nanotechnology environmental, health, and safety issues. pp. 159–220. *In:* Nanotechnology Research Directions for Societal Needs in 2020. Netherlands: Springer.

Okada, A. and A. Usuki. 1995. The chemistry of polymer-clay hybrids. Materials Science and Engineering C 3: 109–115.

Oomen, A.G., K.G. Steinhäuser, E.A.J. Bleeker, F. van Broekhuizen, A. Sips, S. Dekkers, et al. 2018. Risk assessment frameworks for nanomaterials: Scope, link to regulations, applicability, and outline for future directions in view of needed increase in efficiency. NanoImpact 9: 1–13.

Pandey, J.K., K.R. Reddy, A.K. Mohanty and M. Misra. 2014. Handbook of Polymernanocomposites. Processing, Performance and Application. Springer Berlin Heidelberg: Germany.

Park, C., K.E. Wise, J.H. Kang, J. Kim, G. Sauti, S.E. Lowther, et al. 2008. Multifunctional nanotube polymer nanocomposites for aerospace applications: Adhesion between swcnt and polymer matrix. 2008 Annual Meeting of The Adhesion Society: 1–3.

Peyghambarzadeh, S.M., S.H. Hashemabadi, S.M. Hoseini and M. Seifi Jamnani. 2011. Experimental study of heat transfer enhancement using water/ethylene glycol based nanofluids as a new coolant for car radiators. International Communications in Heat and Mass Transfer 38: 1283–1290.

Peyghambarzadeh, S.M., S.H. Hashemabadi, M. Naraki and Y. Vermahmoudi. 2013. Experimental study of overall heat transfer coefficient in the application of dilute nanofluids in the car radiator. Applied Thermal Engineering 52: 8–16.

Privalko, V.P., T.A. Shantalii and E.G. Privalko. 2005. Polyimides Reinforced by a sol-gel derived organosilicon nanophase: Synthesis and structure-property relationships. pp. 63–76. *In:* Z.Z. Klaus Friedrich and Stoyko Fakirov (eds). Polymer Composites. Springer-Verlag, Boston: USA.

Rafiee, M.A., J. Rafiee, Z. Wang, H. Song, Z.-Z. Yu and N. Koratkar. 2009. Enhanced mechanical properties of nanocomposites at low graphene content. ACS Nano 3: 3884–3890.

Rajan, T.P.D., R.M. Pillai and B.C. Pai. 1998. Reinforcement coatings and interfaces in aluminium metal matrix composites. Journal of Materials Science 33: 3491–3503.

Rajendhran, N., S. Palanisamy, P. Periyasamy and R. Venkatachalam. 2018. Enhancing of the tribological characteristics of the lubricant oils using Ni-promoted MoS_2 nanosheets as nano-additives. Tribology International 118: 314–328.

Ramanathan, T., A.A. Abdala, S. Stankovich, D.A. Dikin, M. Herrera-Alonso, R.D. Piner, et al. 2008. Functionalized graphene sheets for polymer nanocomposites. Nature Nanotechnology 3: 327–331.

Ravi, K.R., J. Nampoothiri and B. Raj 2017. Nanocomposites: A gaze through their applications in transport industry. pp. 831–856. *In:* B. Raj, M.V. de Voorde and Y. Mahajan (eds). Nanotechnology for Energy Sustainability. Weinheim, Germany: Wiley-VCH.

Roco, M.C. 2011. The long view of nanotechnology development: The National Nanotechnology Initiative at 10 years. Journal of Nanoparticle Research 13: 427–445.

Rossi, E.M., L. Pylkkänen, A.J. Koivisto, H. Nykäsenoja, H. Wolff, K. Savolainen, et al. 2010. Inhalation exposure to nanosized and fine TiO_2 particles inhibits features of allergic asthma in a murine model. Particle and Fibre Toxicology 7: 1–35.

Satyamkumar, G., S. Brijrajsinh, M. Sulay, T. Ankur and R. Manoj. 2015. Analysis of radiator with different types of nano fluids. Journal of Engineering Research and Studies 6: 1–2.

Sellers, K., N.M.E. Deleebeeck, M. Messiaen, M. Jackson, E.A.J. Bleeker, D.T.H.M. Sijm, et al. 2015. Grouping nanomaterials: A strategy towards grouping and read-across. RIVM Report 2015-0061, Arcadis, Netherlands.

Sequeira, S. 2015. Applications of Nanotechnology in Automobile Industry. Manipal Technologies Limited: 1–3.

Sergent, J.A., V. Paget and S. Chevillard. 2012. Toxicity and genotoxicity of nano-SiO_2 on human epithelial intestinal HT-29 cell line. The Annals of Occupational Hygiene 56: 622–630.

Serrano, E., G. Rus and J. García-Martínez. 2009. Nanotechnology for sustainable energy. Renewable and Sustainable Energy Reviews 13: 2373–2384.

Seubert, C., K. Nietering, M. Nichols, R. Wykoff and S. Bollin. 2012. An overview of the scratch resistance of automotive coatings: Exterior clearcoats and polycarbonate hardcoats. Coatings 2: 221–234.

Shafique, M. and X. Luo. 2019. Nanotechnology in transportation vehicles: An overview of its applications, environmental, health and safety concerns. Materials 12: 11–17.

Shahnazar, S., S. Bagheri and S.B. Abd Hamid. 2016. Enhancing lubricant properties by nanoparticle additives. International Journal of Hydrogen Energy 41: 3153–3170.

Shchukin, D.G. and H. Möhwald. 2007. Self-repairing coatings containing active nanoreservoirs. Small 3: 926–943.

Song, H.-J., Z.-Z. Zhang and X.-H. Men. 2007. Surface-modified carbon nanotubes and the effect of their addition on the tribological behavior of a polyurethane coating. European Polymer Journal 43: 4092–4102.

Soukht Saraee, H., S. Jafarmadar, H. Taghavifar and S.J. Ashrafi. 2015. Reduction of emissions and fuel consumption in a compression ignition engine using nanoparticles. International Journal of Environmental Science and Technology 12: 2245–2252.

Srinivasan, V. and P.G.S. Kumar. 2016. Review on nanoparticles in CI engines with a new and better proposal on stabilisation. International Journal of Innovative Research in Science, Engineering and Technology 5: 1656–1668.

Suresh, S., P. Saravanan, K. Jayamoorthy, S. Ananda Kumar and S. Karthikeyan. 2016. Development of silane grafted ZnO core shell nanoparticles loaded diglycidyl epoxy nanocomposites film for antimicrobial applications. Materials Science and Engineering: C 64 : 286–292.

Tjong, S.C. 2006. Structural and mechanical properties of polymer nanocomposites. Materials Science and Engineering: R: Reports 53: 73–197.

Türk, V., C. Kaiser and S. Schaller. 2008. Invisible but tangible? Societal opportunities and risks of nanotechnologies. Journal of Cleaner Production 16: 1006–1009.

Valsami-Jones, E. and I. Lynch. 2015. How safe are nanomaterials? Science 350: 388–389.

Vivek, P. and Yashwant. 2011. Polymer nanocomposites drive opportunities in the automotive sector. Nanotech Nol. Insights 2: 17–24.

Wang, Z., E. Han and W. Ke. 2006. Effect of nanoparticles on the improvement in fire-resistant and anti-ageing properties of flame-retardant coating. Surface and Coatings Technology 200: 5706–5716.

Wang, Z., K. Zhang, J. Zhao, X. Liu and B. Xing. 2010a. Adsorption and inhibition of butyrylcholinesterase by different engineered nanoparticles. Chemosphere 79: 86–92.

Wang, L., L. Weng, S. Song and Q. Sun. 2010b. Mechanical properties and microstructure of polyetheretherketone–hydroxyapatite nanocomposite materials. Materials Letters 64: 2201–2204.

Wang, J., C. Zhang, Z. Du, H. Li and W. Zou. 2016. Functionalization of MWCNTs with silver nanoparticles decorated polypyrrole and their application in antistatic and thermal conductive epoxy matrix nanocomposite. RSC Advances 6: 31782–31789.

Yang, X.L., C.H. Fan and H.S. Zhu 2002. Photo-induced cytotoxicity of malonic acid [C60]fullerene derivatives and its mechanism. Toxicology *in vitro* 16: 41–46.

Yu, T., K. Greish, L.D. McGill, A. Ray and H. Ghandehari. 2012. Influence of geometry, porosity, and surface characteristics of silica nanoparticles on acute toxicity: Their vasculature effect and tolerance threshold. ACS Nano 6: 2289–2301.

Yu, W.W., E. Chang, J.C. Falkner, J. Zhang, A.M. Al-Somali and C.M. Sayes, et al. 2007. Forming biocompatible and nonaggregated nanocrystals in water using amphiphilic polymers. Journal of the American Chemical Society 129: 2871–2879.

Zheng, X., Y. Su, Y. Chen, R. Wan, M. Li, H. Huang, et al. 2016. Carbon nanotubes affect the toxicity of CuO nanoparticles to denitrification in marine sediments by altering cellular internalization of nanoparticle. Scientific Reports 6: 27748.

Zhou, D., F. Qiu, H. Wang, and Q. Jiang, 2014. Manufacture of nano-sized particle-reinforced metal matrix composites: A review. Acta Metallurgica Sinica (English Letters) 27: 798–805.

Zhou, S., L. Wu, J. Sun and W. Shen. 2002. The change of the properties of acrylic-based polyurethane via addition of nano-silica. Progress in Organic Coatings 45: 33–42.

Chapter 3

Oxide based Nanocomposites for Photocatalytic Applications

T. Vijayaraghavan and Anuradha Ashok*

Functional Materials Laboratory,
PSG Institute of Advanced Studies, Coimbatore-641004, India
ragavan07@gmail.com
anu@psgias.ac.in

1   INTRODUCTION

The majority of current global energy production is by utilizing fossil fuels. Almost 90% of total energy is generated from fossil fuels. Forecast studies reveal that the consumption of energy will increase at a rate of 1.4% per year till 2035 (U.S. EIA, International energy outlook, 2011). Consequently, there will be a huge demand for fossil fuels such as coal and oil. Moreover, the extraction of energy from fossil fuels results in the generation of greenhouse gases such as carbon dioxide, methane and nitrous oxides. These gases create an imbalance in the ongoing carbon cycle in the atmosphere. This imbalance will produce several disastrous effects on the climate such as an increase in the sea level, global warming, etc. Reduction or hydrogenation of CO_2 into environment-friendly non-toxic products and hydrogen generation through water splitting and its utilization for clean energy production are some of the solutions to this problem. Similarly, organic dye removal from industrial wastewater from the textile industry is another major requirement to reduce water pollution. The photocatalytic approach is one of the best eco-friendly methods to address the above issues. This chapter explains the recent research, which has been carried out on various nanocomposites for photocatalytic reduction of CO_2 to useful products and hydrogen generation through water splitting.

*Corresponding Author

2   NEED FOR HYDROGEN ENERGY

The current use of renewable energy resources for energy production is limited due to their poor efficiency, complicated large-scale production, poor stability, etc. On the other hand, the continued increase in demand for oil may create economic disruptions, which could lead to political and military conflicts between nations. There is a dire need for alternative clean energy resources, energy and fuel production technologies and methods for environmental remediation to reduce various types of hazards posed by current industrial activities.

Hydrogen is one of the most powerful energy carriers for future energy generation technologies and it has a great potential to be used as fuel in the industrial and transport sectors. It is a colorless, odorless, and tasteless gas with the highest energy content per unit weight when compared to any other fuel. It is chemically very active and usually exists in combination with other elements such as oxygen in water, carbon in methane, etc. Therefore, it must be broken from its bond with other elements, in order to be used as a fuel.

Hydrogen gas was first isolated by Henry Cavendish in 1766 and it was recognized as a constituent of water by Lavoisier in 1783. The production of hydrogen and oxygen from the electrolysis of water was discovered by Nicholson and Carlisle.

2.1   Hydrogen Production Methods

Hydrogen can be produced by several methods. The following are some of them.

- Steam reforming and coal gasification using fossil fuels
- Electrolysis or photolysis using natural resources like sunlight and water.
- From biomass using micro-organisms

The schematic diagram explaining different hydrogen production methods is shown in Fig. 3.1. Among several methods, steam methane reforming is the most widely used technique for the generation of hydrogen. The efficiency of hydrogen production using this technique is almost 80–90%. Coal gasification is also another way for the production of hydrogen, which is slightly costlier than the steam methane reforming process. Since the availability of coal is vast in all regions of the world, it is predominantly used for electricity generation from thermal power plants and for the production of syn-gases. However, when carbon-based fossil fuels are reformed or gasified to produce hydrogen, the stream of pure carbon dioxide will be formed as a by-product. There are several emerging technologies to isolate this carbon dioxide from the atmosphere by sequestering it in an ocean or geological formations. However, the long-term effects of carbon sequestration and security are not known entirely.

Hydrogen production via biomass conversion involves heating of biomass such as craps, industrial residues and wastes, forestry residues & wastes and municipal wastes. Unfortunately, carbon mono oxide and carbon dioxide are the most common side products from these processes.

The energy production process can be called carbon-free if non-carbon sources such as wind, solar or nuclear power are used or any technology where there is no emission of carbon-based gases. Hydrogen production via electrolysis and/or photolysis will also not emit any carbon dioxide since there is no carbon involved in the splitting of water into hydrogen and oxygen. Among several hydrogen production methods, the photocatalytic approach of producing hydrogen by splitting of water using sunlight is one of the cleaner, non-hazardous, non-carbon emission methods. Several research groups are working to rectify the drawbacks of this method such as complicated synthesis methods, poor efficiencies, etc. The detailed review of the production of hydrogen through photocatalytic water splitting is explained later.

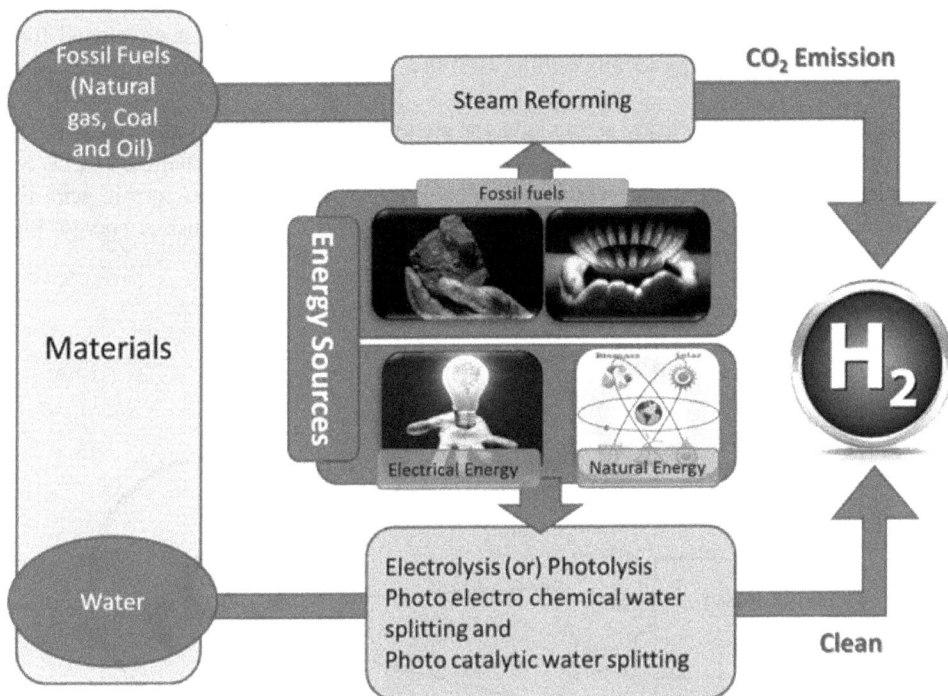

Figure 3.1   Hydrogen production methodologies.

3   NEED FOR CO_2 CONVERSION TO USEFUL FUELS

Extensive utilization of fossil fuels for energy generation and various other industrial and domestic activities not only lead to their faster depletion but also increasing CO_2 level in the atmosphere. Extensive efforts are under progress to minimize the environmental damage due to the increased level of CO_2 in the atmosphere. Most of the methods proposed on CO_2 reduction from fossil fuels such as thermal reforming, hydrothermal conversion, electrochemical reduction, chemical transformation, etc., consume a large amount of thermal/electrical energy for the breakage of the C=O bond. The hydrogenation of CO_2 by using photocatalysis is one of the sustainable methods, which is derived from nature. This will generate several useful products from CO_2 such as formic acid, carbon monoxide, formaldehyde, methanol and methane, based on their combustion-free energy. The detailed mechanism of photocatalytic hydrogenation of CO_2, existing photocatalysts, their advantages/limitations, and recent developments reported on such photocatalysts are discussed in this chapter.

The energy consumption for various industrial and domestic activities mostly depends on fossil fuels due to the fact that humankind has been facing serious problems such as increasing CO_2 levels in the atmosphere, global warming, etc. IPCC (Intergovernmental Panel on Climate Change) has predicted that the CO_2 level will increase upto 590 ppm and the global average temperature may increase upto 1.9 °C by the year 2100. CO_2 is one of the most thermodynamically stable carbon-based compounds. Reducing CO_2 into environmental friendly non-toxic products are one of the solutions to this problem. The methods proposed in recent research articles published on CO_2 reduction from fossil fuels such as thermal reforming, hydrothermal conversion, electrochemical reduction, chemical transformation, etc consume a large amount of thermal/electrical energy for the breakage of the C=O bond. The bond energy of C=O (750 kJ/mol) is higher than C–H (411 kJ/mol) and C–C (336 kJ/mol). So, it is necessary to combine thermal energy with other eco-friendly methods for the efficient hydrogenation of CO_2. Photothermo catalysis under sunlight irradiation is one of the sustainable methods. In this case, the photocatalyst developed for CO_2 reduction produces heat energy locally within itself thus enhancing the overall efficiency of photo conversion.

The reduction of CO_2 by using semiconductor photocatalysis under sunlight irradiation is one of the sustainable methods, which is derived from Mother Nature. The reduction of CO_2 by this way will lead to several products, which are valuable for mankind such as formic acid, carbon monoxide, formaldehyde, methanol and methane, based on their combustion-free energy. Fig. 3.2 shows the useful by-products derived from CO_2.

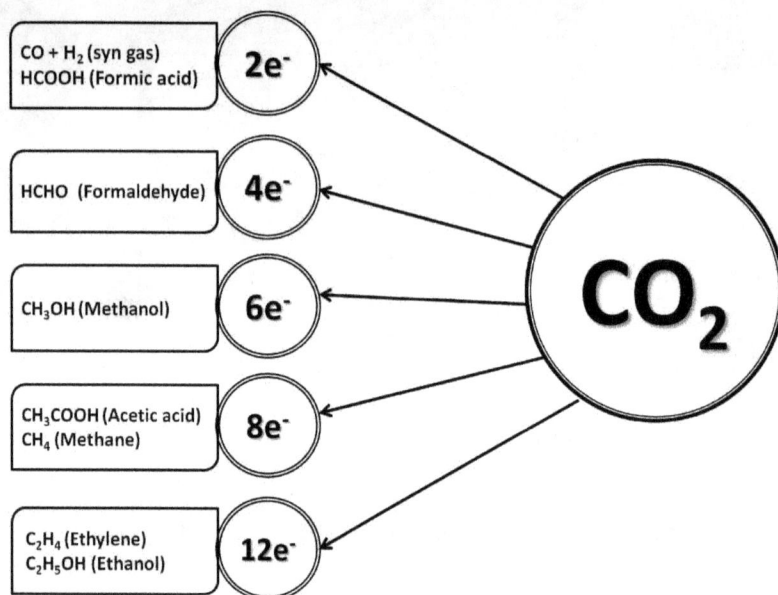

Figure 3.2   By-products derived from CO_2 and the number of proton-coupled electrons required for each of them (circled in green color).

4   NEED FOR TEXTILE DYE REMEDIATION

Dyes are organic compounds that are used in textile industries as coloring agents. After treatment, the unused dyes are discarded along with water, which causes contamination of not only the water body, which they are let out but also the surrounding water reservoirs. These dyes (especially, textile dyes) are engineered to be resistant to all kinds of treatments without fading. They need to sustain either alkaline or acidic environment; withstand washing with soaps and bleaching agents, microbiological fading, are resistant to light and ultraviolet irradiation, etc. Obviously, the better stability of dyes achieved in consumer products worsens the problem they cause in the wastewater stream when subjected to decoloration and degradation.

Chequer et al. 2013 found that 7×10^5 tons of dyes are produced globally and consumed every year. India and China are the major exporters of both textiles and dyes. It was reported that after being used in the industry, 10–50% of the dyes are released to the environment. Due to their high solubility in water, they are very hard to detect. The by-products that are produced from dyes on degradation are carcinogenic.

Among several such artificial dyes, Azo dyes are a class of colored organic compounds bearing the functional group R–N=N–R′, in which R and R′ can be either aryl or alkyl groups. Congo Red (CR) (Sodium salt of benzidinediazobis-1-naphthylarnine-4-sulfonic acid) has been one of the widely used azo dyes in the textile industry. A wide range of methods has been developed for the removal of CR from textile industry effluents to prevent their adverse environmental effects. The conventional treatment methods such as coagulation, flocculation and activated carbon adsorption, etc., are ineffective for the degradation of CR due to the presence of stable, aromatic azo (–N=N–) groups in it (Crini 2006). Several semiconducting catalysts such as TiO_2 (Wahi et al. 2005), Fe_2O_3 (Huayue et al. 2009), ZnO (Ning et al. 2012) doped ZnS and CdS (Erdemoglu et al. 2008) can photo catalytically degrade CR present in the wastewater. However, the major drawback of these catalysts is the requirement of UV light for the efficient degradation of CR. This problem can be overcome by developing a material that can degrade CR under the visible region of sunlight. Therefore, it is necessary to choose a material that can degrade the dye photo chemically by absorbing the visible light.

4.1 Textile Dye Removal Methods

Generally, based on their performance, dyes can be removed by both destructive and non-destructive methods (shown in Fig. 3.3). The majority of the present technologies such as adsorption using activated carbon nanoparticles, sedimentation using nanoparticles, filtration using nano fiber templates, coagulation using nano aerogels, etc., are non-destructive. The disadvantage of these methods is the formation of hazardous by-products at the surface of the catalysts. These by-products need to be degraded again by employing destructive methods such as biodegradation and Advanced Oxidation Processes (AOP). Biodegradation is a time-consuming process whereas AOPs can be done using inorganic materials, with or without light irradiation. Photocatalysis is a method where the dye molecules can be destructed effectively in presence of light. It is one of the environmental friendly, non-hazardous and cheaper ways than other non-irradiation processes.

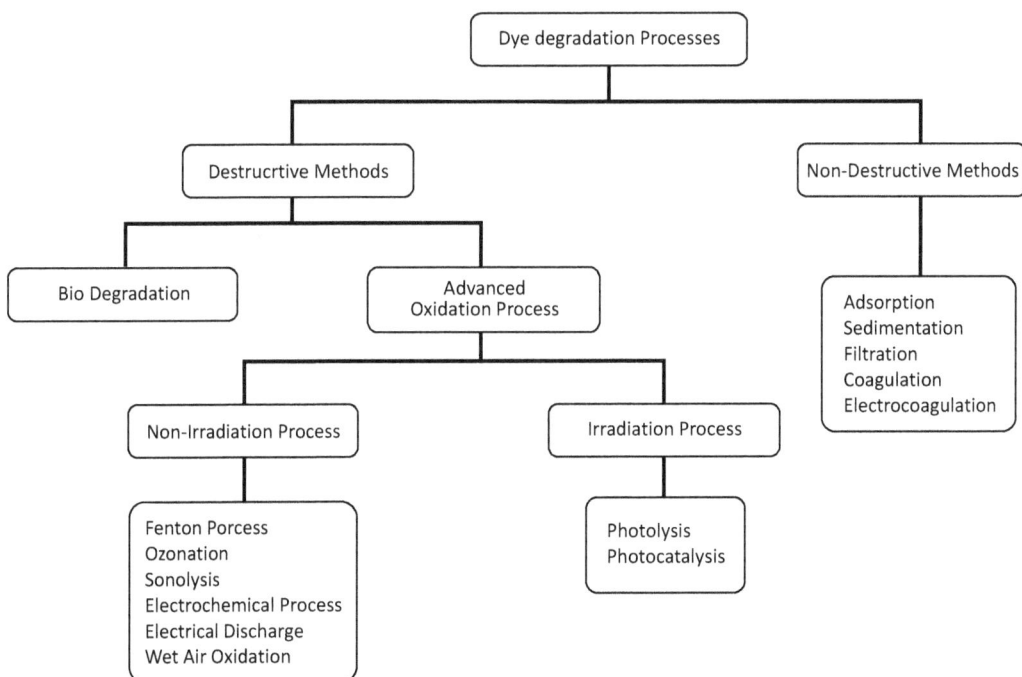

Figure 3.3 Textile dye removal technologies.

The detailed mechanism of photocatalysis, their types, a brief review of the existing photocatalysts, etc., are explained below.

5   BASICS OF PHOTOCATALYTIC PROCESS

The process that uses light energy to initiate or increase the rate/speed of reaction is called photocatalysis. The materials that are used for this process is known as photocatalyst. In this type of reaction, a suitable photocatalyst, when irradiated with sunlight, will initiate redox reactions in the molecules surrounding it. Generally, there are two types of photocatalytic reactions called downhill and uphill reactions. The downhill reactions (lower energy in products than in reactants) initiate with a lower energy input and are spontaneous. They result in a decline of free energy and thus release energy at the end. Photocatalytic dye degradation and removal of organic pollutants are examples of downhill reactions. Whereas uphill reactions (higher energy in products than in reactants) increase free energy and thus require higher energy for the reactions to take place. Photocatalytic water splitting for hydrogen, oxygen evolution and conversion of CO_2 to non-toxic gases are examples of uphill reactions.

5.1   Mechanism of Photocatalytic Water Splitting

Hydrogen production by water splitting (also called artificial photosynthesis), which is a reaction formulated through the motivation from natural photosynthesis has gained a lot of attention. This is one of the uphill reactions with the Gibbs energy of 237 kJ/mol as shown in the following equation (1).

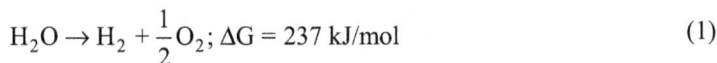

$$H_2O \rightarrow H_2 + \frac{1}{2}O_2; \Delta G = 237 \text{ kJ/mol} \tag{1}$$

Since ΔG is positive, the reaction is non-spontaneous and needs energy equal to or more than 1.23 eV for its initiation. Considering the recombination of photo-excited electron and hole pairs, the optimum bandgap value needed for a photocatalyst for water splitting should be above 2 eV (Bak et al. 2002; Alexander et al. 2008). The energy required for a water-splitting reaction can be renewable (solar) or non-renewable (electricity) sources. Based on the energy input, the hydrogen production process by water splitting can be classified as photocatalytic and photoelectrochemical methods.

In the photocatalytic water splitting method, the photocatalyst along with or without co-catalyst in the form of powder is suspended in a water and electron scavenger medium. It is irradiated under light energy such as sunlight, solar simulator, Mercury lamp, Xenon lamp and Halogen lamp, etc. The amount of hydrogen generated from the photocatalyst and water suspension is measured by gas chromatography. The rate of hydrogen evolution is measured in terms of the number of moles within an hour, per gram of the photocatalyst ($mol/h^{-1}.g^{-1}$).

In the photoelectrochemical water splitting (PEC) method, the photocatalyst material is made as an electrode and connected with a metallic cathodic counter electrode through the external circuit and immersed in an aqueous solution called the electrolyte. When the electrode is exposed to light, photo-induced charges will electrochemically interact with ions in the electrolyte at the solid-liquid interface. Photo-excited holes will react with the aqueous solution and release oxygen at the anodic surface. Similarly, photo-excited electrons will react with proton (H^+ ion) in an aqueous solution and release hydrogen at the cathodic surface.

The detailed mechanism of hydrogen generation through photocatalytic and photoelectrochemical water splitting is explained next.

When the photocatalyst is irradiated under light energy greater than 1.23 eV, photo-induced charges will be generated through excitation from lower energy level to higher energy level (i.e., valence band to conduction band level). Positively charged holes in lower energy levels (i.e., Valence Band-VB) are responsible for the oxidation of water. Similarly, negatively charged electrons, which are at the higher energy level (i.e., Conduction Band-CB) are responsible for the reduction of protons to neutral hydrogen. Instead of participating in these redox reactions, some of the generated charges may recombine thus reducing the efficiency of hydrogen evolution. This can be avoided by using suitable co-catalysts such as Pt, RuO_2, NiO, CuO, Rd, etc.

When the photocatalyst reacts with the water medium in the presence of light, holes will oxidize H_2O to proton and oxide ion immediately. These oxide ions react with the holes having VB energy to produce oxygen. Electrons with CB energy will react with protons and reduce them to hydrogen. Co-catalysts will play a major role in preventing the recombination of photo-induced charges. Electron scavengers such as Methanol, Glycerol, Glucose, etc., are used to supply extra electrons to the CB to achieve better hydrogen production. Figure 3.4 shows the schematic diagram explaining the photocatalytic water splitting mechanism.

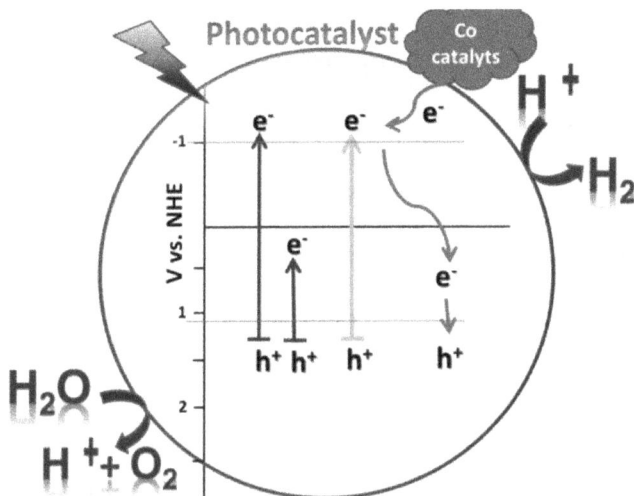

Figure 3.4   Mechanism of photocatalytic water splitting.

5.2   Mechanism of Photoelectro Chemical (PEC) Water Splitting

In this case, the photocatalyst is used in the form of a thin film. Figure 3.5 shows the schematic diagram of the mechanism of PEC reactions.

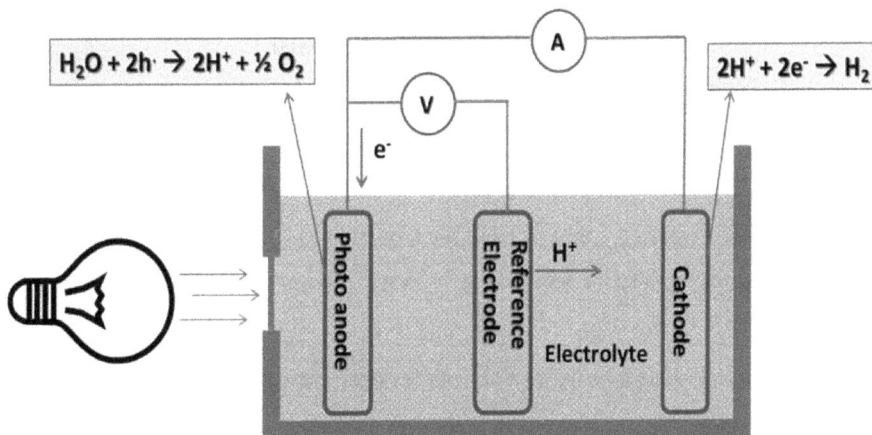

Figure 3.5   Schematic representation of photoelectrochemical water splitting set up showing its mechanism.

When light is irradiated on the PEC system, water will get oxidized at anode and releases proton as shown in equation (2)

$$H_2O + 2h^+ \rightarrow 2H^+ + \frac{1}{2}O_2 \tag{2}$$

Proton produced at the anode will react with electrons from the cathode and releases hydrogen on its surface as shown in equation (3).

$$2H^+ + 2e^- \rightarrow H_2 \tag{3}$$

5.3   Mechanism for Photocatalytic CO_2 Reduction

The mechanism for photocatalytic CO_2 reduction is more complicated than hydrogen generation through water splitting due to several factors such as solubility, by-product formation, etc.

This process involves three steps: (1) adsorption of light and creation of charge, (2) separation of charges and (3) redox reactions at the surface. Figure 3.6 shows the schematic representation of the photocatalytic CO_2 conversion mechanism. When the photon having energy higher than the bandgap of photocatalysts are irradiated on them, charge carriers will be produced. Electrons get excited and reach a higher energy level called the conduction band, whereas holes remain in a lower energy level called the valence band. These electrons are responsible for the reduction and conversion of adsorbed CO_2 to various by-products such as CO, HCOOH, CH_3OH, CH_4, etc., based on the number of electrons available for the reactions (as mentioned in Fig. 3.2). Holes will undergo an oxidation reaction with water and CO_2. The type of co-catalysts used for reduction reaction will decide the formation of the by-product from CO_2. Unlike water-splitting reactions, reduction reactions will contribute more to CO_2 conversion.

The reduction potentials of various by-products are shown in the equations below (1.4–2.1). The single-electron reduction from CO_2 to CO_2^- is not possible due to the large negative reduction potential (i.e., –1.85 V vs SHE (Standard Hydrogen Electrode, which is used for reference on all half- cell potential reactions. The value of SHE is usually 0 at a temperature of 298 K because it acts as a reference for comparison with any other electrode) relative to the conduction band potential of many of the semiconductors. This highly negative reduction potential arises from the change in the hybridization of carbon from sp^2 to sp^3. The reduction potential for the formation of by-products such as HCOOH (–0.665 V), HCHO (–0.485 V), CH_3OH (–0.399 V) and CH_4 (–0.246 V) are smaller and those are positive to the CB edge potentials of many semiconductors. Therefore, it is advisable to undergo proton-coupled electron transfer in CO_2.

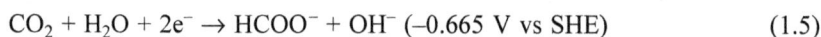

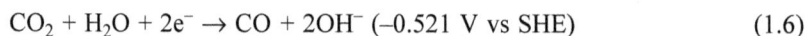

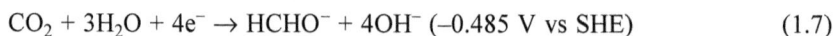

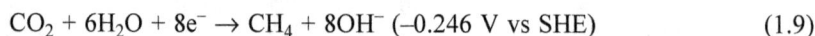

$$CO_2 + e^- \rightarrow CO_2^- \ (-1.85 \text{ V vs SHE}) \tag{1.4}$$

$$CO_2 + H_2O + 2e^- \rightarrow HCOO^- + OH^- \ (-0.665 \text{ V vs SHE}) \tag{1.5}$$

$$CO_2 + H_2O + 2e^- \rightarrow CO + 2OH^- \ (-0.521 \text{ V vs SHE}) \tag{1.6}$$

$$CO_2 + 3H_2O + 4e^- \rightarrow HCHO^- + 4OH^- \ (-0.485 \text{ V vs SHE}) \tag{1.7}$$

$$CO_2 + 5H_2O + 6e^- \rightarrow CH_3OH + 6OH^- \ (-0.399 \text{ V vs SHE}) \tag{1.8}$$

$$CO_2 + 6H_2O + 8e^- \rightarrow CH_4 + 8OH^- \ (-0.246 \text{ V vs SHE}) \tag{1.9}$$

$$2H_2O + 2e^- \rightarrow H_2 + 2OH^- \ (-0.414 \text{ V vs SHE}) \tag{2.0}$$

$$2H_2O \rightarrow O_2 + 4H^+ + 4e^- \ (+0.816 \text{ V vs SHE}) \tag{2.1}$$

The formation energy (i.e., ΔG) of the by-products are in the order of HCOOH < CO < HCHO < CH_3OH < CH_4.

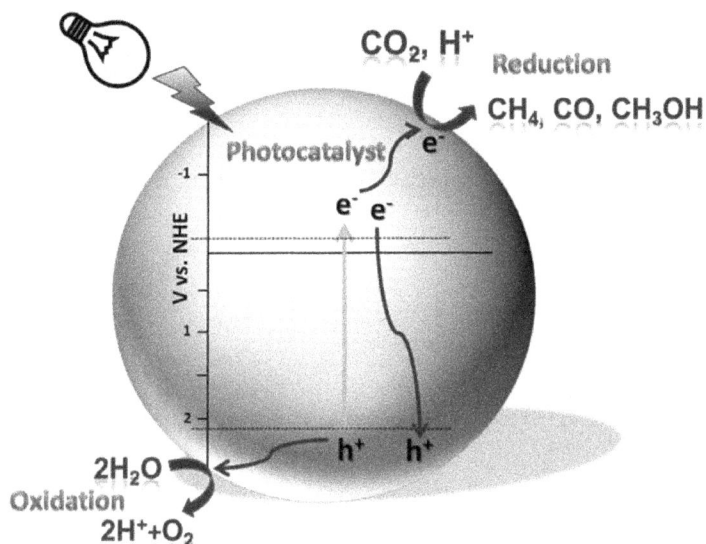

Figure 3.6   Mechanism of photocatalytic CO_2 reduction.

5.4   Mechanism of Photocatalytic Dye Degradation

The mechanism of dye degradation is shown in Fig. 3.7. Dye degradation is a downhill photocatalytic reaction, where the energy involved in the reactants is higher than that of products and thus spontaneous. In this case, partially charged holes are generated at the surface of the photocatalyst. They will react with water under visible light irradiation and produce hydroxyl radicals. Similarly, electrons in photocatalyst will react with water and form super oxide anions. Super oxide anions will react with organic dye molecules in the solution and form hydroperoxyl radical and hydroxyl ions. The hydroxyl ion acts as a strong oxidizing agent to decompose organic dye into non-toxic by products.

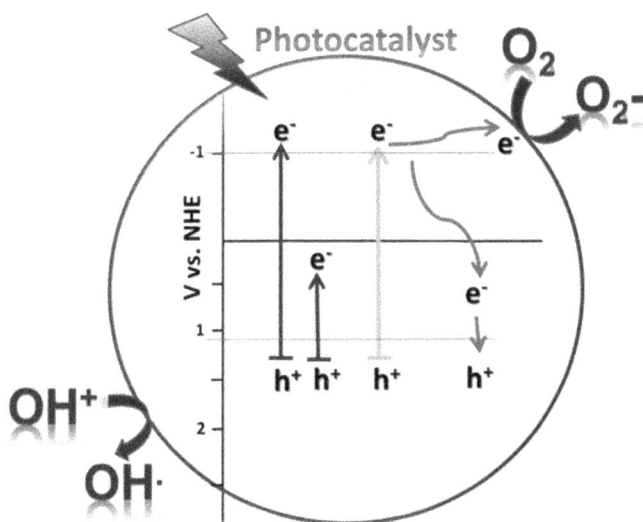

Figure 3.7   Mechanism of formation of active radicals for photocatalytic dye degradation.

6   BASIC REQUIREMENTS FOR PHOTOCATALYTIC REACTIONS

The characteristics required for a material to exhibit photocatalytic reactions are listed below.

1. The theoretical bandgap of the photocatalyst should be greater than 1.23 eV to achieve water splitting, but practically it should be more than 1.7 eV so that recombination of the charges can be reduced. To be active under visible light, it should be lesser than 3.0 eV and for UV light it should be 3.0 eV to 3.9 eV.
2. Photo-induced charges should be separated efficiently to avoid surface or bulk recombination.
3. Larger recombination time.
4. Photo-corrosion resistance.
5. Acceptor level of semiconductor must be more negative than the reduction potential of H^+/H_2 (< 0 eV vs NHE) and donor level must be more positive than oxidation potential of O_2/H_2O (> 1.2 eV vs NHE).
6. Low material and preparation cost.

6.1   Role of Band Edge Potentials

The band edge potentials of a few selected semiconductors are shown in Fig. 3.8. These materials are commonly used for photoelectrochemical water splitting and CO_2 reduction. Redox potentials of key CO_2 reduction reactions are also included. In principle, water splitting and CO_2 reduction can take place on the same semiconductor material if the conduction band energy level is aligned with, or more negative than, the energy level of the targeted CO_2 methanation reaction (−0.24 vs NHE) and the valence band energy level is aligned with, or more positive than, the oxygen evolution reaction energy level (1.23 vs NHE). This is indicated by the position of each material relative to the vertical bar dividing the figure. The materials exhibiting unfavorable band alignment are also included in the figure, as they are commonly used as light-absorbers in photoelectrochemical cells (Ronge et al. 2014).

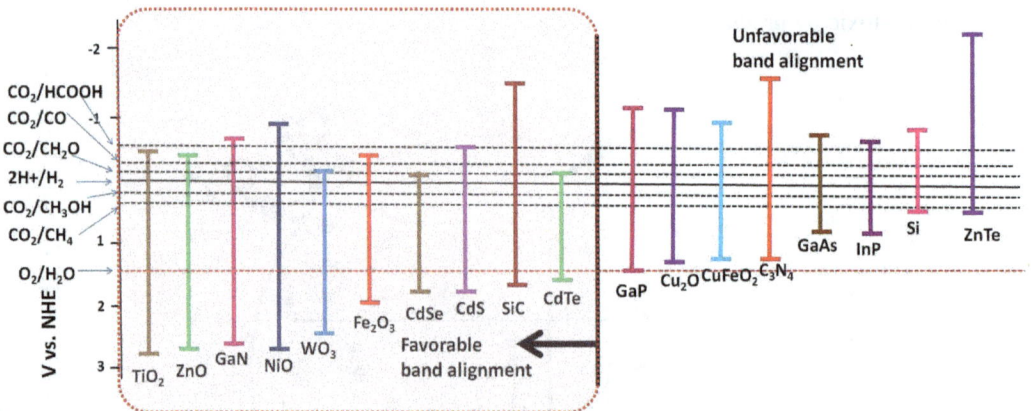

Figure 3.8   The band edge potential diagram of various photocatalysts with different redox potentials of by-products.

6.2   Role of Co-catalysts

The co-catalysts can be used to tune the product selectivity, reduce the activation barrier for the redox reactions and facilitate the separation of charge carriers. Generally, the composite heterogeneous photo redox catalysts are composed of metals (e.g., Cu, Ru or Re) coupled with semiconductors (Eg. P-Si, GaP, GaAs, GaN orTiO$_2$) or other metal co-catalysts like Au or Pt. Table 3.1 shows the list of reported co-catalysts used in photocatalytic reactions. The lead-halide perovskite quantum dots supported on graphene oxide have also been shown to photo catalytically reduce CO$_2$ to CH$_4$ and other products. Yu et al. (Sorcar et al. 2017) demonstrated that the addition of Pt and Cu onto TiO$_2$ photocatalysts yielded CH$_4$ under solar irradiation. The addition of metal oxide co-catalysts, such as NiO and CuO on semiconductors (TiO$_2$) has also increased the photocatalytic activity.

Table 3.1   List of Co-catalysts used in Photocatalytic water splitting experiments

Type of Co-catalysts	Examples	Reference
Noble metals	Pt	Sorcar et al. 2017
	Pd	Sayed et al. 2012
	Ru	Tsuji et al. 2005
	Au	Lin et al. 2019
	Ag	Onsuratoom et al. 2011
Transition metals	Cu	Wu et al. 2019
	Ni	Huang et al. 2013
	Co	Tran et al. 2012
Metal oxides	NiO	Sreethawong et al. 2005
	CuO	Yu et al. 2011
	Cu$_2$O	Xu et al. 2010
Metal sulfides	CuS	Zhang et al. 2013
	MoS$_2$	Zong et al. 2008
	Ni	Wang et al. 2012
	WS$_2$	Zong et al. 2011

6.3   Role of Reaction Medium

The photocatalytic reduction of CO$_2$ can be carried out in the liquid phase and gaseous phase. In the liquid phase,the reduction can be done in a saturated aqueous solution of CO$_2$. But the poor solubility rate of CO$_2$ in water is one of the major drawbacks. The solubility rate of CO$_2$ in water can be improved by adding additives such as NaOH, NaHCO$_3$, Na$_2$CO$_3$. Another factor is the surface adsorption of H$_2$O is preferable over CO$_2$ in the liquid phase, so the reduction of water is favorable. The efficiency of CO$_2$ reduction in the liquid phase depends on the pH of the reaction medium, surface hydroxyl groups, solvent and additives. The increase in the pH increases the rate of reaction. So, it forms different concentrations of species such as CO$_3^{2-}$, HCO$_3^-$ and CO$_2$ at different values of pH. These chemical species having different reduction potentials get adsorbed on the catalyst surface. So, the addition of NaOH can improve the dissolution of CO$_2$ and increases the efficiency of photoreduction of CO$_2$ on TiO$_2$ supporting Cu catalysts. The gas-phase reactions can be carried out with humidified CO$_2$. Xie et al. (Xie et al. 2012) demonstrated both liquid and gas phase reactions with TiO$_2$ and Pt-TiO$_2$. They found that the CH$_4$ production is nearly three times more and hydrogen production is less in the gas phase compared to the liquid phase. Therefore, three times more selectivity in CO$_2$ reduction overwater reduction was observed in the gas phase. The efficiency of CO$_2$ reduction in gas phase reactions depends on surface properties of photocatalysts, CO$_2$–H$_2$O mixture ratio, feed pressure, temperature, etc.

7   REPORTED CATALYSTS FOR PHOTOCATALYTIC REACTIONS

Photocatalysts are typically made of metal oxides, metal sulfides, oxysulfides, oxynitrides and composites. In most cases, metal cations with the highest oxidative states in photocatalysts have

d^0 or d^{10} electronic configuration, while O, S and N show their most negative states. The bottom of the conduction band consists of the d and sp orbitals of the metal cations, while the top of the valence band in metal oxides is composed of O $2p$ orbitals, which are normally located at +3 V (vs NHE) or higher. The valence bands of metal oxy-sulfides and oxynitrides are formed by S $3p$ (and O $2p$) and N $2p$ (and O $2p$), respectively. Some alkali (Li, Na, K, Rb or Cs), alkaline earth (Mg, Ca, Sr and Ba), and transition-metal ions (Y, La or Gd) can construct the crystal structure of layered perovskite and cubic pyrochlore compounds, but do not contribute to the energy band structure of these compounds.

There are many photocatalysts reported in the past five decades. TiO_2 is one of the well-known photocatalyst materials discovered by Fujishima and Honda in 1972. It is one of the widely used photocatalyst materials due to its abundance, lower cost, non-toxicity, and corrosion resistance properties (Inoue et al. 1979; Sayama and Arakawa 1997). Since it is a wider bandgap material, it is less active under visible light irradiation when compared to the UV region. Several modifications on TiO_2 were carried out to improve visible light active photocatalytic properties such as by doping nitrogen (Fu et al. 2006; Li et al. 2009a), sulfur (Yu et al. 2005), carbon (Sakthivel and Kisch 2003; Irie et al. 2003), noble metals, etc. Several other metal oxides such as ZnO (Ramasami et al. 2017), WO_3 (Wilson et al. 1984), Fe_2O_3 (Lingqiao et al. 2018), etc., are also reported as photocatalysts. The details about such photocatalysts are shown in Table 3.2. The list of reported nanocomposite photocatalysts for CO_2 conversion are in Table 3.3.

However, these metal oxides suffer from the drawback of having unfavorable band edge potentials for overall water splitting. Their band edge potentials favor only photo-oxidation and not photoreduction. Other than metal oxides, sulfides (Zong et al. 2008) and nitrides (Leny et al. 2010) also have been reported to be capable of photocatalytic activity. They have a narrower band gap (i.e., 1.9 eV–2.3 eV) with favorable band edge potentials, which are suitable for water splitting. But they are photo-corrosive in nature. To avoid the above drawbacks, recently several combinations of metal oxides/metal sulfides have been used in the form of nanocomposites for water splitting called Z-scheme photocatalysts (ex. TiO_2–CdS (Jang et al. 2008; Park et al. 2008), ZnO–CdS (Wang et al. 2010), WO_3–Cu_2O (Hu et al. 2008)). The nanocomposites of various oxides are explained elaborately later.

Table 3.2 Reported semiconductor photocatalysts for uphill reactions

S. No.	Photocatalysts	Photocatalytic property	Reference
1	TiO_2	Hydrogen evolution – 1497 μ.mol/h.g with Rh co-catalyst under UV light irradiation.	Yamaguti and Sato 1985.
2	N doped TiO_2	Hydrogen evolution – 250 μ.mol/h.g with Pt co-catalyst under 300 W Xenon lamp irradiation.	Yuan et al. 2006
3	S doped TiO_2	Hydrogen evolution – 0.5 μ.mol/h.g with Pt co-catalyst under 500 W Xenon lamp irradiation.	Fang et al. 2010
4	N, S doped TiO_2	Hydrogen evolution – 9.97 μ.mol/h.g with Pt co-catalyst under 300 W Xenon lamp irradiation.	Nishijima et al. 2007
5	CdS	Hydrogen evolution – 5400 μ.mol/h.g with MoS_2 co-catalyst under Visible light irradiation.	Zong et al. 2008
6	ZnO	Oxygen evolution – 140 μ.mol/h.g without co-catalyst under 250 W Hg lamp irradiation.	Ramasami et al. 2017
7	Fe_2O_3	Oxygen evolution – 5 μ.mol/h.g without co-catalyst under 500 W Xenon lamp irradiation.	Lingqiao et al. 2018
8	Ta_2O_5	Hydrogen evolution – 1154 μ.mol/h.g with NiO co-catalyst under 400 W Hg lamp irradiation.	Yoshiko et al. 2001
9	WO_3	Oxygen evolution – 67 μ.mol/h.g with RuO_2 co-catalyst under 500 W Xenon lamp irradiation.	Wilson et al. 1984

7.1 TiO₂ based Nanocomposites for Photocatalytic Reactions

TiO_2 has received more attention because it is stable, corrosion-resistant, nontoxic, abundant and less expensive. However, its practical and economical applications are limited by two factors. First, the photon-to-hydrogen efficiency of TiO_2 is too low due to the rapid recombination of photo generated electrons and holes as well as the fast back-reaction of H_2 and O_2 to H_2O. Second, it is inactive under visible light irradiation due to its large bandgap, which impedes its usage as a solar energy harvesting photocatalyst. In order to overcome these deficiencies, many bulk surfaces and operating environment modifications have been conducted.

Photocatalytic activity for water splitting was effectively enhanced in TiO_2 doping with metal ions. Chae et al. (2009) reported that Ga doped TiO_2 powder could split pure water stoichiometrically under UV irradiation, whereas pure TiO_2 did not show any activity. The Ni^{2+} doping enhanced the photoactivity of the TiO_2 for hydrogen production from an aqueous methanol solution (Jing et al. 2005). Sn/Eu co-doped TiO_2 exhibited high activity for hydrogen generation with a quantum efficiency of ~40% with Pd as the co-catalyst under irradiation from a fluorescent lamp (Sasikala et al. 2008). Zalas and La studied the effect of lanthanide doping on the photocatalytic activity of TiO_2 (Parida and Sahu 2008). The best performance for hydrogen production from an aqueous methanol solution was obtained for the TiO_2 containing 0.5 mol% of Gd oxide as the dopant. The UV-driven photocatalytic activity of TiO_2 was also improved by combining it with a second oxide semiconductor. All the mixed oxides with hetero-phase structures, SnO_2/TiO_2 (Sasikala et al. 2009), ZrO_2/TiO_2 (Yuan et al. 2009), Cu_xO/TiO_2 (Xu et al. 2010; Xu and Sun 2009; Choi et al. 2007), Ag_xO/TiO_2 (Park and Kang 2008; Park et al. 2008; Lalitha et al. 2010) and $MTiO_3/TiO_2$ (M=Ca, Sr, Ba) (Zielinska et al. 2008) displayed higher rates of photocatalytic hydrogen evolution from aqueous solutions containing electron donors other than TiO_2 alone. With Pt as a co-catalyst, the Ti/B binary oxide stoichiometrically decomposes pure water under UV irradiation (Moon et al. 1998; 2000). When TiO_2 nanoclusters were dispersed in the mesoporous structures of MCM-41 and MCM-48, the formed Ti-MCM-41 (Liu et al. 2003) and Ti-MCM-48 (Zhao et al. 2010) showed much higher photocatalytic activity for hydrogen evolution under UV irradiation than bulk TiO_2.

Li et al. (2012) reported that the nitrogen-doped mesoporous TiO_2 photocatalyst showed good visible-light absorption and enhanced activity for the hydrogenation of CO_2 to CH_4 under visible light irradiation. Ong et al. (Ong et al. 2015) synthesized anatase nitrogen-doped TiO_2 nanoparticles with exposed (001) facets, showing photocatalytic activity under visible light for the hydrogenation of CO_2 to CH_4, which is 11-fold higher than that of pure TiO_2 under visible light irradiation. Zhang et al. (Zhang et al. 2011) successfully prepared 10% iodine doped TiO_2 samples with a mixture of anatase and brookite phases exhibiting a high CO_2 hydrogenation activity (2.4 $\mu mol.g^{-1}.h^{-1}$) under visible light irradiation. The doping of transient metal ions such as V, Cr, Mn, Fe and Ni is another effective strategy to extend the light adsorption range of TiO_2. The incorporation of an optimal weight percentage (1.5 wt.%) Ni into the TiO_2 matrix does not only extends optical absorption in the visible light range but also enhances the hydrogenation of CO_2. The different concentrations of Cr, V and Co, metals-doped TiO_2 show obvious red shifts of absorption edges and increased light absorption in the visible light region compared to pure TiO_2. The 0.5 wt.% V doped TiO_2 sample showed a maximum acetaldehyde formation rate of 11.13 $\mu mol.g^{-1}.h^{-1}$ in vapor-phase hydrogenation of CO_2 under visible light irradiation. The metal and non-metal non-metal co-doped TiO_2 were also found to be efficient in the hydrogenation of CO_2 under visible light irradiation. The effects of N and Ni-doped on TiO_2 in the hydrogenation of CO_2 photocatalytic reduction showed an optimal methanol yield of 3.59 $\mu mol.g^{-1}.h^{-1}$ under visible light irradiation. The doped N and Ni could enhance the light response of TiO_2 and improve the photocatalytic activity of TiO_2 by acting as a probable electron trapper to ensure good separation of electron-hole pairs, respectively.

Table 3.3 List of Nanocomposite photocatalysts for CO_2 reduction along with their by-products and yield

Material	Preparation method	Reaction conditions	By product and Yield ($\mu mol/g.h$)
Boron doped $SrTiO_3$	Ball milling	Liquid phase reaction under 500 W Xenon lamp	CO – 21 μmol CH_4 – 14 μmol and O_2 – 35 μmol.
$Ca_xTi_yO_3$	Hydrothermal	Liquid phase reaction under 6 W UV lamp	CH_4 – 17 $\mu mol/g$ and O_2 – 34 μmol.
3D leaf like $SrTiO_3$	Chemical method	Liquid phase reaction under 300 W Xenon lamp	CO – 3 $\mu mol/g$ and CH_4 – 0.29 $\mu mol/g$
$LaFeO_3$-TiO_2	Chemical method	Liquid phase reaction under 300 W Xenon lamp	CO – 18.75 $\mu mol/g$ and CH_4 – 13.75 $\mu mol/g$
Pt- $SrTi_{0.98}Co_{0.02}O_3$	Chemical method	Liquid phase reaction under 300 W Xenon lamp	CH_4 – 63.6 $\mu mol/g$
CuO loaded $SrTiO_3$ nanorod thin film	Hydrothermal method	Liquid phase reaction under UV lamp	CO – 1.4 $\mu mol/g$
$ZnFe_2O_4$/Ag/TiO_2	Solvothermal and Physical mixing methods	Gas phase reaction under 200 W Hg lamp	CO – 1025 $\mu mol/g$, CH_4– 132 $\mu mol/g$, CH_3OH – 31 $\mu mol/g$ and C_2H_6 – 19 $\mu mol/g$
$ZnFe_2O_4$/TiO_2	Chemical method	Liquid phase reaction under 500 W Xenon lamp	CH_3OH – 141.22 $\mu mol/g$
$M_{0.33}WO_3$ (M = K, Rb, Cs)	Solvothermal method	Liquid phase reaction under 500 W Xenon lamp	CH_3OH – 17.5 $\mu mol/g.h$
Vacancy Bi_2WO_6	Hydrothermal	Liquid phase reaction under 500 W Xenon lamp	CO – 40 $\mu mol/g.h$
Cu-TiO_2	Chemical method	Liquid phase reaction under 250 W Xenon lamp	Oxalic acid – 65.6 $\mu mol/g$, Acetic acid – 12 $\mu mol/g$ and Methanol – 0.8 $\mu mol/g$
Ni doped $InTaO_4$	Sol-gel method	Liquid phase reaction under 20 W white LED	Methanol – 200 $\mu mol/g$
P-25 TiO_2	Sol-gel method	Liquid phase reaction under 250 W Hg lamp	Methanol – 914 $\mu mol/g$ and Methane – 0.7 $\mu mol/g$
Co(II) Phathalocyanine with Ni/NiO core shell	Chemical method	Liquid phase reaction under 20 W white LED	Methanol – 3641.2 $\mu mol/g$
Oxygen vacancy with CeO_2	Solution combustion method	Liquid phase reaction under 300 W Xenon lamp	Methanol – 0.702 $\mu mol/g.h$
rGO/$InVO_4$/Fe_2O_3	Chemical method	Liquid phase reaction under 20 W white LED	Methanol – 16.9 $mmol/g$
g-C_3N_4/$FeWO_4$	Hydrothermal method	Liquid phase reaction under 300 W Xenon lamp	CO – 6 $\mu mol/g.h$
rGO-CdS on porous alumina support	Chemical method	Gas phase reaction under sunlight irradiation	Methanol – 153.8 $\mu mol/g.h$
Au-S-TiO_2	Chemical method	Liquid phase reaction under 20 W white LED	Methanol – 74.3 $\mu L/g$ and H_2 – 57.01 $\mu mol/g.h$
$InVO_4$	Chemical method	Liquid phase reaction under 20 W white LED	CO – 520 ppm and H_2–236.41 ppm
Cu_3SnS_4	Chemical method	Gas phase reaction under solar simulator irradiation	Methanol – 14 $\mu mol/g.h$
g-C_3N_4/NiAl-LDH	Chemical method	Liquid phase reaction under 300 W Xenon lamp	Methanol – 2 $\mu mol/g$ and H_2 – 0.45 $\mu mol/g$, O_2 – 1.2 $\mu mol/g$
CNT-TiO_2	Chemical method	Liquid phase reaction under 20 W white LED	CH_4 – 2360 $\mu mol/g.h$, H_2 – 3246 $\mu mol/g.h$ and HCOOH – 1520 $\mu mol/g.h$

Recently several studies have reported the use of plasmonic Au or Ag nanoparticles decorated TiO_2 materials for photocatalytic hydrogenation of CO_2. Hou et al. reported that TiO-Au material catalyzed reduction of CO_2 and water vapor over a wide range of wavelengths (Hou et al. 2011). When they used the light of wavelength 532 nm to catalyze the photoreduction of aqueous CO_2, methane was the only product. In this case, a 24-fold enhancement was observed because of the strong electric fields created by the surface plasmon resonance of the Au nanoparticles, which excited electron-hole pairs locally in the TiO_2 at a rate several orders of magnitude higher than the normal incident light. When the photon energy was higher to excite d-band electronic transitions of Au nanoparticles in the UV range, different products including C_2H_6, CH_3OH and HCHO were observed. Plasmonic Ag nanoparticles deposited on TiO_2 nanowire films were used for photocatalytic hydrogenation of CO_2 to CH_3OH under visible light irradiation. The results indicated that the CH_3OH yield reached 8.3 $\mu mol.cm^{-1}$ over the optimal TiO_2-Ag sample due to the charge transfer property of Ag nanoparticles and the efficient light utilization based on the overlapped visible light-harvesting of Ag nanoparticles and TiO_2 films. Liu et al. (2014) prepared a 2.5 wt.% Ag-TiO_2 composite, which exhibited a methanol production rate of 135.1 $\mu mol.g^{-1}.h^{-1}$ under UV and visible light irradiation. The rate was 9.4 times higher than the pure TiO_2. Compared with Au or Ag plasmonic nanoparticles, Cu nanoparticles are reported to exhibit the Surface Plasmon Resonance (SPR) effect for visible light-harvesting properties. Liu et al. (2015) used plasmonic Cu nanoparticle modified TiO_2 nanoflower films for hydrogenation of CO_2 to CH_3OH. The CH_3OH production rate was 1.8 $\mu mol.g^{-1}.h^{-1}$ over 0.5 wt.% Cu-TiO_2 under UV and visible light irradiation, which is six times higher than the pure TiO_2 film. The commercial P25-TiO_2 decorated with Au-Cu alloy nanoparticles (Au : Cu = 1.2) exhibited a CH_4 production rate of 2000 $\mu mol.g^{-1}.h^{-1}$ under-stimulated sunlight (Neatu et al. 2014). The monometallic Au/TiO_2 and Cu/TiO_2 samples exhibited a preferred hydrogen evolution or lower activity to methane formation, respectively. FT-IR spectroscopy and chemical analysis revealed a mechanism, in which the role of Au was to respond to visible light and that of Cu was to bind to CO. So, through rational design of nanocomposite, one could harvest visible light through the SPR effect to enhance the activities of semiconductors for the photocatalytic reduction of CO_2. It is widely known that combining TiO_2 with visible-light responsive photocatalysts to form hetero junctions could greatly improve the performance of photocatalysts for CO_2 reduction on TiO_2 under visible light irradiation.

The combination of TiO_2 with graphene has been proven to be a promising way to improve the photocatalytic reduction of CO_2. When combining TiO_2 with graphene the excited electrons of TiO_2, which are transferred from the conduction band of graphene react with adsorbed CO_2 and H_2O to produce CH_3OH and O_2 respectively. Solvent exfoliated graphene (Liang et al. 2011) showed enhanced photocatalytic activity than conventional graphene for the photocatalytic reduction of CO_2 to CH_4, with up to seven-fold improvement compared to pure TiO_2 under visible light irradiation. This is due to its superior electrical mobility, which facilitates the diffusion of photoinduced electrons to reactive sites. Liang et al. (2012) found that carbon nanomaterial dimensionality is a key factor in determining the spectral response and reaction specificity of graphene titania nanosheet composite photocatalysts. The 2D-2D graphene titania nanosheet composites had superior electronic coupling compared to the 1D-2D carbon nanotube-titania nanosheet composites leading to greater enhancement factors for CO_2 photoreduction under UV irradiation. Cu-Pt bimetallic doped on TiO_2 nanotubes showed four times enhanced CO_2 reduction to CH_4, C_2H_4 and C_3H_6 in presence of H_2O (Yasuhiro et al. 2013). They have been attempting to mimic natural photosynthesis by making three-dimensional leaf-like architectures using titanates (Han et al. 2013) and reported a reasonable 3 $\mu mol/g$ of CO and 0.3 $\mu mol/g$ of CH_4 conversion from CO_2. Jingjing et al. (2017) showed that the boron-doped $SrTiO_3$ exhibited four times increased conversion of CO_2 to useful products such as CO, CH_4 and oxygen. Bi_2WO_6 with oxygen vacancies also showed (Xin et al. 2016) photoreduction of CO_2 to CO. Tahir et al. (Muhammad 2020) showed that the $ZnFe_2O_4$/Ag/TiO_2 nanocomposite exhibited excellent formation of various byproducts such as CO, C_2H_4, CH_3OH and CH_4 using this indigenous photocatalytic reactor. They also developed a monolith photoreactor (Tahir and Amin 2013) that enhances CO_2 reduction than the liquid phase commercial reactors.

Many white titanates are known to work as efficient photocatalysts for water splitting under UV irradiation. Shibata et al. (Shibata et al. 1987) reported that the layered titanates, $Na_2Ti_3O_7$, $K_2Ti_2O_5$ and $K_2Ti_4O_9$ were active in photocatalytic H_2 evolution from aqueous methanol solutions even without the presence of Pt co-catalyst. These layered titanates, consisting of titanium oxide layers and interlayers, can be modified using ion-exchange reactions (Shibata et al. 1987; Allen et al. 2010). Of the materials studied, the H^+-exchanged $K_2Ti_2O_5$ exhibited a high activity with a quantum yield of up to 10%. After being pillared with SiO_2 in the interlayers, $K_2Ti_4O_9$ showed enhanced photocatalytic activity for H_2 evolution from CH_3OH/H_2O mixtures (Machida et al. 2000).

7.2   Copper-based Nanocomposites for Photocatalytic Reactions

Cu based semiconductors including copper oxide (CuO), cuprous oxide (Cu_2O) and cuprous sulfide (Cu_2S) with band gaps of 1.7 eV, 2.2 eV and 1.2 eV, are of particular interest due to their efficient visible light-harvesting, suitable conduction band states and good selectivity towards hydrogenation of CO_2. But the narrow band gap leads to poor photocatalytic activity because of the fast recombination of photo-induced carriers. So, incorporating a suitable n-type semiconductor with a Cu-based semiconductor to form a p-n junction, can efficiently increase the lifetime of the charge carriers by establishing an electric field at the interface with a potential difference between the two sides. This is considered to be an effective strategy to promote the performance of hydrogenation of CO_2. The iron oxide cuprous oxide nanocomposites could show CO yield of 1.67 $\mu mol.g^{-1}.h^{-1}$ under visible light irradiation in presence of water vapor (Kim et al. 2017). The ZnO–CuO nanocomposites showed the highest CO yield of 1.98 $\mu mol.g^{-1}.h^{-1}$ under UV Visible irradiation (Wang et al. 2015). In this composite, the CuO was surface engineered with ZnO islands using a few pulsed cycles of atomic layer deposition. The detailed investigations revealed that the formation of the p-n hetero junction as the contact surface, which significantly influenced the migration behavior of charge carriers. The PL studies also confirmed a very high density of defects on these ZnO islands, which could enhance the lifetime of photogenerated electrons. Gusain et al. (2016) synthesized CuO nanoparticles covered with rGO nanosheets via chemical synthesis. This composite showed excellent photocatalytic activity and stability for transforming CO into CH_3OH at the rate of 1228 $\mu mol.g^{-1}.h^{-1}$, which is seven times higher than that of the pure CuO sample under visible light irradiation. The yield of CH_3OH remained almost the same after six cycles of reaction, which ruled out the possibility of surface carbon contamination. The improved photoactivity of rGO-CuO was explained to be the strong interaction between rGO and CuO, which facilitated rGO to efficiently capture and transfer photoelectrons generated from CuO to the surface catalytic sites for reduction of CO_2.

7.3   Tungsten based Nanocomposites for Photocatalytic Reactions

WO_3 is one of the narrower bandgap materials around 2.5–2.8 eV and shows good light absorption up to the NIR region. It also has a better electron transport mobility than TiO_2. The crystal structure of WO_3 also affects its charge separation, optical absorption, redox capability and electrical conductivity. Huang et al. (2015) proved that single-crystal WO_3 nanosheet of 4–5 nm thickness showed enhanced CO_2 to CH_4 conversion rate of 1.14 $\mu mol.g^{-1}.h^{-1}$. Xie et al. (Xie et al. 2012) synthesized rectangular sheets like WO_3 with predominant (002) facets by controlling acid and showed CO_2 to CH_4 conversion rate of 0.34 $\mu mol.g^{-1}.h^{-1}$. Kadowaki (Kadowaki et al. 2007) found that $PbWO_4$ incorporating a WO_4 tetrahedron showed high and stable photocatalytic activity for the overall splitting of water. A stoichiometric quantity of H_2 and O_2 was produced under

UV irradiation when RuO_2 was loaded onto the metal oxide. Kudo and co-workers have extensively investigated the photocatalytic activities of tungstates. Under UV irradiation, $Na_2W_4O_{13}$ (Kudo and Kato 1997) and $Bi_2W_2O_9$ (Kudo 1999) with layered structures, were active for photocatalytic hydrogen (Pt as co-catalyst) and oxygen evolution in the presence of suitable sacrificial reagents.

7.4 Metal Sulfide-based Nanocomposites for Photocatalytic Reactions

Metal sulfide systems are attractive photocatalysts for the water-splitting reaction because of their narrower band gaps, which allow the absorption of visible light. Metal sulfide photocatalysts are unstable in the water-oxidation reaction under visible light because sulfur has more oxidation rate, thereby causing photo degradation of the photocatalyst. A common method for reducing the photo corrosion of the sulfides under irradiation is the use of suitable sacrificial reagents. The photo corrosion of sulfide-based catalysts may be effectively suppressed by using a Na_2S/Na_2SO_3 salt mixture as an electron donor. Among the available sulfide semiconductors, CdS with a wurtzite structure is probably the best-studied metal sulfide photocatalyst. Due to its relatively narrow band gap (2.4 eV) and it can absorb visible light efficiently (Kamat 2002). The photocatalytic activity was studied by combining CdS with other semiconductors with different energy levels: TiO_2 (Fujii et al. 1998), ZnO (Spanhel et al. 1987), or CdO (Navvaro et al. 2008). In these composite systems, the photogenerated electrons move from CdS to the surface of surrounding oxide semiconductor particles, while photogenerated holes remain in CdS. This charge-carrier separation stops charge recombination, thereby improving the photocatalytic activity of CdS.

ZnS is interesting as a semiconductor for combining with CdS in the composite. CdS and ZnS make a continuous series of solid solutions ($Cd_{1-x}Zn_xS$) in which metal atoms are mutually substituted in the same crystal lattice (Fedorov et al. 1993; Nayeem et al. 2001). Del Valle et al. (Del Valle et al., 2009) investigated the photophysical and photocatalytic properties of $Cd_{1-x}Zn_xS$ solid solutions with different Zn concentrations ($0.2 < x < 0.35$). The solid solution between CdS and ZnS showed a blue shift of the absorption edge with increasing Zn concentration. Photocatalytic activity of these samples increases gradually when the Zn concentration increases from 0.2 to 0.3. The change in activity for H_2 production of these samples is due to the modification of the energy level of the conduction band as the concentration of Zn increases in the solid-solution photocatalyst.

ZnS is another sulfide semiconductor investigated for photochemical water splitting. ZnS is unable to split water under visible light because of its wider band gap (3.66 eV), which restricts light absorption to the UV region. Studies have been carried out to improve the visible light sensitivity of ZnS-based photocatalysts. One of the strategies for inducing a visible light response in ZnS is the chemical doping of ZnS with metal ions (Cu^{2+} (Kudo and Hijii 1999), Ni^{2+} (Kudo 2000) and Pb^{2+} (Kudo et al. 2007). The doped ZnS materials can absorb visible light as a result of the transitions from M+ (M=Cu, Ni, Pb) levels to the conduction band of ZnS. These doped-ZnS photocatalysts showed high photocatalytic activity under visible light for H_2 production from aqueous solutions using SO_3^{2-}/S^{2-} as electron donor reagents.

7.5 g-C_3N_4 based Nanocomposites for Photocatalytic Reactions

Graphitic carbon nitride is a two-dimensional conjugated polymer, which is a promising metal-free, visible light responsive photocatalyst. The conduction band potential of g-C_3N_4 is located at 1.3 V, which is high enough to drive the multi-electron transfer reactions to produce CO, HCOOH, CH_3OH and other hydrocarbon fuels. At the same time, H_2O can act as both H source and hole consumption agent due to its higher oxidation potential than the g-C_3N_4 valence band (+1.4 V). This material can be easily synthesized by thermal polymerization of abundant nitrogen-rich precursors, such as urea, melamine, dicyanide, cyanamide, thiourea, ammonium thiocyanate, which provides a great possibility for practical application of g-C_3N_4. The hetero junction of g-C_3N_4/Fe_2O_3 hybrid showed enhanced photoreduction of CO_2 to CO at the rate of 27.2 $\mu mol.g^{-1}.h^{-1}$ without co-catalyst and sacrificial reagent, which is 2.2 times higher than that bare g-C_3N_4 (10.3 $\mu mol.g^{-1}.h^{-1}$) (Jiang et al. 2018). Li et al. (2016) hybridized mesoporous structured CeO_2 with g-C_3N_4 to form heterojunction photocatalyst for CO_2 reduction. The optimal CeO_2-g-C_3N_4 with 3 wt.% CeO_2 showed the highest rate of CH_4 conversion (13.88 $\mu mol.g^{-1}.h^{-1}$). When g-C_3N_4 was combined with 42 wt.% SnO_2 the CO_2 reduction rate reached 22.7 $\mu mol.g^{-1}.h^{-1}$, which was four and five times higher than those of g-C_3N_4 and P25 respectively. This is due to the formation of a heterojunction structure between two components, which efficiently promoted the separation of electron-hole pairs by a direct Z scheme mechanism to enhance the photocatalytic activity (He et al. 2015). Boron and phosphorus co-doped g-C_3N_4 coupled with SnO_2 exhibited an enhanced visible light activity for CO_2 conversion to CH_4 with a yield of 37.5 $\mu mol.g^{-1}.h^{-1}$ from CO_2-containing water (Raziq et al. 2017). Zhou et al. (Zhou et al. 2014) reported the coupling of g-C_3N_4 with n-TiO_2 and showed an enhanced activity CO_2 conversion rate of 12.2 $\mu mol.g^{-1}.h^{-1}$. Wang et al. synthesized novel 2D MnO_2-g-C_3N_4 heterojunction photocatalyst for enhanced CO_2 to CO conversion efficiency (Wang et al. 2017). Metal oxide frameworks have attracted intense attention due to their extremely high surface area, and availability in both inorganic and organic nature. g-C_3N_4 nanotubes integrated ZIF-8-Zn showed enhanced CH_3OH production efficiency by a factor of more than three times than that of the bare g-C_3N_4 (Shi et al. 2015). g-C_3N_4 decorated with carbon materials such as carbon dots, reduced graphene oxides have attracted great attention as an effective photocatalyst. Ong et al. (2017) reported that the enhanced CH_4 conversion rate of 29.23 $\mu mol.g^{-1}.h^{-1}$ after 10 hours of irradiation under visible light, which is 3.6 times higher than the bare g-C_3N_4. Ong et al. (Ong et al. 2015) constructed a 2D heterojunction nanocomposite containing g-C_3N_4-rGO with different ratios of rGO contents that exhibit remarkably enhanced activity than the bare g-C_3N_4. The optimized sample (g-C_3N_4-15 wt.% rGO) showed enhanced CH_4 evolution of 13.93 $\mu mol.g^{-1}.h^{-1}$in 10 h with a quantum efficiency of 0.56% and high stability. According to them the reason for this behavior was the charge modification at the surface of g-C_3N_4 extended the 2D/2D interlamination region and facilitated the separation of photo-induced charge carriers, thus improving photoreduction performance.

7.6 Metal-Organic Frameworks based Nano Composites

Metal-Organic Frameworks (MOF) are solid-state compounds consisting of metal ions or clusters coordinated to organic ligands and are demonstrably suitable for such catalytic reactions. Initial reports have shown the successful integration of known homogeneous CO_2 photoreduction catalysts into the backbone of a solid-state framework and demonstrated the photocatalytic CO_2 reduction activity of the obtained compounds. This concept has been expanded to integrate plasmonic metal clusters, metal oxides and photosensitizers into MOF architectures, thereby yielding new, high surface area, porous materials with tunable optoelectronic properties and catalytic activity. Reaction products were primarily formic acid, formate and CO; however, this concept could potentially be applied to design photoredox catalysts with high selectivity for CH_4.

7.7 Complex Oxide Structures based Photocatalysts

A combination of two or more metal oxides combined together to form a definite structure is called complex oxide structures. There are several types of complex oxide structures namely Perovskites, Spinels, Brownmillerites, etc. They are reported to be useful in several applications such as ionic conductors, dielectrics, piezoelectrics, thermoelectrics, magnetism, multiferroics, photocatalysis, etc. In this work, three complex oxides: Perovskites, Spinels and Brownmillerites are discussed for photocatalysis applications.

7.7.1 Perovskite-based Complex Oxide Nano Composites

Perovskite oxides (ABO_3) are one of the promising materials for photocatalytic applications due to their adaptation to modification through various degrees of tilting in octahedral sites, which in turn influences their band structures, electron-hole transport properties, etc. The advantages of perovskites over other binary oxides in photocatalytic properties are, (1) More optimum band edge potentials for photocatalytic activity, specifically, increased conduction band potential, and (2) Band structures can be adjusted by using appropriate A and/or B site cations in their lattice (Kanhere et al. 2012). Several perovskite materials have been reported as photocatalysts for hydrogen evolution through water splitting and/or organic pollutant removal (shown in Table 3.4).

Table 3.4 List of reported perovskites with different photocatalytic properties

Type	Photocatalyst	Photocatalytic property	References
Titanates	$BaTiO_3$	Hydrogen evolution rate of 30.8 $\mu mol.h^{-1}.g^{-1}$	Maeda 2014
	$CaTiO_3$	Hydrogen evolution rate of 670 $\mu mol.h^{-1}.g^{-1}$	Zhang et al. 2010
	$SrTiO_3$	Hydrogen evolution rate of 28 $\mu mol.h^{-1}.g^{-1}$	Townsend et al. 2012
	$MnTiO_3$	Rhodamine B degradation of 64% at 240 min	Dong et al. 2013
	$FeTiO_3$	Removal of 2-propanal	Kim et al. 2009
	$CoTiO_3$	Oxygen evolution rate of 64.8 $\mu mol.h^{-1}.g^{-1}$	Qu et al. 2014
	$NiTiO_3$	Nitrobenzene degradation of 98% at 240 min	Qu et al. 2012
	$PdTiO_3$	Methylene Blue degradation of 95% at 300 min	Li et al. 2012
	$CdTiO_3$	Rhodamine 6G degradation of 97% at 300 min	Hassan et al. 2014
	$LaTiO_3$	Congo–Red dye decolorization of 81% at 60 min	Bradha et al. 2015
Tantalates	$NaTaO_3$	Hydrogen evolution rate of 0.86 $\mu mol.h^{-1}.g^{-1}$	Kanhere et al. 2012
	$AgTaO_3$	Hydrogen evolution rate of 76 $\mu mol.h^{-1}.g^{-1}$	Ni et al. 2013
Niobates	$KNbO_3$	Hydrogen evolution rate of 5.3 $\mu mol.h^{-1}.g^{-1}$	Wang et al. 2013
	$NaNbO_3$	2-propanal degradation to acetone	Shi et al. 2009
	$AgNbO_3$	Oxygen evolution rate of 9.9 $\mu mol.h^{-1}.g^{-1}$	Li et al. 2009b
Vanadates	$AgVO_3$	Rhodamine B degradation of 64% at 150 min	Xu et al. 2012
Ferrites	$LaFeO_3$	Hydrogen evolution rate of 3315 $\mu mol.h^{-1}.g^{-1}$	Tijare et al. 2014
	$BiFeO_3$	Methylene Blue degradation of 94% at 240 min	Mohan 2014
	$GaFeO_3$	Hydrogen evolution rate of 289 $\mu mol.h^{-1}.g^{-1}$	Dhanasekaran et al. 2012
	$YFeO_3$	Rhodamine B degradation of 98% at 120 min	Tang et al. 2011
	$PrFeO_3$	Hydrogen evolution rate of 2847 $\mu mol.h^{-1}.g^{-1}$	Tijare et al. 2014
	$AlFeO_3$	Methyl Orange degradation of 98% at 40 minutes	Yuan et al. 2006
Others	$NaBiO_3$	Methylene Blue degradation of 99% at 40 minutes	Takei et al. 2011
	$LaCoO_3$	Methyl Orange degradation of 98% at 50 hours	Sun et al. 2010
	$LaNiO_3$	Methyl Orange degradation of 70% at 90 minutes	Tang et al. 2011
	$LaCrO_3$	Hydrogen evolution rate of 8.2 $\mu mol.h^{-1}.g^{-1}$	Shi et al. 2011

Titanate perovskites are highly reported photocatalysts due to their wider bandgap as well as improved conduction band potential than widely used photocatalyst TiO_2 (Kavan et al. 1996). They favor higher electron-hole recombination rates due to their wider bandgap. In order to reduce the recombination rate, it is necessary to use co-catalysts along with these photocatalysts. Among several titanates, $CaTiO_3$ is reported to exhibit a higher hydrogen evolution rate in presence of NiO as a co-catalyst. It was enhanced eight times after Cu doping in its Ti site as $CaTi_{0.98}Cu_{0.02}O_3$ (Zhang et al. 2010). $SrTiO_3$ is another important photocatalyst having a bandgap of > 3.2 eV. Several attempts were made to improve its photocatalytic activity by doping Mn, Ru, Rh and Ir in its Ti site (Konta et al. 2004) and making it a Z-scheme photocatalyst like Rh-SrTiO$_3$ with BiVO$_4$, AgNbO$_3$, WO$_3$, etc. (Sasaki et al. 2009). The disadvantages of these titanate perovskites are that they show higher photocatalytic efficiency under UV irradiation than visible light irradiation, expensive co-catalysts and preparation methods.

Niobates and Tantalates are also highly reported wider bandgap photocatalysts. They are more active under UV light irradiation than visible light. Recently, a solid solution of KNbO$_3$-BaNiNbO$_3$ perovskites showed ferroelectricity as well as photocatalytic properties (Grinberg et al. 2013). Doping several elements in NaTaO$_3$ such as La-N (Zhao et al. 2011) and La-Fe (Kanhere et al. 2012) improved its photocatalytic activity. The disadvantages of these perovskite photocatalysts are expensive preparation methods, availability and expensive co-catalysts. AgVO$_4$ and BiVO$_4$ are the reported photocatalyst among vanadate perovskites and the cost involved in its synthesis is high.

Among several perovskite oxides, niobates and ferrites are the only photocatalysts, which are active under visible light due to their narrower bandgap (i.e. > 2.6 eV). Their lower conduction band potential makes them less effective for hydrogen evolution through water splitting than organic pollutant removal. Therefore, they are used for degradation of several dyes such as Rhodamine B, Methylene Blue, Methyl Orange, Congo Red, etc., and degradation of organic pollutants like Phenol, 2-Propanal and conversion of methane, etc. The conduction band edges of these perovskites are made more suitable for photocatalytic water splitting by doping suitable elements or making as a composite for Z-scheme photocatalysts.

Among all, iron-based perovskites are earth-abundant, inexpensive photocatalysts having narrower bandgaps (Kanhere et al. 2012). Several iron-based perovskites such as LaFeO$_3$, BiFeO$_3$, GaFeO$_3$, etc., are reported as photocatalysts for hydrogen evolution as well as organic pollutants removal. LaFeO$_3$ is one of the highly reported photocatalysts compared to other ferrite perovskites for hydrogen evolution and organic pollutants removal.

7.7.1.1 Iron-based Perovskite Complex Oxides Nanocomposites

Fe$_2$O$_3$ is one of the non-toxic, earth-abundant, less expensive, narrow bandgap semiconductor photocatalyst. However, due to its higher valence band potential, it is hygroscopic in nature. This can be resolved by making it a complex oxide material with better chemical stability. Fortunately, the electronic structure of Fe in Fe$_2$O$_3$ has partially filled d orbitals, which supports Fe atom to form bonds easily with other atoms. A detailed explanation of complex oxide structures is given later.

LaFeO$_3$ is one of the highly reported perovskite photocatalysts for hydrogen evolution and organic pollutants removal due to its narrower bandgap and suitable band-edge potentials when compared to other ferrites like BiFeO$_3$, GaFeO$_3$, etc. However, its photocatalytic efficiency is found to be lower when compared to some of the other complex oxide photocatalysts. This was found to be due to the recombination of electron-hole pairs during the reaction as well as higher valence band potential and low conduction band potential. It was found that point defects in parent crystals act as recombination barriers during photocatalytic reactions. Like all perovskites, LaFeO$_3$ has the adaptability to incorporate suitable elements as dopants for property enhancement. The creation of point defects can be achieved by substituting suitable alkaline/transition metal ions in the A site or B site, or both in perovskites.

Doping aliovalent cation in the La site of LaFeO$_3$ leads to stabilization of multiple valence states of Fe as well as the formation of structural defects (Pecchi et al. 2011). Similarly doping at Fe site with lower valent cation will create charge imbalance. Based on the number of dopant ions the charge compensation can be achieved by a change of Fe valence state from Fe^{3+} to Fe^{4+} and the formation of oxygen vacancies (Parrino et al. 2016).

Several attempts to increase the photocatalytic performance of LaFeO$_3$ (LFO) were carried out by doping different elements. Table 3.5 shows the list of doped LFO and Table 3.6 shows LFO-in the form of composites with their enhanced photocatalytic properties. The reports show that composite LFO is more effective in photocatalytic hydrogen evolution as well as dye degradation

Table 3.5 List of reported doped LaFeO$_3$ used for photocatalytic applications

Compound	Preparation	Studies	Results
La$_{1-x}$Li$_x$FeO$_3$ (Li et al. 2006)	Sol-gel method	Photocatalytic Dye degradation	Methylene Blue degradation-46% at 300 minutes
Boron doped LaFeO$_3$ (Haitao et al. 2015)	Sol-gel method	Photocatalytic activity	Phenol degradation
LaFe$_{1-x}$Cu$_x$O$_3$ (Parrino et al. 2016)	Auto combustion	Photocatalytic activity	2-Propanol degradation
LaFe$_{1-x}$Cu$_x$O$_3$ (Juanjuan et al. 2015)	Citric acid complexing method	Photocatalytic water splitting	Hydrogen evolution of 343 μmol.h^{-1}.g^{-1}.
LaFe$_{0.9}$Cu$_{0.1}$O$_{3-\delta}$ (Qi et al. 2015)	Sol-gel method	Photo electrochemical water splitting	Photocurrent density-0.99 mA cm^2
LaFe$_{1-x}$Pd$_x$O$_3$ (Arnim et al. 2010)	Amorphous citrate-sol-gel method	Catalytic Activity	Methane combustion
La$_{0.9}$Ca$_{0.1}$FeO$_3$ (Fa-tang et al. 2010)	Micro emulsion method	Photocatalytic Dye degradation	Methylene Blue degradation-78% at 90 minutes
La$_{1-x}$Ca$_x$FeO$_3$ (Pecchi et al. 2011)	Citrate method	Catalytic Activity	Methane combustion
LaFe$_{0.9}$Mn$_{0.1}$O$_{3-\delta}$ (Wei et al. 2013)	Sol-gel method	Photocatalytic Dye degradation	Methyl Orange degradation-95% at 240 minutes
LaFe$_{0.9}$Mn$_{0.1}$O$_{3-\delta}$ (Qi Peng et al. 2015)	Sol-gel method	Photo electrochemical water splitting	Photocurrent density-0.52 mA/cm^2
LaFe$_{0.9}$Co$_{0.1}$O$_{3-\delta}$ (Qi et al. 2015)	Sol-gel method	Photo electrochemical water splitting	Photocurrent density-0.85 mA/cm^2
LaFe$_{1-x}$Zn$_x$O$_3$ (Shuhua et al. 2009)	sol-gel auto-combustion	Photocatalytic Dye degradation	Methylene Blue degradation-75% at 150 minutes
La$_{1-x}$Ce$_x$FeO$_3$ (Xiang-Ping et al. 2013)	Sol-gel method	Catalytic Activity	Methane combustion
La$_{1-x}$Sr$_x$FeO$_{3-\delta}$ (Isabella et al. 2013)	Sol-gel method	Photo electrochemical water splitting	Photocurrent density-2.9 mA.cm^2
LaFe$_{1-x}$Mg$_x$O$_3$ (Li et al. 2015)	Sol-gel method	Photocatalytic Dye degradation	Methyl Orange degradation-75% at 180 minutes
Ruthenium doped LaFeO$_3$ (Iervolino, et al. 2017)	Solution combustion method	Photocatalytic water splitting	Hydrogen evolution of 545 μmol.h^{-1}.g^{-1}

than doped LFO. Tijare et al. 2014 showed the highest rate of photocatalytic hydrogen evolution of 3315 μmol.h^{-1}.g^{-1} with Pt co-catalyst under 400 W Tungsten lamp in undoped LFO than others. They claimed that the reason for the higher hydrogen evolution rate is the smaller particle size of LFO, which reduces the migration of electron-hole pairs for recombination. Parida et al. 2010 prepared LFO by the sol-gel method and obtained photocatalytic hydrogen evolution of 1296 μmol.h^{-1}.g^{-1} without any co-catalysts under 125 W Hg-lamp. In addition, several attempts were made by them to increase its efficiency without using noble metal co-catalyst, such as RGO-LFO, Zn-Fe double hydroxide-LFO and g-C$_3$N$_4$-LFO Z-scheme nanocomposites. However, no further improvement was observed due to increased recombination rate rather than separation. Though undoped LFO photocatalyst showed higher percentage degradation of dyes such as methyl orange, Methylene blue and Rhodamine B, etc., the reaction was completed after a longer duration (i.e. 180 minutes). An attempt was made by Vijayaraghavan et al. 2020 to improve the photocatalytic efficiency of LFO in terms of better hydrogen evolution, dye degradation, and also faster rate of reaction by doping at La site by Ba, Ca, Sr and Fe site by Mg. The defects created by the dopants are expected to generate new active sites, which can also help to reduce the recombination during the photocatalytic reaction.

Table 3.6 List of reported $LaFeO_3$ composites for photocatalytic applications

Compound	Preparation	Studies	Results
Ag + $LaFeO_3$ (Yongchun et al. 2017)	Polyacrylamide-gel method	Photocatalytic Dye degradation	Rhodamine B degradation-91% at 360 minutes
RGO + $LaFeO_3$ (Xiao et al. 2015)	Sol-gel method	Photocatalytic Dye degradation	Methylene Blue degradation-98% at 70 minutes
RGO + $LaFeO_3$ (Acharya et al. 2020)	Sol-gel method	Photocatalytic water splitting	Hydrogen evolution of 316.24 $\mu mol.h^{-1}.g^{-1}$
g-C_3N_4 + $LaFeO_3$ (Acharya et al., 2017)	Co-precipitation method	Photocatalytic Dye degradation	Methylene Blue degradation-95% at 15 minutes with EDTA
g-C_3N_4 + $LaFeO_3$ (Acharya et al. 2017)	Co-precipitation method	Photocatalytic water splitting	Hydrogen evolution of 158 $\mu mol.h^{-1}.g^{-1}$
g-C_3N_4 + $LaFeO_3$ (Ke et al. 2017)	Sol-gel method	Photocatalytic water splitting	Hydrogen evolution of 1152 $\mu mol.h^{-1}.g^{-1}$
Zn-Fe + $LaFeO_3$ (Saumyaprava and Parida 2017)	Sol-gel method	Photocatalytic Dye degradation	Rhodamine B degradation-96% at 60 minutes.
Zn-Fe + $LaFeO_3$ (Saumyaprava and Parida 2017)	Sol-gel method	Photocatalytic water splitting	Hydrogen evolution of 963 $\mu mol.h^{-1}.g^{-1}$
TiO_2 + $LaFeO_3$ (Dhinesh et al. 2016)	Sol-gel method	Photocatalytic Dye degradation	Methyl orange degradation-85% at 180 minutes
$SrTiO_3$ + $LaFeO_3$ (Qian et al. 2016)	Ultrasonic spray pyrolysis method	Photocatalytic activity	NO_x removal
Ag_3PO_4 + $LaFeO_3$ (Yang et al. 2016)	*In-situ* precipitation method	Photo electrochemical water splitting	Photocurrent density-0.14 $mA.cm^2$
Co-Pi + $LaFeO_3$ (Peng et al. 2016)	Sol-gel + doctor blade method	Photo electrochemical water splitting	Photocurrent density-0.5 $mA.cm^2$
MMT + $LaFeO_3$ (Kang et al. 2016)	Sol-gel method	Photocatalytic Dye degradation	Rhodamine B degradation-99 % at 90 mins.

7.7.2   Spinel Complex Oxide-based Nano Composites

Spinels are a type of complex oxides possessing the general formula AB_2O_4, where A and B represent metal cations. These materials have a unique crystal structure where the two cations occupy tetrahedral and octahedral sites in an FCC lattice made up of oxygen atoms. Due to their stable crystal structure and good magnetic properties, they have been widely used in electronic devices and information storage systems. Several reports reveal that they exhibit various photocatalytic properties such as oxidative dehydration of hydrocarbons (Gibson and Hightower 1976), decomposition of alcohols and hydrogen peroxide (Manova et al. 2004), oxidation of carbon monoxide (Paldey et al. 2005) and degradation of dyes (Mathur et al. 2005; Wahi et al. 2005; Crini 2006; Erdemoglu et al. 2008; Huayue et al. 2009). Table 3.7 shows various spinel oxides used for different photocatalytic applications. Almost all spinel oxides exhibit a narrower bandgap except aluminates (MAl_2O_4, M = Co, Cu, Zn, Mg). Most of them are used as photocatalysts for the degradation of dyes due to their narrower bandgap and higher valence band potential. Among several spinel oxides, iron-based spinel oxides are highly reported photocatalysts due to their earth abundance and simple preparation methods.

Table 3.7 List of Spinel oxides reported for various photocatalytic properties

Photocatalysts	Photocatalytic property	Reference
$CuMn_2O_4$	PEC hydrogen evolution of 0.6 mA $cm^{-3}mg^{-1}h^{-1}$	Bessekhouad and Trari 2002
$NiCo_2O_4$	Degradation of Methylene Blue – 60%	Bai Cui et al. 2009
$CuCo_2O_4$	Degradation of Methyl Orange – 99%	Nithya et al. 2017
$ZrO_2+ZnCo_2O_4$	Degradation of 2–chloro phenol – 99%	Mousavi et al. 2016
$MnCo_2O_4$	Photocatalytic CO_2 reduction	Sibo et al. 2015
$CuGa_2O_4$	Hydrogen evolution rate of 1662 $\mu mol.h^{-1}.g^{-1}$	Karuppasamy et al. 2008
$RuO_2+ CuGa_{2-x}Fe_xO_4$	Hydrogen evolution rate of 4861 $\mu mol.h^{-1}.g^{-1}$	Rashid et al. 2004
$CuAl_2O_4$	Degradation of MO – 60%	Hassanzadeh-Tabrizi et al. 2016
$Co(Al_{1-x}Ga_x)_2O_4$	Degradation of MO – 72%	Kyureon et al. 2015
$MgAl_2O_4$	Degradation of MB– 99%	Fa-tang et al. 2011
$MgAl_2O_4$	Degradation of MB – 88%	Rahman and Jayaganthan 2015
$ZnAl_2O_4$	Degradation of Eosin B – 99%	Archana et al. 2018
$ZnCr_2O_4$	Degradation of MB – 87%	Cheng and Lian 2008
$ZnCr_2O_4$	Degradation of Eosin Y – 97%	Mousavi. et al. 2016
$CuCr_2O_4$	Degradation of MB – 98%	Yuan et al. 2014
$BaCr_2O_4$	Hydrogen evolution rate of 38 $\mu mol.h^{-1}.g^{-1}$	Defa et al. 2003
$BaFe_2O_4$	Hydrogen evolution rate of 47 $\mu mol.h^{-1}.g^{-1}$	Borse et al. 2011
$BaFe_2O_4$	Degradation of Methyl Orange – 65%	Yang et al. 2007
$CaFe_2O_4$	Hydrogen evolution rate of 6.94 $\mu mol.h^{-1}.g^{-1}$	Dom et al. 2017
$CaFe_2O_4 + MgFe_2O_4$	Hydrogen evolution rate of 82.7 $\mu mol.h^{-1}.g^{-1}$	Kim et al. 2009
$MgFe_2O_4$	PEC hydrogen evolution of 0.6 $cm^3mg^{-1}h^{-1}$	Zazoua et al. 2014
$NiFe_2O_4$	Degradation of Congo Red – 65%	Lixia et al. 2012
$CuFe_2O_4$	Degradation of Congo Red – 92%	Niyaz 2011
$MnFe_2O_4$	Degradation of Methylene Blue – 99%	Niyaz 2013a
$ZnFe_2O_4$	Hydrogen evolution rate of 134 $\mu mol.h^{-1}.g^{-1}$	Niyaz 2013b

7.7.2.1 Iron-based Spinel Complex Oxide Nano Composites

Most of the spinel ferrites are semiconductors with narrower bandgaps that enable visible light absorption. Their band edge potentials are suitable for either reduction of protons or/and oxidation of water (Xiaobo et al. 2010). Several spinel ferrites possess band edge potentials that satisfy the condition for photocatalytic reactions such as hydrogen evolution through water splitting (Table 3.8). These ferrites can be used in two forms for photocatalytic water splitting reactions: (1) in the form of photoelectrodes for photoelectrochemical water splitting reactions (2) as powder suspended in a mixture of water and other electron scavengers (Matsumoto et al. 1987).

Among all ferrites, $CaFe_2O_4$ is one of the promising *p*-type photocatalysts for hydrogen generation through water splitting and dye degradation applications. However, it has the disadvantage of low photoresponse due to the recombination of photogenerated carriers present in it. The addition of a co-catalyst lowers the over the potential for H_2 and O_2 evolution reactions and separates the electron-hole pairs at the surface of the photocatalyst (Jingrun et al. 2014). Several attempts have been made on $CaFe_2O_4$ to enhance its hydrogen evolution activity photo catalytically as a colloidal suspension (e.g. $Pt/CaFe_2O_4/MgFe_2O_4/RuO_2$ nanocomposite (Kim et al. 2009) or photo electrochemically by making it as a cathode and noble metal Pt (Ida et al. 2011; Ida et al. 2010), TiO_2 (Ida et al. 2010), ZnO (Ida et al. 2011) separately as an anode in a two or three-electrode system. Vijayaraghavan et al. 2018 reported an improved photocatalytic activity in $CaFe_2O_4$ without adding any external co-catalyst.

Table 3.8 List of spinel ferrites used as photocatalysts for water splitting reactions.

Applications	Mode of usage	Photocatalysts	References
Photoelectrochemical water splitting	Photo electrode	$CaFe_2O_4$ $NiFe_2O_4$, $ZnFe_2O_4$ $CoFe_2O_4$ $MgFe_2O_4$ $CaFe_2O_4$–$MgFe_2O_4$ composite	Kim et al. 2005; Rekhila et al. 2013; Borse et al. 2009; Boumaza et al. 2010; Dom, et al. 2012; Yang et al. 2013; Zazoua et al. 2014; Kim et al. 2009
Photocatalytic water splitting	Powder along with water and electron scavengers	$CaFe_2O_4$, $BaFe_2O_4$ $ZnFe_2O_4$	Shintaro et al. 2012; Borse et al. 2011; Jang et al. 2009; Vijayaraghavan et al. 2018

7.7.3 Brownmillerites Complex Oxides based Nanocomposites

Brownmillerites are oxides with a distorted structure formed due to the excess loss of oxygen from regular perovskite structures (Yang et al. 2006). Excess oxygen loss from BO_6 octahedron will transform it into BO_4 tetrahedron (Bertaut et al. 1959; Liu et al. 1996). There are few Brownmillerite structures reported as photocatalysts such as $Sr_2Co_2O_5$ (Saib et al. 2017), $Ca_2Fe_2O_5$ (Suchita et al. 2016; Sutka et al. 2018), $Sr_2Fe_2O_5$ (Saib et al. 2017). Yang et al. 2006 studied the difference in photocatalytic properties of perovskites and brownmillerites. They claimed that the reduced/modified BO_4 tetrahedron in brownmillerites consists of lesser amounts of electron-hole pairs than the BO_6 octahedron in perovskites. So perovskites show higher photocatalytic properties than brownmillerites. Recently, Sutka et al. 2018 showed enhanced photocatalytic property of $Ca_2Fe_2O_5$/Fe_2O_3 composite synthesized by solid-state method without any co-catalysts. Vijayaraghavan et al. (Vijayaraghavan et al. 2018) found that the doping of Mg in calcium ferrite resulted in a 19.6 times higher hydrogen evolution rate than undoped calcium ferrite due to the co-operative assistance of each phase present in the nanocomposite. The reasons for the higher hydrogen evolution capability of 1000 °C calcined Mg-doped calcium ferrite in this study could be suitable band edge potentials and presence of a combination of secondary oxides, like n-type Fe_2O_3 and $MgFe_2O_4$, p-type spinel phase $CaFe_2O_4$ and brownmillerite phase $Ca_2Fe_2O_5$, which assists each other and support the enhanced activity.

8   FUTURE CONSIDERATIONS

The photocatalytic reactions not only depend on the bandgap (i.e.) visible light activity of photocatalysts, but also particle size, surface area, crystallinity, carrier mobility and defects. Exploration of novel materials is important, in order to satisfy all the above factors and also chemical stability under different environmental conditions, eco-friendly and low material and preparation cost. Finding novel, cost-effective, heterostructures with easily available materials will improve the charge separation as well as quantum efficiency during photocatalytic reactions. Proper reactor design is very important for all photocatalytic reactions under different conditions. Gas-phase reactors are upcoming new technologies that also report higher quantum yields. *In-situ* techniques should be focused more to understand the mechanism during reduction reactions.

9   CONCLUSION

Photocatalytic hydrogen generation and CO_2 reduction are some of the promising green energy technologies. The present techniques are not suitable for industrial level usage due to several reasons such as lower conversion efficiency, long reaction time, complications in experimental methods, etc. It is expected that the material should give at least 10% of quantum efficiency in the visible region of the solar spectrum. The only way to make these processes industrially viable is through globalized collaboration on finding novel materials and optimizing process parameters. If successful, this initiative would ease the burden on the environment to a significant extent and improve life on this planet.

■ References

Archana, C., M. Akbar and M.M. Shaikh. 2018. Facile synthesis of phase pure $ZnAl_2O_4$ nanoparticles for effective photocatalytic degradation of organic dyes. Materials Science & Engineering B 227: 136–144.

Acharya, S., S. Mansingh and K.M. Parida. 2017. Enhanced photocatalytic activity of $g-C_3N_4-LaFeO_3$ for water reduction reaction through mediator free Z-scheme mechanism. Inorganic Chemistry Frontiers 4: 1022–1032.

Acharya, S., D.K. Padhi and K.M. Parida. 2020. Visible light driven $LaFeO_3$ nano sphere/RGO composite photocatalysts for efficient water decomposition reaction. Catalysis Today 353: 220–231.

Alexander, B.D., P.J. Kulesza, L. Rutkowska and R. Solarska. 2008. Metal oxide photoanodes for solar hydrogen production. Journal of Materials Chemistry 18: 2298–303.

Allen, M.R., A. Thibert, E.M. Sabio, N.D. Browning, D.S. Larsen and F.E. Osterloh. 2010. Evolution of physical and photocatalytic properties in the layered titanates $A_2Ti_4O_9$ (A = K, H) and in nanosheets derived by chemical exfoliation. Chemistry of Materials 22: 1220–1228.

Arnim, E., M. Peter, W. Alexander, H. Paul, S. Olga, F. Rento, et al. 2010. The effect of the state of Pd on methane combustion in Pd-doped $LaFeO_3$. Journal of Physical Chemistry C 114: 4584–4594.

Bai Cui, H. Lin, Y.-Z. Liu, J.-B. Li, P. Sun, X.-C. Zhao and C.-J. Liu. 2009. Photophysical and photocatalytic properties of core-ring structured $NiCo_2O_4$ nanoplatelets. Journal of Physical Chemistry C 113: 14083–14087.

Bak, T., J. Nowotny, M. Rekas and C.C. Sorrell. 2002. Photo-electrochemical hydrogen generation from water using solar energy. Materials-related aspects. International Journal of Hydrogen Energy 27: 991–1022.

Bertaut, E.F., P. Blum and A. Sagnieres. 1959. Structure du ferrite bicalcique et de la brownmillerite. Acta Crystallographica 12: 149–159.

Bessekhouad, Y. and M. Trari. 2002. Photocatalytic hydrogen production from suspension of spinel powders AMn_2O_4 (A = Cu and Zn). International Journal of Hydrogen Energy 27: 357–362.

Borse, P.H., J.S. Jang, S.J. Hong, J.S. Lee, J.H. Jung, T.E. Hong, et al. 2009. Photocatalytic hydrogen generation from water-methanol mixtures using nanocrystalline $ZnFe_2O_4$ under visible light irradiation. Journal of Korean Physical Society 55: 1472–1477.

Borse, P.H., C.R. Cho, K.T. Lim, Y.J. Lee, T.E. Hong, J.S. Bae, et al. 2011. Synthesis of barium ferrite for visible light photocatalysis applications. Journal of Korean Physical Society 58: 1672–1676.

Boumaza, S., A. Boudjemaa, A. Bouguelia, R. Bouarab and M. Trari. 2010. Visible light induced hydrogen evolution on new hetero-system $ZnFe_2O_4/SrTiO_3$. Applied Energy 87(7): 2230–2236

Bradha, M. and A. Ashok. 2014. Total conductivity in Sc-doped $LaTiO_3 + \delta$ perovskites. Ionics 20: 1343–1350.

Bradha, M., T. Vijayaraghavan, S.P. Suriyaraj, A. Anuradha and R. Selvakumar. 2015. Synthesis of photocatalytic $La_{(1-x)}A_xTiO_{3.5-\delta}$ (A=Ba, Sr, Ca) nano perovskites and their application for photocatalytic oxidation of congo red dye in aqueous solution. Journal of Rare Earths 33(2): 160–167.

Chequer, F.M.D., G.A.R. de Oliveira, E.R.A. Ferraz, J.C. Cardoso, M.V.B. Zanoni and D.P. de Oliveira. 2013. Textile dyes: Dyeing process and environmental impact. pp. 154–176. In: Melih Günay (ed.). Eco-Friendly Textile Dyeing and Finishing. London, United Kingdom: IntechOpen. [Online]. Available: https://www.intechopen.com/chapters/41411 doi: 10.5772/53659.

Chae. J., J. Lee, J. Jeong and H. Kang. 2009. Hydrogen production from photo splitting of water using the ga-incorporated TiO_2 prepared by a solvothermal method and their characteristics. Bulletin of the Korean Chemical Society 30: 302–308.

Cheng, P. and G. Lian. 2008. Optical and photocatalytic properties of spinel $ZnCr_2O_4$ nanoparticles synthesized by a hydrothermal route. Journal of American Ceramic Society 91: 2388–2390.

Choi, H.J. and M. Kang. 2007. Hydrogen production from methanol/water decomposition in a liquid photosystem using the anatase structure of Cu loaded TiO_2. International Journal of Hydrogen Energy 2007 32: 3841–3848.

Crini, G. 2006. Non-conventional low-cost adsorbents for dye removal: A review. Bioresource Technology 97(9): 1061–1085.

Defa, W., Z. Zhigang and Y. Jinhua. 2003. A new spinel-type photocatalyst $BaCr_2O_4$ for H_2 evolution under UV and visible light irradiation. Chemical Physics Letters 373: 191–196.

Del Valle. F., A. Ishikawaa, K. Domena, J.A. Villoria de La Mano, M.C. Sánchez-Sánchez, I.D. González, S. Herreras, et al. 2009. Influence of Zn concentration in the activity of $Cd_{1-x}Zn_xS$ solid solutions for water splitting under visible light. Catalysis Today 143: 51–56.

Dhanasekaran, P. and N.M. Gupta. 2012. Factors affecting the production of H_2 by water splitting over a novel visible-light-driven photocatalyst $GaFeO_3$. International Journal of Hydrogen Energy 37: 4897–4907.

Dhinesh, K.R. and R. Jayavel. 2014. Facile hydrothermal synthesis and characterization of $LaFeO_3$ nanospheres for visible light photocatalytic applications. Journal of Materials Science: Materials in Electronics 25: 3953–3961.

Dhinesh, K.R., R. Thangappan and R. Jayavel 2016. Synthesis and characterization of $LaFeO_3/TiO_2$ nanocomposites for visible light photocatalytic activity. Journal of Physics and Chemistry of Solids 101: 25–39.

Dom, R., R. Subasri, K. Radha and P.H. Borse. 2011. Synthesis of solar active nanocrystalline ferrite, MFe_2O_4 (M: Ca, Zn, Mg) photocatalyst by microwave irradiation. Solid State Communications 151: 470–473.

Dom, R., R. Subasri, N.Y. Hebalkar, A.S. Chary and P.H. Borse. 2012. Synthesis of a hydrogen producing nanocrystalline $ZnFe_2O_4$ visible light photocatalyst using a rapid microwave irradiation method. RSC Advances 2: 12782–12791.

Dom, R., H.G. Kim and P.H. Borse. 2017. Photo chemical hydrogen generation from orthorhombic $CaFe_2O_4$ nanoparticles synthesized by different methods. Chemistry Select 2: 2556–2564.

Dong, W., D. Wang, L. Jiang, H. Zhu, H. Huang, J. Li, et al. 2013. Synthesis of F doping $MnTiO_3$ nanodiscs and their photocatalytic property under visible light. Materials Letters 98: 265–268.

Erdemoglu, S., S.K. Aksu, F. Sayılkan, B. Izgi, M. Asiltürk, H. Sayılkan, et al. 2008. Photocatalytic degradation of congo red by hydrothermally synthesized nanocrystalline TiO_2 and identification of degradation products by LC–MS. Journal of Hazardous Materials 155: 469–472.

Fang, J., F. Shi, J. Bu, J. Ding, S. Xu, J. Bao, et al. 2010. Bifunctional N-doped mesoporous TiO_2 photocatalysts. Journal of Physical Chemistry C 112(46): 18150–18156.

Fa-tang, L., L. Ying, L. Rui-hong, S. Zhi-min, Z. Di-shun and K. Cheng-guang. 2010. Preparation of Ca-doped $LaFeO_3$ nanopowders in a reverse microemulsion and their visible light photocatalytic activity. Materials Letters 64: 223–225.

Fa-tang, L., Z. Ye, L. Ying, H. Ying-juan, L. Rui-hong and Z. Di-shun. 2011. Solution combustion synthesis and visible light-induced photocatalytic activity of mixed amorphous and crystalline $MgAl_2O_4$ nanopowders. Chemical Engineering Journal 173: 750–759.

Fedorov, V.A., V.A. Ganshing and Y.N. Norkeshko. 1993. Solid-state phase diagram of the zinc sulfide-cadmium sulfide system. Material Research Bulletin 28: 59–66.

Fu, H., L. Zhang, S. Zhang, Y. Zhu and J. Zhao. 2006. Electron spin resonance spin-trapping detection of radical intermediates in N-doped TiO_2-assisted photodegradation of 4-chlorophenol. Journal of Physical Chemistry B 110: 3061–3065.

Fujii, H.M., K. Ohtaki and H.A. Eguchi. 1998. Preparation and photocatalytic activities of a semiconductor composite of CdS embedded in a TiO_2 gel as a stable oxide semiconducting matrix. Journal of Molecular Catalysis A 129: 61–68.

Fujishima, A. and K. Honda. 1972. Electrochemical photolysis of water at a semiconductor electrode. Nature 238: 37–38.

Gibson, M.A. and J.W. Hightower. 1976. Oxidative dehydrogenation of butenes over magnesium ferrite kinetic and mechanistic studies. Journal of Catalysis 41: 420–430.

Grinberg, I., D.V. West, M. Torres, G. Gou, D.M. Stein, L. Wu, et al. 2013. Perovskite oxides for visible-light-absorbing ferroelectric and photovoltaic materials. Nature 503: 509–512.

Gusain, R., P. Kumar, O.P. Sharma, S.L. Jain and O.P. Khatri. 2016. Reduced graphene oxide–CuO nanocomposites for photocatalytic conversion of CO_2 into methanol under visible light irradiation. Applied Catalysis B: Environmental 181: 352–362.

Haitao, W., H. Ruisheng, Z. Tingting, L. Chun and J.Y. WanwanMeng. 2015. A novel efficient boron-doped $LaFeO_3$ photocatalyst with large specific surface area for phenol degradation under simulated sunlight. Crystal Engineering Communications 17: 3859–3865.

Han, Z., G. Jianjun, L. Peng, F. Tongxiang, D. Zhang and Y. Jinhua. 2013. Leaf-architectured 3D hierarchical artificial photosynthetic system of perovskite titanates towards CO_2 photoreduction into hydrocarbon fuels. Scientific Reports 3: 1667.

Hassan, M.A., T. Amna and M.S. Khil. 2014. Synthesis of high aspect ratio $CdTiO_3$ nanofibers via electrospinning: Characterization and photocatalytic activity. Ceramics Internationa l40: 423–427.

Hassanzadeh-Tabrizi, S.A., Reza Pournajaf, Amin Moradi-Faradonbeh and Sayedkayvan Sadeghinejad. 2016. Nanostructured $CuAl_2O_4$: Co-precipitation synthesis, optical and photocatalytic properties. Ceramics International 42 (12): 14121–14125.

He, Y., L. Zhang, M. Fan, X. Wang, M.L. Walbridge, Q. Nong, et al. 2015. Z-scheme SnO_{2-x}/g-C_3N_4 composite as an efficient photocatalyst for dye degradation and photocatalytic CO_2 reduction. Solar Energy Materials and Solar Cells 137: 175–184.

Hou, W., W.H. Hung, P. Pavaskar, A. Goeppert, M. Aykol and S.B. Cronin. 2011. Photocatalytic conversion of CO_2 to hydrocarbon fuels via plasmon-enhanced absorption and metallic interband transitions. ACS Catalysis 1: 929–936.

Hu, C.C., J.N. Nian and H. Teng. 2008. Electrodeposited p-type Cu_2O as photocatalyst for H_2 evolution from water reduction in the presence of WO_3. Solar Energy Materials and Solar Cells 92(9): 1071–1073.

Huang, Z.F., J. Song, L. Pan, X. Zhang, L. Wang and J.J. Zou. 2015. Tungsten oxides for photocatalysis, electrochemistry, and phototherapy. Advanced Materials 27: 5309–5327.

Huang, L., X.L. Wang, J.H. Yang, G. Liu, J.F. Han and C. Li. 2013. Dual cocatalysts loaded type I CdS/ZnS core/shell nanocrystals as effective and stable photocatalysts for H_2 evolution. Journal of Physical Chemistry C 117: 11584.

Huayue, Z., R. Jiang, L. Xiao, Y. Chang, Y. Guan, X. Li, et al. 2009. Photocatalytic decolorization and degradation of congo red on innovative crosslinked chitosan/nano-CdS composite catalyst under visible light irradiation. Journal of Hazardous Materials 169: 933–940.

Ida, S., K. Yamada, T. Matsunaga, H. Hagiwara, Y. Matsumoto and T. Ishihara. 2010. Preparation of p-type $CaFe_2O_4$ photocathodes for producing hydrogen from water. Journal of American Chemical Society 132: 17343–17345.

Ida, S., K. Yamada, T. Matsunaga, H. Hagiwara, T. Ishihara, T. Taniguchi, et al. 2011. Photoelectrochemical hydrogen production from water using p-type $CaFe_2O_4$ and n-type ZnO. Electrochemistry 79: 797–800.

Iervolino, G., V. Vaiano, D. Sannino, L. Rizzo and P. Ciambelli. 2015. Production of hydrogen from glucose by $LaFeO_3$ based photocatalytic process during water treatment. International Journal of Hydrogen Energy 41 (2): 959–966.

Iervolino, G., V. Vaiano, D. Sannino, L. Rizzo and V. Palma. 2017. Enhanced photocatalytic hydrogen production from glucose aqueous matrices on Ru-doped $LaFeO_3$ Applied Catalysis B, Environmental 207: 182–194.

Inoue, T., A. Fujishima, S. Konishi and K. Honda. 1979. Photoelectrocatalytic reduction of carbon dioxide in aqueous suspensions of semiconductor powders. Nature 277: 637–638.

Irie, H., Y. Watanabe and K. Hashimoto. 2003. Carbon-doped anatase TiO_2 powders as a visible-light sensitive photocatalyst. Chemistry Letters 32: 772–773.

Isabella, N.S., F. Fontana, R. Passalacqua, C. Ampelli, S. Perathoner, G. Centi, et al. 2013. Photoelectrochemical properties of doped lanthanum orthoferrites. Electrochimica Acta 109: 710–715.

Jang, J.S., H.G. Kim, U.A. Joshi, J.W. Jang, and J.S. Lee. 2008. Fabrication of CdS nanowires decorated with TiO_2 nanoparticles for photocatalytic hydrogen production under visible light irradiation. International Journal of Hydrogen Energy 33: 5975–5980.

Jang, J.S., S.J. Hong, J.S. Lee, P.H. Borse, Ok-S. Jung, T.E. Hong, et al. 2009. Synthesis of zinc ferrite and its photocatalytic application under visible light. Journal of the Korean Physical Society 54: 204–208.

Jiang. Z., W. Wan, H. Li, S. Yuan, H. Zhao and P.K. Wong. 2018. A hierarchical Z-scheme α-Fe_2O_3/g-C_3N_4 Hybrid for Enhanced Photocatalytic CO_2 Reduction. Advanced Materials 30: 1706108.

Jing, D., Y. Zhang and L. Guo. 2005. Study on the synthesis of Ni doped mesoporous TiO_2 and its photocatalytic activity for hydrogen evolution in aqueous methanol solution. Chemical Physics Letters 415: 74–78.

Jingjing, S., R. Fazal, H. Muhammad, Z. Wei, Q. Yang, W. Guofeng, et al. 2017. Improved charge separation and surface activation via boron-doped layered polyhedron $SrTiO_3$ for co-catalyst free photocatalytic CO_2 conversion. Applied Catalysis B: Environmental 219: 10–17.

Jingrun, R., Z. Jun, Y. Jiaguo, J. Mietek and Q. Shi Zhang. 2014. Earth-abundant cocatalysts for semiconductor-based photocatalytic water splitting. Chemical Society Reviews 43: 7787–7812.

Juanjuan, L., P. Xinwei, X. Yingrui, J. Lishan, Y. Xiaodong and F. Weiping. 2015. Synergetic effect of copper species as cocatalyst on $LaFeO_3$ for enhanced visible-light photocatalytic hydrogen evolution. International Journal of Hydrogen Energy 40: 13918–13925.

Kadowaki, H., N. Saito, H. Nishiyama, H. Kobayashi, Y. Shimodaira and Y. Inoue. 2007. Overall splitting of water by RuO_2-Loaded $PbWO_4$ Photocatalyst with $d^{10}s^2$-d0 Configuration. Journal of Physical Chemistry C 111: 439.

Kamat, P.V. 2002. Photophysical, photochemical and photocatalytic aspects of metal nanoparticles. Journal of Physical Chemistry B 106(32): 7729–7747.

Kang, P., F. Liangjie, Y. Huaming and Q. Jing. 2016. Perovskite $LaFeO_3$/montmorillonite nanocomposites: Synthesis, interface characteristics and enhanced photocatalytic activity. Scientific Reports 6: 19723.

Kanhere, P., J. Nisar, Y. Tang, B. Pathak, R. Ahuja, J. Zheng and Z. Chen. 2012. Electronic Structure, Optical Properties, and Photocatalytic Activities of $LaFeO_3$-$NaTaO_3$ Solid Solution. Journal of Physical Chemistry C 116: 22767–22773.

Kanhere, P.D., J. Zheng and Z. Chen. 2012. Site specific optical and photocatalytic properties of Bi-doped $NaTaO_3$. Journal Physical Chemistry C 115(23): 11846–11853.

Karuppasamy, G., B. Jin-Ook, M.L. Sang, S. Esakkiappan, M. Sang-Jin and K. Ki-jeong. 2008. Visible light active pristine and Fe_3+ doped $CuGa_2O_4$ spinel photocatalysts for solar hydrogen production. International Journal of Hydrogen Energy 33: 2646–2652.

Kavan, L., M. Grätzel, S. Gilbert, C. Klemenz and H. Scheel. 1996. Electrochemical and photoelectrochemical investigation of single-crystal anatase. Journal of American Chemical Society 118: 6716–6723.

Ke, X. and F. Jian. 2017. Superior photocatalytic performance of $LaFeO_3$/g-C_3N_4 heterojunction nanocomposites under visible light irradiation. RSC Advances 7: 45369–45376.

Kim, H.G., P.H. Borse, W. Choi and J.S. Lee. 2005. Photocatalytic nanodiodes for visible-light photocatalysis. Angewandte Chemie 44: 4585–4589.

Kim, H.G., P.H. Borse, J.S. Jang, J.O.S. Jeong, Y.J. Suh and J.S. Lee. 2009. Fabrication of $CaFe_2O_4$/$MgFe_2O_4$ bulk heterojunction for enhanced visible light photocatalysis. Chemical Communications 13: 5889–5891.

Kim, Y.J., B. Gao, S.Y. Han, M.H. Jung, A.K. Chakraborty, T. Ko, et al. 2009. Heterojunction of $FeTiO_3$ nanodisc and TiO_2 nanoparticle for a novel visible light photocatalyst. Journal of Physical Chemistry C 113(44): 19179–19184.

Kim, H.R., A. Razzaq and C.A. Grimes. 2017. Heterojunction p-n-p Cu_2O/S-TiO_2/CuO: Synthesis and application to photocatalytic conversion of CO_2 to methane. Journal of CO_2 Utilization 20: 91–96.

Konta, R., T. Ishii, H. Kato and A. Kudo. 2004. Photocatalytic activities of noble metal ion doped SrTiO3 under visible light irradiation. Journal of Physical Chemistry B 108 (26): 8992–8995

Kudo, A. and H. Kato. 1997. Photocatalytic activities of $Na_2W_4O_{13}$ with layered structure. Chemistry Letters 26(5): 421–422.

Kudo, A. and S. Hijii. 1999. H_2 or O_2 Evolution from aqueous solutions on layered oxide photocatalysts consisting of Bi^{3+} with $6s^2$ configuration and d^0 transition metal ions. Chemistry Letters 28(10): 1103–1104.

Kudo, M.S. 1999. Photocatalytic H_2 evolution under visible light irradiation on $Zn_{1-x}Cu_xS$ solid solution. Chemistry Letters 58: 241–243.

Kudo, M.S. 2000. Photocatalytic H_2 evolution under visible light irradiation on Ni-doped ZnS photocatalyst. Chemical Communications 15: 1371–1372.

Kudo, A., R. Niishiro, A. Iwase and H. Kato. 2007. Effects of doping of metal cations on morphology, activity, and visible light response of photocatalysts. Chemical Physics 339: 104–110.

Kyureon, L., A.R. Daniel, D. Gordana and R.N. Nathan. 2015. Synthesis, optical, and photocatalytic properties of cobalt mixed-metal spinel oxides $Co(Al_{1-x}Gax)_2O_4$. Journal of Materials Chemistry A 3: 8115–8122.

Lalitha, K., J.K. Reddy, M.V.P. Sharma, V.D. Kumari and M. Subrahmanyam. 2010. Continuous hydrogen production activity over finely dispersed Ag_2O/TiO_2 catalysts from methanol: water mixtures under solar irradiation: a structure–activity correlation. International Journal of Hydrogen Energy 35: 3991–4001.

Leny, Y., Y. Jae-Hun, W. Xinchen, M. Kazuhiko, T. Tsuyoshi, A. Markus, et al. 2010. Highly active tantalum(v) nitridenanoparticles prepared from a mesoporous carbon nitride template for photocatalytic hydrogen evolution under visible light irradiation. Journal of Materials Chemistry A 20: 4295–4298.

Li, H., S. Guifang, L. Kuan, L. Yong and G. Faming. 2006. Preparation, characterization and investigation of catalytic activity of Li-doped $LaFeO_3$ nanoparticles. Journal of Sol-gel Science and Technology 40: 9–14.

Li, G., J.C. Yu, D. Zhang, X. Hu and W.M. Lau. 2009a. A mesoporous TiO_2–xNx photocatalyst prepared by sonication pretreatment and in situ pyrolysis. Separation and Purification Technology 67(2): 152–157.

Li, G., S. Yan, Z. Wang, X. Wang, Z. Li, J. Ye and Z. Zou. 2009b. Synthesis and visible light photocatalytic property of polyhedron-shaped $AgNbO_3$. Dalton Transactions 40: 8519–8524.

Li, L., Y. Zhang, A.M. Schultz, X. Liu, P.A. Salvador and G.S. Rohrer. 2012. Visible light photochemical activity of heterostructured $PbTiO_3$–TiO_2 core-shell particles. Catalysis Science and Technology 2: 1945–1952.

Li, X., Z. Zongjin, L. Wei and P. Huiqi. 2012. Photocatalytic reduction of CO_2 over noble metal-loaded and nitrogen-doped mesoporous TiO_2 Applied Catalysis A: General. 429–430: 31–38.

Li, J.K., X. Liu, L. Huang and Y. Wang. 2015. Preparation of high specific surface area micro/meso-porous SiOC ceramics by the low temperature phase separation method. Journal of Inorganic Materials 30: 1223–1227.

Li, M., L. Zhang, M. Wu, Y. Du, X. Fan, M. Wang, L. Zhang, et al. 2016. Mesostructured CeO_2/g-C_3N_4 nanocomposites: Remarkably enhanced photocatalytic activity for CO_2 reduction by mutual component activations. Nano Energy 19: 145–155.

Liang, Y.T., B.K. Vijayan, K.A. Gray and M.C. Hersam. 2011. Minimizing graphene defects enhances titania nanocomposite-based photocatalytic reduction of CO_2 for improved solar fuel production. Nano Letters 11: 2865–2870.

Liang. Y.T., B.K. Vijayan, O. Lyandres, K.A. Gray and M.C. Hersam. 2012. Effect of dimensionality on the photocatalytic behavior of carbon–titania nanosheet composites: Charge transfer at nanomaterial interfaces. Journal of Physical Chemistry Letters 3: 1760–1765.

Lingqiao, K., Y. Junqing, L. Ping and L. Shengzhong. 2018. Fe_2O_3/C–C_3N_4-Based tight heterojunction for boosting visible-light-driven photocatalytic water oxidation. ACS Sustainable Chemical Engineering 6(8): 10436–10444.

Lin, J., J. Hu, C. Qiu, H. Huang, L. Chen, Y. Xie, et al. 2019. *In situ* hydrothermal etching fabrication of $CaTiO_3$ on TiO_2 nanosheets with heterojunction effects to enhance CO_2 adsorption and photocatalytic reduction. Catalysis Science & Technology 9: 336–346.

Liu, L.M., T.H. Lee, Y.L. Qiu, Y.L. Yang and A.J. Jacobson. 1996. A thermogravimetric study of the phase diagram of strontium cobalt iron oxide, $SrCo_{0.8}Fe_{0.2}O_{3-\delta}$. Materials Research Bulletin 31(1): 29–35.

Liu, S.H., H.P. Wang, Y.J. Huang, Y.M. Sun, K.S. Lin and M.C. Hsiao. 2003. Photodecomposition of water catalyzed by Zr- and Ti-MCM-41. Energy Sources 25: 591–596.

Liu. E., L. Kang, F. Wu, T. Sun, X. Hu, Y. Yang, et al. 2014. Photocatalytic reduction of CO_2 into methanol over Ag/TiO_2 nanocomposites enhanced by surface plasmon resonance. Plasmonics 9: 61–70.

Liu, E.L., L. Qi, J. Bian, Y. Chen, X. Hu, J. Fan, et al. 2015. A facile strategy to fabricate plasmonic Cu modified TiO_2 nano-flower films for photocatalytic reduction of CO_2 to methanol. Materials Research Bulletin 68: 203–209.

Lixia, W., L. Jianchen, W. Yingqi, Z. Lijun and J. Qing. 2012. Adsorption capability for congo red on nanocrystalline MFe_2O_4 (M = Mn, Fe, Co, Ni) spinel ferrites. Chemical Engineering Journal 181–182: 72–79.

Machida, M., X.W. Ma, H. Taniguchi, J. Yabunaka and T. Kijima. 2000. Pillaring and photocatalytic property of partially substituted layered titanates, $Na_2Ti_{3-x}M_xO_7$ and $K_2Ti_{4-x}M_xO_9$ (M = Mn, Fe, Co, Ni, Cu). Journal of Molecular Catalysis A : Chemistry 155: 131–142.

Maeda, K. 2014. Rhodium-doped barium titanate perovskite as a stable p-type semiconductor photocatalyst for hydrogen evolution under visible light. ACS Applied Materials and Interfaces 6: 2167–2173.

Manova, E., T. Tsoncheva, D. Paneva, I. Mitov, K. Tenchev and L. Petrov. 2004. Mechanochemically synthesized nano-dimensional iron–cobalt spinel oxides as catalysts for methanol decomposition. Applied Catalysis A 277: 119–127.

Mathur, N., P. Bhatnagar and P. Bakre. 2005. Assessing mutagenicity of textile dyes from pali (rajasthan) using ames bioassay. Applied Ecology and Environmental Research 4: 111–118.

Matsumoto, Y., M. Omae, K. Sugiyama and E. Sato. 1987. New photocathode materials for hydrogen evolution: calcium iron oxide ($CaFe_2O_4$) and strontium iron oxide ($Sr_7Fe_{10}O_{22}$). Journal of Physical Chemistry 91: 577–581.

Mohan, S., B. Subramanian, I. Bhaumik, P.K. Gupta and S.N. Jaisankar. 2014. Nanostructured $Bi_{(1-x)}Gd_{(x)}$ FeO_3 – a multiferroic photocatalyst on its sunlight driven photocatalytic activity. RSC Advances 4: 16871–16878.

Moon, S.C., H. Mametsuka, E. Suzuki and A. Masakazu. 1998. Stoichiometric decomposition of pure water over Pt-loaded Ti/B binary oxide under UV-irradiation. Chemistry Letters 27: 117–118.

Moon, S.C., H. Mametsuka, S. Tabata and E. Suzuki. 2000. Photocatalytic production of hydrogen from water using TiO_2 and B/TiO_2 Catal Today 58: 12–132.

Mousavi, Z., F. Soofivand, M. Esmaeili-Zare, M. Salavati-Niasari and S.B. Bagheri. 2016. $ZnCr_2O_4$ Nanoparticles: Facile synthesis, characterization and photocatalytic properties. Scientific Reports 6: 20071.

Muhammad, T. 2020. Well-designed $ZnFe_2O_4/Ag/TiO_2$ nanorods heterojunction with Ag as electron mediator for photocatalytic CO_2 reduction to fuels under UV/visible light. Journal of CO_2 Utilization 37: 134–146.

Navarro, R.M., F. Del Valle and J.L.G. Fierro. 2008. Photocatalytic hydrogen evolution from CdS–ZnO–CdO systems under visible light irradiation: Effect of thermal treatment and presence of Pt and Ru cocatalysts. International Journal of Hydrogen Energy 33: 4265–4273.

Nayeem, A., G. Vajralingam, K. Yadaiah, P. Mahesh and M. Nagabhooshanam. 2001. Synthesis and characterization of cd1-xznxs:cu crystals by co-precipitation method. International Journal of Modern Physics B 15: 2387 – 2407.

Neaţu, S.J.A. Maciá-Agulló, P. Concepción and H. Garcia. 2014. Gold–copper nanoalloys supported on TiO_2 as photocatalysts for CO_2 reduction by water. Journal of the American Chemical Society 136: 15969–15976.

Ni, L., M. Tanabe and H. Irie. 2013. A visible-light-induced overall water-splitting photocatalyst: Conduction-band-controlled silver tantalite. Chemical communications 49: 10094–10096.

Nishijima, K., T. Kamai, N. Murakami, T. Tsubota and T. Ohno. 2007. Photocatalytic hydrogen or oxygen evolution from water over S- or N-doped under visible light. International Journal of Photoenergy 2008: 173943.

Nithya, J.S.M., J.Y. Do and M. Kang. 2017. Fabrication of flower-like copper cobaltite/graphitic-carbon nitride ($CuCo_2O_4/g\text{-}C_3N_4$) composite with superior photocatalytic activity. Journal of Industrial Engineering Chemistry 57: 405–415.

Niyaz, M.M. 2011. Photocatalytic ozonation of dyes using copper ferrite nanoparticle prepared by co-precipitation method. Desalination 279: 332–337.

Niyaz, M.M. 2013a. Preparation of PVA-chitosan blend nanofiber and its dye removal ability from colored wastewater. Desalination and Water Treatment 51: 1–7.

Niyaz, M.M. 2013b. Zinc ferrite nanoparticle as a magnetic catalyst: Synthesis and dye degradation. Materials Research Bulletin 48(10): 4255–4260.

Ong, W.J, L.L. Tan, S.P. Chai, S.T. Yong and A.R. Mohamed. 2015. Surface charge modification via protonation of graphitic carbon nitride (g-C3N4) for electrostatic self-assembly construction of 2D/2D reduced graphene oxide ($rGO)/g\text{-}C_3N_4$ nanostructures toward enhanced photocatalytic reduction of carbon dioxide to methane. Nano Energy 13: 757–770.

Ong, W.J., L.K. Putri, Y.C. Tan, L.L. Tan, N. Li, Y.H. Ng, et al. 2017. Unravelling charge carrier dynamics in protonated g-C_3N_4 interfaced with carbon nanodots as co-catalysts toward enhanced photocatalytic CO_2 reduction: A combined experimental and first-principles DFT study. Nano Research 10: 1673–1696.

Onsuratoom, S., T. Puangpetch and S. Chavadej. 2011. Comparative investigation of hydrogen production over Ag-, Ni-, and Cu-loaded mesoporous-assembled TiO_2–ZrO_2 mixed oxide nanocrystal photocatalysts. Chemical Engineering Journal 173: 667–675.

Paldey, S., S. Gedevanishvili, W. Zhang and F. Rasouli. 2005. Evaluation of a spinel based pigment system as a CO oxidation catalyst. Applied Catalysis B 56: 241–250.

Parida K.M. and N. Sahu. 2008. Visible light induced photocatalytic activity of rare earth titania nanocomposites. Journal of Molecular Catalysis A 287(1-2): 151–158.

Parida, K.M., K.H. Reddy, S. Martha, D.P. Das and N. Biswal. 2010. Fabrication of nanocrystalline $LaFeO_3$: An efficient sol-gel auto-combustion assisted visible light responsive photocatalyst for water decomposition. International Journal of Hydrogen Energy 35: 12161– 12168.

Park, J.W. and M. Kang. 2007. Synthesis and characterization of Ag_xO, and hydrogen production from methanol photodecomposition over the mixture of Ag_xO and TiO_2. International Journal of Hydrogen Energy 32: 4840–4846.

Park, M.S. and M. Kang. 2008. The preparation of the anatase and rutile forms of Ag–TiO_2 and hydrogen production from methanol/water decomposition. Materials Letters 62: 183–187.

Park, H., W. Choi and M.R. Hoffmann. 2008. Effects of the preparation method of the ternary CdS/TiO_2/Pt hybrid photocatalysts on visible light-induced hydrogen production. Journal of Materials Chemistry 18: 2379–2385.

Parrino, F., E. Garcia-Lopez, G. Marci, L. Palmisano, V. Fedlice, I. NataliSora, et al. 2016. Cu-substituted lanthanum ferrite perovskites: Preparation, characterization and photocatalytic activity in gas-solid regime under simulated solar light irradiation. Journal of Alloys and Compounds 682: 686–694.

Pecchi, G., M.G. Jiliberto, A. Buljan and E. Delgado. 2011. Relation between defects and catalytic activity of calcium doped $LaFeO_3$ perovskite. Journal of Solid State Ionics 187: 27–32.

Peng, Q., J. Wang, Y.W. Wen, B. Shan and R. Chen. 2016. Surface modification of $LaFeO_3$ by Co-Pi electrochemical deposition as an efficient photoanode under visible light. RSC Advances 6: 26192–26198.

Qi, P., S. Bin, W. Yanwei and C. Rong. 2015. Enhanced charge transport of $LaFeO_3$ via transition metal (Mn, Co, Cu) doping for visible light photoelectrochemical water oxidation. International Journal of Hydrogen Energy 40: 15423–15431.

Qian, Z., Y. Huanga, S. Peng, Y. Zhang, Z. Shen, J. Cao, et al. 2016. Perovskite $LaFeO_3$-$SrTiO_3$ composite for synergistically enhanced NO removal under visible light excitation. Applied Catalysis B, Environmental 204: 346–357.

Qianyi, Z., L. Ying, A. Erik, M.G.J. Ackerman and L. Hailong. 2011. Visible light responsive iodine-doped TiO_2 for photocatalytic reduction of CO_2 to fuels. Applied Catalysis A : General 400: 195–202.

Qu, Y., W. Zhou, Z. Ren, S. Du, X. Meng, G. Tian, K. Pan, G. Wang and H. Fu. 2012. Facile preparation of porous $NiTiO_3$nanorods with enhanced visible-light-driven photocatalytic performance. Journal of Materials Chemistry 22: 16471–16476.

Qu, Y., W. Zhou and H. Fu. 2014. Porous cobalt titanate nanorod: A new candidate for visible light-driven photocatalytic water oxidation. ChemcatChem 6: 265–270.

Rahman, A. and R. Jayaganthan. 2015. Study of photocatalyst magnesium aluminate spinel nanoparticles. Journal of Nanostructure in Chemistry 5: 147–151.

Ramasami A.K., T.N. Ravishankar, G. Nagaraju, T. Ramakrishnappa, S.R. Teixeira and R.G. Balakrishna. 2017. Gel-combustion-synthesized ZnO nanoparticles for visible light-assisted photocatalytic hydrogen generation. R. Bulletin of Materials Science 40(2): 345–354.

Rashid, J., M.A. Barakat, R.M. Mohamed and I.A. Ibrahim, I.A. 2014. Enhancement of photocatalytic activity of zinc/cobalt spinel oxides by doping with ZrO_2 for visible light photocatalytic degradation of 2-chlorophenol in wastewater. Journal of Photochemistry & Photobiology A: Chemistry 284: 1–7.

Raziq, F., Y. Qu, M. Humayun, A. Zada, H. Yu and L. Jing. 2017. Synthesis of SnO_2/B-P co-doped g-C_3N_4 nanocomposites as efficient cocatalyst-free visible-light photocatalysts for CO_2 conversion and pollutant degradation. Applied Catalysis B: Environmental 201: 486–494.

Rekhila, G., Y. Bessekhouad and M. Trari. 2013. Visible light hydrogen production on the novel ferrite $NiFe_2O_4$. International Journal of Hydrogen Energy 38(15): 6335–6343.

Rongé, J., T. Bosserez, D. Martel, C. Nervi, L. Boarino, F. Taulelle, et al. 2014. Monolithic cells for solar fuels. Chemical Society Reviews 43: 7963–7981.

Saib, F., M. Mekiria, B. Bellal, M. Chibane and M. Trari. 2017. Photoelectrochemical properties of the brownmillerite $Sr_2Fe_2O_5$: Application to electrochemical oxygen evolution. Russion journal of physical Chemistry A 91(8): 1562–1570.

Sakthivel, S. and H. Kisch. 2003. Daylight photocatalysis by carbon-modified titanium dioxide. Angewendte Chemie 42(40): 4908–4911.

Sasaki, Y., H. Nemoto, K. Saito and A. Kudo. 2009. Solar water splitting using powdered photocatalysts driven by Z-schematic interparticle electron transfer without an electron mediator. Journal of Physical Chemistry C 113: 17536–17542.

Sasikala, R., V. Sudarsan, C. Sudakar, R. Naik, T. Sakuntala and S.R. Bharadwaj. 2008. Enhanced photocatalytic hydrogen evolution over nanometer sized Sn and Eu doped titanium oxide. International Journal of Hydrogen Energy 33: 4966–4973.

Sasikala, R., A. Shirole, V. Sudarsan, T. Sakuntala, C. Sudakar, R. Naik, et al. 2009. Highly dispersed phase of SnO_2 on TiO_2 nanoparticles synthesized by polyol-mediated route: Photocatalytic activity for hydrogen generation. International Journal of Hydrogen Energy 34: 3621–3630

Saumitra, N.T., V.J. Meenal, S.P. Priyanka, A.M. Priti, S.R. Sadhana and K.L. Nitin. 2012. Photocatalytic hydrogen generation through water splitting on nano-crystalline $LaFeO_3$ perovskite. International Journal of Hydrogen Energy 37: 10451–10456.

Saumyaprava, A. and K. Parida 2017. A visible light-driven $Zn/Cr–LaFeO_3$ nanocomposite with enhanced photocatalytic activity towards H_2 production and RhB degradation. Chemistry Select 2: 10239–10248.

Sayama, K. and H. Arakawa. 1997. Effect of carbonate salt addition on the photocatalytic decomposition of liquid water over $Pt–TiO_2$ catalyst. Journal of the Chemical Society, Faraday Transactions 93: 1647–1654.

Sayed, F.N., O.D. Jayakumar, R. Sasikala, R.M. Kadam, S.R. Bharadwaj, L. Kienle, et al. 2012. Photochemical hydrogen generation using nitrogen-doped $TiO_2–Pd$ nanoparticles: Facile synthesis and effect of Ti^{3+} incorporation. Journal of Physical Chemistry C 116: 12462–12467.

Shi, H., X. Li, H. Iwai, Z. Zou and J. Ye. 2009. 2-Propanol photodegradation over nitrogen-doped $NaNbO_3$ powders under visible-light irradiation. Journal of Physics and Chemisty of Solids 70: 931–935.

Shi, J., J. Ye, Z. Zhou, M. Li and L. Guo. 2011. Hydrothermal synthesis of $Na_{0.5}La_{0.5}TiO_3–LaCrO_3$ solid-solution single-crystal nanocubes for visible-light-driven photocatalytic H_2 evolution. Chemistry: A European Journal 17: 7858–7867.

Shi, L., T. Wang, H. Zhang, K. Chang and J. Ye. 2015. Electrostatic self-assembly of nanosized carbon nitride nanosheet onto a zirconium metal–organic framework for enhanced photocatalytic CO_2 reduction. Advanced Functional Materials 25: 5360–5367.

Shibata, M., A. Kudo, A. Tanaka, K. Domen, K. Maruya and T. Ohishi. 1987. Photocatalytic activities of layered titanium compounds and their derivatives for H_2 evolution from aqueous methanol solution. Chemistry Letters 16: 1017–1018.

Shintaro, I., K. Yamada, M. Matsuka, H. Hagiwara and T. Ishihara. 2012. Photoelectrochemical hydrogen production from water using p-type and n-type oxide semiconductor electrodes. Electrochimica Acta 82: 397–401.

Shuhua, D., X. Kejing and T. Guishan. 2009. Photocatalytic activities of $LaFe_{1-x}Zn_xO_3$ nanocrystals prepared by sol–gel auto-combustion method. Journal of Materials Science 44: 2548–2552.

Sibo, W., H. Yidong and W. Xinchen. 2015. Development of a stable $MnCo_2O_4$ cocatalyst for photocatalytic CO_2 reduction with visible light. ACS Applied Materials and Interfaces 7(7): 4327–4335.

Spanhel, L., H. Weller and A. Henglein. 1987. Photochemistry of semiconductor colloids. 22. Electron ejection from illuminated cadmium sulfide into attached titanium and zinc oxide particles. Journal of the American Chemical Society 109: 6632–6635.

Sorcar, S., Y. Hwang, C.A. Grimes and S.-I. In. 2017. Highly enhanced and stable activity of defect-induced titania nanoparticles for solar light-driven CO_2 reduction into CH_4. Materials Today 20: 507–515.

Sreethawong, T., Y. Suzuki and S. Yoshikawa. 2005. Photocatalytic evolution of hydrogen over mesoporous TiO_2 supported NiO photocatalyst prepared by single-step sol-gel process with surfactant template. International Journal of Hydrogen Energy 30: 1053–1062.

Suchita, D., B. Gopal, K. Baskar and S. Shubra. 2016. Synthesis and characterization of polycrystalline brownmillerite cobalt doped $Ca_2Fe_2O_5$ AIP Conference Proceedings 1731: 140032.

Sun, M., Y. Jiang, F. Li, M. Xia, B. Xue and D. Liu. 2010. Dye degradation activity and stability of perovskite-type $LaCoO_{3-x}$ ($x = 0{\sim}0.075$). Materials Transactions 51: 2208–2214.

Sutka, A., M. Vanags, U. Joost, K. Smits, J. Ruza, J. Locs, et al. 2018. Aqueous synthesis of Z-scheme photocatalyst powders and thin-film photoanodes from earth abundant elements. Journal of Environmental Chemical Engineering 6(2): 2606–2615.

Tahir, M. and N.S. Amin. 2013. Photocatalytic CO_2 reduction with H_2O vapors using montmorillonite/TiO_2 supported microchannel monolith photoreactor. Chemical Engineering Journal 230: 314–327.

Takei, T., R. Haramoto, Q. Dong, N. Kumada, Y. Yonesaki, N. Kinomura, et al. 2011. Photocatalytic activities of various pentavalent bismuthates under visible light irradiation. Journal of Solid State Chemistry 184: 2017–2022.

Tang, P., H. Chen, F. Cao and G. Pan. 2011. Magnetically recoverable and visible-light-driven nanocrystalline $YFeO_3$ photocatalysts. Catalysis Science & Technology 1: 1145–1148.

Tang, P., H. Sun, F. Cao, J. Yang, S. Ni and H. Chen. 2011. Visible-light driven $LaNiO_3$ nanosized photocatalysts prepared by a Sol-Gel process. Advanced Materials Research 279: 83–87.

Thirumalairajan, S., K. Girija, I. Ganesh, D. Mangalaraj, C. Viswanathan, A. Balamurugan, et al. 2012. Controlled synthesis of perovskite $LaFeO_3$ microsphere composed of nanoparticles via self-assembly process and their associated photocatalytic activity. Chemical Engineering Journal 209: 420–428.

Thirumalairajan, S., K. Girija, N.Y. Hebalkar, D. Mangalaraj and N. Ponpandian. 2013. Shape evolution of perovskite $LaFeO_3$ nanostructures: a systematic investigation of growth mechanism, properties and morphology dependent photocatalytic activities. RSC Advances 3: 7549–7561.

Tijare, S.N., S. Bakardjieva, J. Subrt, M.V. Joshi, S.S. Rayalu, S. Hishita, et al. 2014. Synthesis and visible light photocatalytic activity of nanocrystalline $PrFeO_3$ perovskite for hydrogen generation in ethanol-water system. Journal of Chemical Sciences 126: 517–525.

Townsend, T.K., N.D. Browning and F.E. Osterloh. 2012. Overall photocatalytic water splitting with NiOx–$SrTiO_3$ – A revised mechanism. Energy and Environmental Sciences 5: 9543–9550.

Tran, P.D., L.F. Xi, S.K. Batabyal, L.H. Wong, J. Barber and J.S.C. Loo. 2012. Enhancing the photocatalytic efficiency of TiO_2 nanopowders for H_2 production by using non-noble transition metal co-catalysts. Physical Chemistry Chemical Physics 14: 11596–11599.

Tsuji, A.K. 2003. H_2 evolution from aqueous sulfite solutions under visible-light irradiation over Pb and halogen-codoped ZnS photocatalysts. Journal of Photochemistry and Photobiology A A 156: 249–252.

Tsuji, I., H. Kato and A. Kudo. 2005. Visible-Light-Induced H_2 Evolution from an Aqueous Solution Containing Sulfide and Sulfite over a $ZnS–CuInS_2–AgInS_2$ Solid-Solution Photocatalyst. Angewandte Chemie International Edition 44: 3565.

U.S. Energy Information Administration, International energy outlook, 2011 (IEO 2011), projections for world energy markets through 2035, 2011.

Vijayaraghavan, T., N. Lakshmana Reddy, M.V. Shankar, S. Vadivel and A. Ashok. 2018. A co-catalyst free, eco-friendly, novel visible light absorbing iron based complex oxide nanocomposites for enhanced photocatalytic hydrogen evolution. International Journal of Hydrogen Energy 43(31): 14417–14426.

Vijayaraghavan, T., M. Bradha, P. Babu, K.M. Parida, G. Ramadoss, S. Vadivel, et al. 2020. Influence of secondary oxide phases in enhancing the photocatalytic properties of alkaline earth elements doped LaFeO3 nanocomposites. Journal of Physics and Chemistry of Solids 140: 109377

Wahi, R.K., W.W. Yu, Y. Liu, M.L. Mejia, J.C. Falkner, W. Nolte, et al. 2005. Photodegradation of congo red catalyzed by nanosized TiO_2. Journal of Molecular Catalysis A: Chemistry 242: 48–56.

Wang, X., G. Liu, G. Lu and H. Cheng. 2010. Stable photocatalytic hydrogen evolution from water over ZnO–CdS core–shell nanorods. International Journal of Hydrogen Energy 35: 8199–8205.

Wang, J., B. Li, J.Z. Chen, N. Li, J.F. Zheng, J.H. Zhao, et al. 2012. Enhanced photocatalytic H-2-production activity of CdxZn1-xS nanocrystals by surface loading MS (M = Ni, Co, Cu) species. Applied Surface Science 259: 118–123.

Wang, R., Y. Zhu, Y. Qiu, C.F. Leung, J. He, G. Liu, et al. 2013. Synthesis of nitrogen-doped $KNbO_3$ nanocubes with high photocatalytic activity for water splitting and degradation of organic pollutants under visible light. Chemical Engineering Journal 226: 123–130.

Wang, W.N., F. Wu, Y. Myung, D.M. Niedzwiedzki, H.S. Im, J. Park, P. Banerjee and P. Biswas. 2015. Surface engineered CuO nanowires with ZnO islands for CO_2 photoreduction. ACS Applied Materials and Interfaces 7: 5685–5692.

Wang, M, M. Shen, L. Zhang, J. Tian, X. Jin, Y. Zhou, et al. 2017. 2D-2D MnO_2/g-C_3N_4 heterojunction photocatalyst: In-situ synthesis and enhanced CO_2 reduction activity. Carbon 120: 23–31.

Wee-Jun, O., T. Lling-Lling, C. Siang-Piao, Y. Siek-Ting and R.M. Abdul. 2014. Self-assembly of nitrogen-doped TiO_2 with exposed {001} facets on a graphene scaffold as photo-active hybrid nanostructures for reduction of carbon dioxide to methane. Nano Research 7: 1528–1547.

Wei, Z.X., Y. Wang, J.-P. Liu, C.-M. Xiao, W.-W. Zeng and S.-B. Ye. 2013. Synthesis, magnetization, and photocatalytic activity of $LaFeO_3$ and $LaFe_{0.9}Mn_{0.1}O_{3-\delta}$. Journal of Materials Science. 48: 1117–1126.

Wilson, E., D. Jean, B. Enrico and G. Michael. 1984. Visible-light-induced oxygen generation from aqueous dispersions of tungsten(VI) oxide. Journal of Physical Chemistry 88: 4001–4006.

Wu, D., J.Z. Ma, Y. Bao, W.Z. Cui, T. Hu, J. Yang, et al. 2017. Fabrication of porous $Ag/TiO_2/Au$ coatings with excellent multipactor suppression. Scientific Reports 7: 43749.

Wu, X., Y. Li, G. Zhang, H. Chen, J. Li and K. Wang, et al. 2019. Photocatalytic CO_2 conversion of $M_{0.33}WO_3$ directly from the air with high selectivity: Insight into full spectrum-induced reaction mechanism. Journal of the American Chemical Society 141: 5267–5274.

Xiang-Ping, X., Z. Lei-Hong, T. Bo-Tao, L. Jia-Jian, H. Xin, L. Tie, et al. 2013. Catalytic combustion of methane on $La_{1-x}Ce_xFeO_3$ oxides. Applied Surface Science 276: 328–332.

Xiao, R., Y. Haitao, G. Sai, Z. Jun, Y. Tianzhong, Z. Xiangqun, et al. 2015. Controlled growth of $LaFeO_3$ nanoparticles on reduced graphene oxide for highly efficient Photocatalysis. Nanoscale 8: 752–756.

Xiaobo, C., S. Shaohua, G. Liejin and S.M. Samuel. 2010. Semiconductor-based photocatalytic hydrogen generation. Chemical Reviews 110: 6503–6570.

Xie, Y.P., G. Liu, L. Yin and H.M. Cheng. 2012. Crystal facet-dependent photocatalytic oxidation and reduction reactivity of monoclinic WO_3 for solar energy conversion. Journal of Materials Chemistry 22: 6746–6752.

Xin, Y.K., Y.C. Yen, C. Siang-Piao, K.S. Ai and R.M. Abdul. 2016. Oxygen vacancy induced Bi_2WO_6 for the realization of photocatalytic CO_2 reduction over the full solar spectrum: from the UV to the NIR region. Chemical Communications 52: 14242–14245.

Xu, S. and D.D. Sun. 2009. Significant improvement of photocatalytic hydrogen generation rate over TiO_2 with deposited CuO. International Journal of Hydrogen Energy 34: 6096–6104.

Xu, S., J. Ng, X. Zhang, H. Bai and D.D. Sun. 2010. Fabrication and comparison of highly efficient Cu incorporated TiO_2 photocatalyst for hydrogen generation from water. International Journal of Hydrogen Energy 35: 5254–5261.

Xu, J., C. Hu, Y. Xi, B. Wan, C. Zhang and Y. Zhang. 2012. Synthesis and visible light photocatalytic activity of β-$AgVO_3$ nanowires. Solid State Sciences 14: 535–539.

Yamaguti, K. and S. Sato. 1985. Photolysis of water over metallized powdered titanium dioxide. Journal of the Chemical Society, Faraday Transactions I 81: 1237–1246.

Yang Y., Y. Sun and Y. Jiang. 2006. Structure and photocatalytic property of perovskite and perovskite-related compounds. Materials Chemistry and Physics 96(2–3): 234–239.

Yang, Y., Y. Jiang, Y. Wang, Y. Sun, L. Liu and J. Zhang. 2007. Influences of sintering atmosphere on the formation and photocatalytic property of $BaFe_2O_4$. Materials Chemistry and Physics 105: 154–156.

Yang, H., Y. Mao, M. Li, P. Liu and Y. Tong. 2013. Electrochemical synthesis of $CoFe_2O_4$ porous nanosheets for visible light driven photoelectrochemical applications. New Journal of Chemistry 37: 2965–2968.

Yang, J., R. Hu, W. Menga and Y. Dua. 2016. A novel p-$LaFeO_3$/n-Ag_3PO_4 heterojunction photocatalyst for phenol degradation under visible light irradiation. Chemical Communications 52: 2620–2623.

Yasuhiro, S., S. Hirokatsu, S. Yoshitsune, I. Satoshi and H. Takayuki. 2013. Pt–Cu bimetallic alloy nanoparticles supported on anatase TiO_2: Highly active catalysts for aerobic oxidation driven by visible light. ACS Nano 7(10): 9287–9297.

Yongchun, Y., Y. Hua, L. Ruishan and W. Xiangxian. 2017. Enhanced photocatalytic performance and mechanism of Ag-decorated LaFeO$_3$ nanoparticles. Journal of Sol-Gel Science and Technology 82: 509–518.

Yoshiko, T., N.K. Junko, T. Tsuyoshi, L. Daling and D. Kazunari. 2001. Mesoporous tantalum oxide. 1. Characterization and photocatalytic activity for the overall water decomposition. Chemistry of Materials 13: 1194–1199.

Yu, J.C., W. Ho, J. Yu, H. Yip, P.K. Wong and J. Zhao. 2005. Efficient visible-light-induced photocatalytic disinfection on sulfur-doped nanocrystalline titania. Journal of Environmental science and Technology 39: 1175–1179.

Yu, J.G., L.F. Qi and M. Jaroniec. 2010. Hydrogen Production by Photocatalytic Water Splitting over Pt/TiO$_2$ nanosheets with exposed (001) facets. The Journal of Physical Chemistry C 114: 13118–13125.

Yu, J.G., Y. Hai and M. Jaroniec. 2011. Photocatalytic hydrogen production over CuO-modified titania. Journal of Colloids and Interface Science. 357: 223–228.

Yuan, J., M. Chen, J. Shi, and W. Shangguan 2006. Preparations and photocatalytic hydrogen evolution of N-doped TiO$_2$ from urea and titanium tetrachloride. International Journal of Hydrogen Energy 31: 1326–1331.

Yuan, Q., Y. Liu, L.-L. Li, Z.-X. Li, C.-J. Fang, W.-T. Duan, et al. 2009. Highly ordered mesoporous titania–zirconia photocatalyst for applications in degradation of rhodamine-B and hydrogen evolution. Microporous and Mesoporous Materials 124: 169–178.

Yuan, W., X. Liu and L. Li. 2014. Synthesis, characterization and photocatalytic activity of cubic-like CuCr$_2$O$_4$ for dye degradation under visible light irradiation. Applied Surface Science 319: 350–357.

Yuan, Z., Y. Wang, Y. Sun, J. Wang, L. Bie and Y. Duan. 2006. Sunlight-activated AlFeO$_3$/TiO$_2$ photocatalyst. Science in China Series B Chemistry 49: 67–74.

Zazoua, H., A. Boudjemaa, R. Chebout and K. Bachari. 2014. Enhanced photocatalytic hydrogen production under visible light over a material based on magnesium ferrite derived from layered double hydroxides (LDHs). International Journal of Energy Research 38: 2010–2018.

Zhang, H., G. Chen, Y. Li and Y. Teng. 2010. Electronic structure and photocatalytic properties of copper-doped CaTiO$_3$. International Journal of Hydrogen Energy 35: 2713–2716.

Zhang, Q. Y. Li, E.A. Ackerman, M. Gajdardziska-Josifovska and H. Li. 2011. Visible light responsive iodine-doped TiO$_2$ for photocatalytic reduction of CO$_2$ to fuels. Applied Catalysis A: General 400(1-2): 195–202.

Zhang, L.J., T.F. Jiang, S. Li, Y.C. Lu, L.L. Wang, X.Q. Zhang, et al. 2013. Enhancement of photocatalytic H$_2$ evolution on Zn$_{0.8}$Cd$_{0.2}$S loaded with CuS as cocatalyst and its photogenerated charge transfer properties. Dalton Transactions 42: 12998.

Zhao, D., S. Budhi, A. Rodriguez and R.T. Koodali. 2010. Rapid and facile synthesis of Ti-MCM-48 mesoporous material and the photocatalytic performance for hydrogen evolution. International Journal of Hydrogen Energy 35: 5276–5283.

Zhao, Z., R. Li, Z. Li and Z. Zou. 2011. Photocatalytic activity of La–N-codoped NaTaO$_3$ for H$_2$ evolution from water under visible-light irradiation. Journal of Physics D: Applied Physics 44 (16): 165401.

Zhou. S., Y. Liu, J. Li, Y. Wang, G. Jiang, Z. Zhao, et al. 2014. Facile in situ synthesis of graphitic carbon nitride (g-C$_3$N$_4$)-N-TiO$_2$ heterojunction as an efficient photocatalyst for the selective photoreduction of CO$_2$ to CO. Applied Catalysis B: Environmental 158–159: 20–29.

Zielinska, B., E. Borowiak-Palen and R.J. Kalenczuk. 2008. Photocatalytic hydrogen generation over alkaline-earth titanates in the presence of electron donors. International Journal of Hydrogen Energy 33: 1797–1802.

Zong, X., H.J. Yan, G.P. Wu, G.J. Ma, F.Y. Wen, L. Wang, et al. 2008. Enhancement of photocatalytic H$_2$ evolution on CdS by loading MoS$_2$ as cocatalyst under visible light irradiation. Journal of the American Chemical Society 130: 7176.

Zong, X., J.F. Han, G.J. Ma, H.J. Yan, G.P. Wu, and C. Li. 2011. Photocatalytic H$_2$ Evolution on CdS loaded with WS$_2$ as Cocatalyst under visible light irradiation. Journal of Physical Chemistry C 115: 12202.

Chapter 4

Advanced Nanocomposites as Cathode Material for Li-ion Batteries

A.S. Kornyushchenko*

Sumy State University, Laboratory of Vacuum Nanotechnologies,
2, Rimsky-Korsakov Str., 40007 Sumy, Ukraine
a.kornyushchenko@mss.sumdu.edu.ua

Ahalapitiya H. Jayatissa

University of Toledo, Mechanical Industrial and
Manufacturing Engineering Department, Mail Stop 312, 43606, Toledo, Ohio, USA
ahalapitiya.jayatissa@utoledo.edu

1 INTRODUCTION

Lithium-ion batteries are most widely used among rechargeable batteries due to their outstanding properties, such as high specific energy, high energy density, low self-discharge rate, good cycling performance and long lifetime (Linden and Reddy 2002; Yoshio et al. 2009; Pistoia 2014; Yazami 2014). They are smaller and lighter than other battery chemistries, such as nickel-cadmium, nickel-metal hydride and lead-acid batteries, which makes them especially attractive for vehicle applications where both size and weight should be minimized. Besides that, Li cells operate at much higher voltages than the rest of the secondary cells, 3.7 V as compared to 1.2 for NiMH and NiCd. This means in order to generate the voltage 3.7 V, three nickel metal hydrate or nickel-cadmium batteries should be used together. It demonstrates a capacity of around 150 mAh/g and power over 200 Wh/kg (Linden and Reddy 2002; Yoshio et al. 2009).

The first-generation of commercialized Li-ion batteries used $LiCoO_2$ as the cathode material and graphite as the anode material (Yuan et al. 2011; Ozoemena and Chen 2016). However, new emerging applications in the field of future energy storage and electrical vehicles, require even better performances from the Li-ion cell in terms of energy density, charge-discharge rate, power and price. That is why, to improve the characteristics of existing Li-ion batteries, different electrode materials have been extensively investigated with the aim to find a novel electrode material with a suitable crystal structure that is able to sustain numerous repetitive charge-

---

*Corresponding Author

discharge cycles without changes of its inner structure. A large amount of different materials have been tested for possible application such as cathode or anode over the past decades. The conducted researches have shown, that one component electrode material alone is not sufficient to satisfy continuously growing requirements for high capacity, high energy density. Therefore, new advanced nanocomposites are needed to satisfy the demand of the modern Li-ion batteries market. Nanocomposites are beneficial because they provide unique benefits of combining two or more functional components having different properties (Demirocak et al. 2017; Liu et al. 2018).

The main constituent elements of the battery cell are the anode, cathode, separator, electrolyte and current collector. Transitional metal oxides are used, as a rule, as a traditional material for the positive electrode (cathode) and graphite as a negative electrode (anode) (Linden and Reddy 2002; Yoshio et al. 2009; Yuan et al. 2011; Pistoia 2014; Yazami 2014; Ozoemena and Chen 2016). The schematic representation of the working principle of the Li-ion battery cell and its main functional components are given in Fig. 4.1. During the discharge process, the anode material gives up electrons, and simultaneously Li^+ ions are released from the negative electrode. In other words, at the anode surface oxidation reaction takes place, because the anode loses electrons. Ions of Li^+ travel from the anode to the cathode surface through the electrolyte, at the same time electrons move to the cathode through the load. The positive electrode accepts electrons coming from the external circuit and a reduction reaction occurs at the surface. Li^+ ions reaching the cathode material meet the electrons and intercalate, or in other words, build into the positive electrode surface. During the charging process, the reactions on the electrodes are exactly the opposite, in this case, the cathode releases electrons or oxidizes and the anode gains electrons or reduces. Electrolyte does not participate in overall chemical reactions on the cell and serves exclusively as a medium for Li^+ ions transport between the electrodes (Linden and Reddy 2002; Yoshio et al. 2009; Yuan et al. 2011; Pistoia 2014; Yazami 2014; Ozoemena and Chen 2016). Current collectors distribute current flowing out and in electrode material, usually, copper is used as the current collector on the anode side, and aluminum as positive electrode current collector.

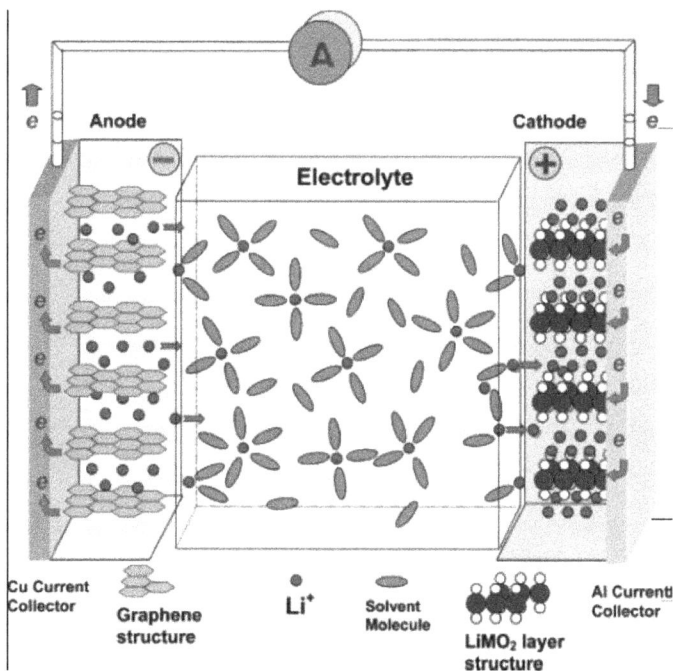

Figure 4.1 Schematic representation of Li-ion cell working principle. (Reprinted with permission from MDPI Open Excess Publisher. *Source:* Demirocak et al. 2017. A Review on Nanocomposite Materials for Rechargeable Li-ion Batteries. *Applied Sciences* 7(2017): 731–757.)

In order for a material to be used as the electrode in a Li-ion battery, it should satisfy the following requirements (Yuan et al. 2011; Julien et al. 2014; Ozoemena and Chen 2016; Demirocak et al. 2017; Liu et al. 2018):

- first, the material should have an open crystalline structure with available vacancies sites and diffusion pathways for lithium ions insertion-extraction and unimpeded motion through the host material without changes in the active material structure;
- secondly, the material should have high electronic conductivity, and be able to give up electrons to the external circuit and to accept electrons from the external circuit;
- the material should have a free energy reaction with lithium which results in the high voltage;
- the material must be unsolvable in the electrolyte;
- the surface area of the electrode should be as high as possible because lithium ions enter the electrode through the surface, as a result, an increase in the surface area will increase the rate of lithium insertion-extraction.

2   CATHODE MATERIAL FOR LI-ION BATTERIES

It is known, that among all active components of LIBs, cathode material has a significant impact on battery capacity, cycle life, safety and cost structure. That is why a large amount of research is devoted to the development of new alternatives to traditional cathode materials. The majority of commercially available cathodes utilize the intercalation mechanism of lithium ions insertion and extraction, such materials have a structure with diffusional channels available for Li-ions motion (Chen et al. 2019; Wang et al. 2019a; Nourhan and Nageg 2020; Divakaran et al. 2021; Zhao et al. 2021). In this case, lithium ions can be reversibly inserted and removed from the electrode. It is very important to note that the electrode material itself does not participate in the electrochemical reaction, lithium ions are stored in the vacancies within the electrode material without changing the structure of the electrode, and they can be easily released from the electrode upon cycling. There are three types of the intercalation electrode structure: layered ($LiCoO_2$, $LiNiO_2$, $LiMnO_2$) (Chen et al. 2019; Wang et al. 2019a; Nourhan and Nageg 2020; Divakaran et al. 2021; Zhao et al. 2021) with two-dimensional diffusion channels, spinel with three-dimensional channels ($LiMn_2O_4$) and olivine with one-dimensional channels ($LiFePO_4$), see Fig. 4.2.

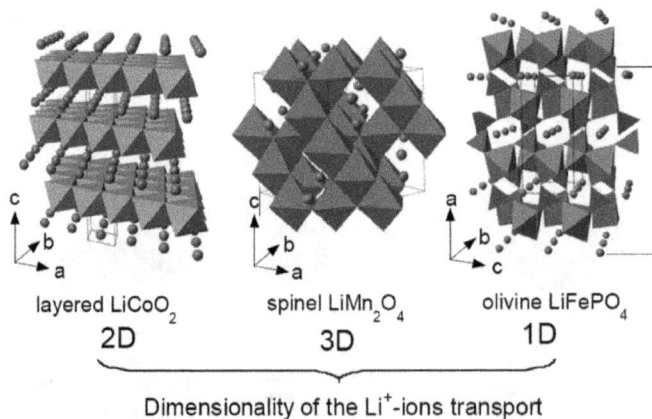

layered $LiCoO_2$ spinel $LiMn_2O_4$ olivine $LiFePO_4$
        2D   3D   1D

Dimensionality of the Li^+-ions transport

Figure 4.2   Crystal structure of the three lithium-insertion compounds in which the Li^+ ions are mobile through the 2-D (layered), 3-D (spinel) and 1-D (olivine) frameworks. (Reprinted with permission from MDPI Open Excess Publisher. *Source*: Julien et al. 2014. Comparative Issues of Cathode Materials for Li-Ion Batteries. *Inorganics* 2(2014): 132–154.)

The surface area of the cathode plays a very important role in the performance (Jiang et al. 2006; Liu et al. 2006; Uddin et al. 2017; Cao et al. 2019). The higher the surface area, the higher will be the charge-discharge rate of the battery and as a result higher the battery cell power due to enlarged electrolyte/electrode contact area and a decrease in both Li^+ ions transport and the electron conduction distances. The increase in the surface area can be achieved by reducing the size of particles comprising the electrode to a nanolevel of about 1–100 nm (Uddin et al. 2017).

2.1 Cathode Materials with Layered Crystal Structure

The first layered intercalation material and the most famous one used in commercial Li-ion battery cells is $LiCoO_2$ (LCO) (Johnston et al. 1958), it provides the capacity value 140 mAh/g and the nominal voltage 3.9 V. The attempts to increase the cell voltage have not been successful, because of the electrode structural transformations and a capacity fade at the voltages higher than 4.2 V. Under such high voltages, the electrode degrades quickly upon the charging-discharging cycles. Another important limitation of the cobalt oxide-based cathode is associated with the fact that only half of Li^+ ions can be reversibly extracted from the electrode at the voltages below 4.5 V. If one wants to extract all Li^+ ions and to obtain higher cell capacities, voltages more than 4.5 V should be applied. However, as it was already mentioned under such high voltages, the electrode structural changes are inevitable because of cobalt loss. It has been shown, that $LiCoO_2$ is stable upon cycling under the voltage range from 3.7 to 4.2 V, and the electrode has the cycleability 500 charge-discharge cycles (Kalluri et al. 2017; Chen et al. 2019; Lyu et al. 2020a). One method used to improve the life cycle of the cathode is a deposition of a protecting layer on the surface of $LiCoO_2$. Such a protective layer can be a metal oxide, among them, Al_2O_3, SnO_2, ZrO_2, TiO_2, MgO (Hwang et al. 2010; Hudaya et al. 2014; Jian et al. 2018; Wang et al. 2002; 2019b; 2020a), metal phosphate, for example, $AlPO_4$ (Kim et al. 2004; Cho et al. 2005), metal fluoride coatings, AlF_3 and LaF_3 (Sun et al. 2006; Yang et al. 2011). The protective layer prevents the structural transformation of the electrode during the cycling and acts as a barrier decreasing the interaction of the active material with the electrolyte, inhibiting in such a way cobalt dissolution in the electrolyte. After a great amount of research was done to solve the stability problem, the upper cut-off voltage of the modern commercial LIBs using the modified LCO cathode has been increased up to 4.5 V during charging, and the reversible capacity of 185 mAh g^{-1} has been achieved (Chen et al. 2019).

Another drawback of $LiCoO_2$ material is the high price of cobalt and its toxicity. Besides lithium cobalt oxides, manganese, chromium, iron oxides belong to the family of oxide materials having a layered structure. These materials have been investigated as an alternative to LCO positive electrodes; however, researchers have not achieved satisfactory battery characteristics in terms of capacity, voltage and life cycle (Tuccillo et al. 2020). With an aim to reduce the price of $LiCoO_2$ and to make the electrode material more environmentally friendly, new composite materials have been developed by partial or full substitution of Co with Ni, Zr, Mn, and Al elements (Itou and Ukyo 2005; Sivajee Ganesh et al. 2018; Wang et al. 2015a; 2020b; Ding et al. 2021). If to replace all cobalt with nickel, lithium nickel oxide ($LiNiO_2$ or LNO) will be formed. This material also has a layered structure that is capable of lithium intercalation and deintercalation. The LNO-based cell operates under around the same voltage range as in the case of $LiCoO_2$ cathode material; however, the capacity is higher in this case (Choi et al. 2019; Taha and El-Molla 2020). As a result, $LiNiO_2$ electrode has higher energy density for the same amount of material and as a result, it is a very promising chemistry for the high-energy Li-ion batteries applications. Besides, Ni is an attractive alternative to Co due to its cheapness and abundance in nature and lower toxicity. From one side, nickel oxide looks like a very promising cathode material, however, on the other hand, it is very hard to sensitize stoichiometric $LiNiO_2$. Nickel has a weaker chemical bond with oxygen as compared to cobalt binding energy in $LiCoO_2$, as a result, nickel tends to occupy lithium positions during the cycling, causing structural changes of the electrode. Moreover, LNO suffers from being less thermally stable than LCO, providing a higher probability of combustion during the cell overcharging. Doping with Ti, Sb, Al, Mg, TiO_2, ZnO, etc. (Cui et al. 2011; Muto et al. 2012; Youp Song et al. 2014; ShinKo et al. 2017; Cao et al. 2020) of the nickel-based electrode can stabilize the structure, but in this case, the specific capacity decreases, because the dopant elements do not participate in the electrochemical reactions with lithium. The stability of the material can also be improved by partial substitution of Ni with Co because it belongs to the same family of layered materials. For example, it has been shown that use of $LiNi_{1/2}Co_{1/2}O_2$ (LNC) allowed to improve the stability of the electrode and to increase the capacity up to 180 mAh/g^{-1} (Belharouak et al. 2003; Shang et al. 2016).

Alternative chemistry is to replace some nickel with cobalt and some with aluminum, this composition is called NCA, and is composed of 80% Ni, 15% Co, and 5% Al ($LiNi_{0.8}Co_{0.15}Al_{0.05}O_2$) (Ryu et al. 2020; Nie et al. 2021). Nie et al. 2021 developed NCA composite material with 4% Al content (NC-A4) having advanced electrochemical properties. The Li-ion cell assembled with NC-A4 delivers a remarkably high initial discharge specific capacity of 196.3 mAh g^{-1}. The capacity value after 100 cycles at 1.0°C was 170.7 mAh g^{-1}, which corresponds to a 95.5% capacity retention coefficient. Even at the high current density value of 5.0°C, it also has an enhanced capacity rate value of 162.1 mAh g^{-1}. Trivalent doping with Al ions provides a significant improvement in the structural stability of the composite that can maintain a stable lattice structure during the lithiation/delithiation process and as a result, enhances its electrochemical properties. Therefore, nickel-reach composite material stabilized with aluminium (NCA) has outstanding structural and thermal stability as compared with other layered compounds, and it is currently used in commercial batteries for Tesla automobiles.

One of the most popular substitution chemistry nowadays is replacing some nickel with cobalt and some with manganese. In this case, one ends up with the composition $LiNi_{1/3}Mn_{1/3}Co_{1/3}O_2$, this material is called NMC or NCM. In this case, nickel is responsible for the heightened electrode voltage and manganese together with cobalt makes the electrode more thermally and structurally stable. Many automotive industries are currently using NMC cathode material in their battery packs. The NMC electrode has a capacity value of 200 mAh/g at the same cell voltage as NCA, and the cycle-ability of such electrode is more than 3000 charge-discharge cycles (Lyu et al. 2020b; Stephan 2020; Wood et al. 2020). Therefore, partial substitution of manganese instead of aluminum is very promising in terms of high capacity values. NMC and NCA are currently superior among the cathode materials on the market. In traditional NMC chemistry, there is an equal amount of nickel, manganese and cobalt. Recently many research groups have been working on the optimization of NMC cathode composition by varying the nickel, manganese and cobalt content in the compound. Among the different NMC compositions, Ni-rich chemistry offers an increase in the energy density of the cathode, Mn-rich compositions provide a better cycle life and thermal stability, Co-reach compound is favorable for improving the rate performance. On the other hand, nickel-rich NMC composition suffers from a structural degradation that causes a decrease of the cycle life, manganese-rich chemistry deteriorating from a reduced capacity, the cobalt-reach composition has high price and environmental issues.

Ni-rich NMC materials also experience from Ni/Li disordering during the cycling process, resulting in Ni and Li exchange in the octahedral sites (Fu et al. 2014; Zheng et al. 2019). This effect becomes a critical issue especially at high cut-off voltages leading to Li-ions diffusivity suppression and resulting in a deterioration of the cycling stability and overall fade of the cathode performance. Numerous research efforts have been devoted to improving the electrochemical performance of NCM composites by reduction of cationic intermixing (Fu et al. 2014; Zheng et al. 2017; Zheng et al. 2019; Zha et al. 2021). In their work, Zha et al. 2021 described that the Ni/Li disordering was suppressed by partial substitution of Co with Fe into the Ni-rich NMC cathode. The new cathode material ($LiNi_{0.8}Co_{0.07}Fe_{0.03}Mn_{0.1}O_2$ (Fe$_3$-NCM871)) has proved to have remarkably enhanced electrochemical characteristics, it delivered the initial capacity of 207.5 mAh g^{-1} at the 0.1 C rate in the voltage range 2.8 and 4.3 V, while the NMC811 cathode has a value of 188.7 mAh g^{-1}. It still provides a high capacity value was 145.8 mAh g^{-1} even at a high rate of 5 C, while the NCM811 cathode is 120.1 mAh g^{-1}. Moreover, it maintains better cycle stability performance than NMC cathode and reaches 80% State-Of-Health (SOH) only after 400 cycles, which is higher than that of the traditional NCM (nearly 200 cycles) at C/2 between 2.8 and 4.3 V. In Fig. 4.3 electrochemical characteristics of Fe_3-NCM871 cathode are given in comparison with NMC811 cathode.

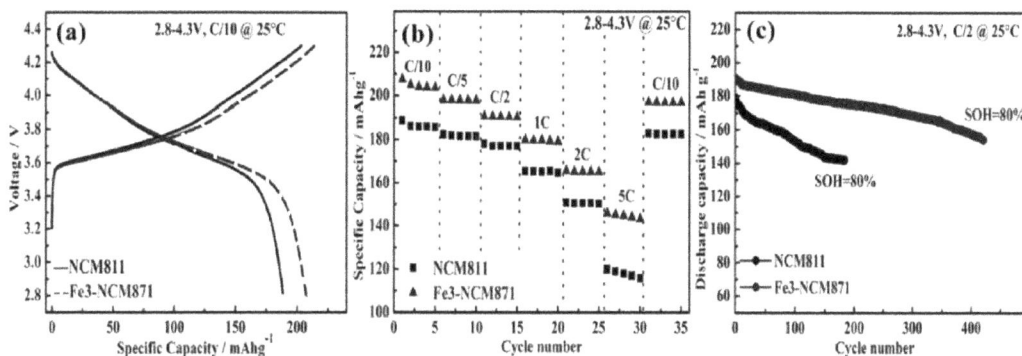

Figure 4.3 Electrochemical performances of NCM811 and Fe3-NCM871 cathodes: (a) initial discharge curves. (b) Rate performances. (c) Long-term cycling performance. (Reprinted with permission from Copyright Elsevier. *Source*: Zha et.al. 2021. High performance layered $LiNi_{0.8}Co_{0.07}Fe_{0.03}Mn_{0.1}O_2$ cathode materials for Li-ion battery. *Chemical Engineering Journal* 409(2021): 128343).

Another strategy used to achieve improved performance of NMC and to balance advantages and drawbacks of individual Ni, Co, Mn components has been implemented by means of gradient core-shell materials development (Manthiram et al. 2016; Ma et al. 2019a; Primer 2019; Hua et al. 2020; Zheng et al. 2020; Phattharasupakun et al. 2021). In the case of concentration gradient chemistry, the elemental composition changes from the center of the particle to the outside edge. There are different types of core-shell compound structures, typically they contain a Ni-rich high-energy core combined with an Mn-rich outer shell for improved safety (Fig. 4.4(a)) (Manthiram et al. 2016; Ma et al. 2019a; Zheng et al. 2020; Phattharasupakun et al. 2021). The transition from a more nickel-rich composition to a more manganese-rich can be sharp (Fig. 4.4(a)) or gradual (Fig. 4.4(b, c)) (Manthiram et al. 2016). Hua et al. 2020 demonstrated that the core-shell NCM cathode has integrated the electrochemical properties of the Ni-rich core and the Mn-rich shell, they explain such behavior by a synergistic effect of the two-layered phases in the core-shell morphology on the electrochemical performance of NCM cathode materials. It was shown, that the Mn-rich shell dominates the potential curve at the initial stage of the charging process and the Ni-rich NCM core is responsible for higher capacity value. The developed cathode delivered a high capacity value, retaining approximately 69% of its initial capacity even after a high current density cycling at rates from 0.1 C to 20 C. In their work, Phattharasupakun et al. 2021 have developed the core-shell Ni-rich NMC-nanocarbon cathode for the high-performance Li-ion batteries. The battery prototype with the core-shell NMC-nanocarbon cathode has shown to possess higher specific capacity and energy as well as excellent cycling stability compared to the pristine NMC. Thus, the core-shell architectures have shown to be promising cathode candidates for the next-generation lithium-ion battery applications. However, more investigations are required in order to clearly understand the intercalation/deintercalation mechanism of lithium ions into/from the core-shell-architectured cathode materials. The question of whether both phases are intercalated/deintercalated subsequently or simultaneously has not been answered yet.

Lithium-reach Layered Oxide compounds (LLOs) are alternatives to the traditional nickel-rich ternary cathode materials. LLOs have a complex structure that depending on a formation method and a composition can be either a single-phase solid solution or a nanocomposite consisting of trigonal $LiTMO_2$ phase (here TM is collection of metals, most often Ni, Co, Mn, Fe, Cr, etc.) and monoclinic Li_2MnO_3 phase (Manthiram et al. 2016; Zheng et al. 2020; Çetin et al. 2020a). It has been shown that the content of the different transition metals in the composition of the LLOs composite drastically affects the electrochemical performance of the cathodes (Roziera and Tarascon 2015; Pan et al. 2018; Çetin et al. 2020a). Therefore, numerous research works have been devoted to the problem of determining the favorable content of TMs in the cathode composition for the high-power and high-energy-density batteries applications (Roziera and Tarascon 2015; Pan et al. 2018; Çetin et al. 2020a).

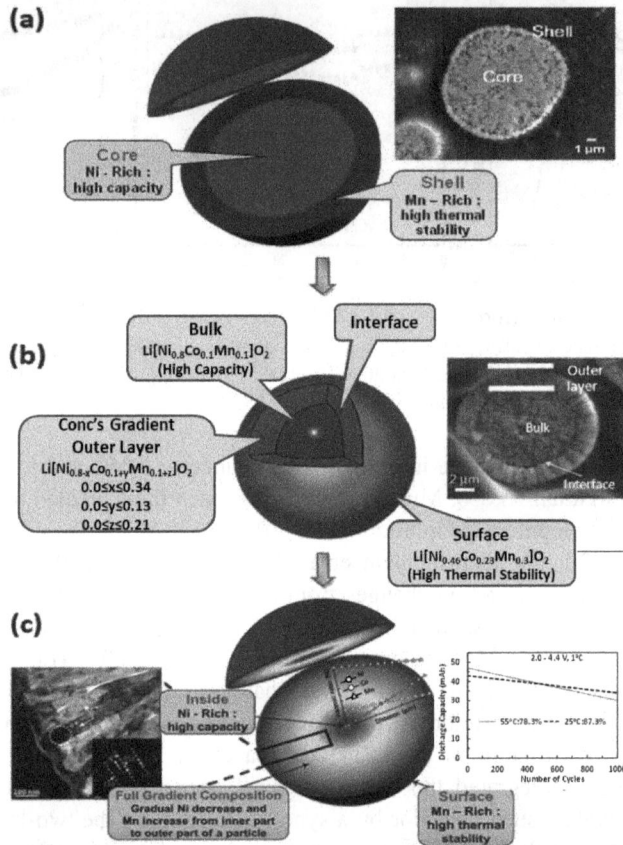

Figure 4.4 Schematic representation of core-shell particles: (a) sharp concentration gradient between Ni-rich core and Mn-rich shell and corresponding SEM image; (b) Ni-rich core surrounded by concentration-gradient outer Mn-rich layer together with SEM image; (c) Full concentration gradient with the nickel concentration decreasing from the center towards the outer layer and the manganese concentration increasing from outer layer toward the center and TEM image of a full concentration gradient particle. (Reprinted with permission from Wiley Online Library. *Source*: A. Manthiram et al. 2016. Nickel-Rich and Lithium-Rich Layered Oxide Cathodes: Progress and Perspectives. *Advanced Energy Materials* 6(2016): 1501010).

Nowadays LLOs attract significant interest of researchers due to even higher capacity and energy density values as compared to NMC and NCA materials. LLOs are capable of delivering a high specific capacity up to 400 mAh g^{-1} and have a maximum energy density higher than 1000 Wh kg^{-1} (Manthiram et al. 2016; Pan et al. 2018; Çetin et al. 2020a; Zheng et al. 2020). Because LLOs can deliver extremely high energy densities, they have great potential for applications in high-energy Li-ion cells and many researchers consider them as next-generation cathode materials for LIBs (Roziera and Tarascon 2015; Pan et al. 2018). Furthermore, this cathode material is safe, environment-friendly, due to its high content of Mn, and have a low cost. However, the practical application of LLOs is still limited and additional investigations are required in order to eliminate the materials undesirable properties, such as an irreversible oxygen loss from the lattice during charging-discharging cycles, high initial capacity loss during the cycling, a poor rate capability, side reactions on the cathode-electrolyte interface, the voltage decay during prolonged cycling. With an aim to overcome these issues different strategies have been proposed, among them surface modification with protective coatings (Ma et al. 2016; Zhang et al. 2018a; Yang et al. 2019a, b; Wang et al. 2020c; Xu et al. 2020a). In their work Yang et al. 2019 described that Li-rich Mn-based cathode material was coated with a layer of $Li_4V_2Mn(PO_4)_4$. Such surface modification allowed to reduce the Li-ion diffusion energy barrier, restrain the dissolution of the transition metals and provided efficient diffusion pathways for Li-ions. The composite material delivered the discharge capacity of 300 mAh g^{-1} with a high initial coulombic efficiency (84.2%) and excellent cycling stability (capacity retention of 78.1% after 200 cycles at 1 C). Xu et al. 2020 discussed how Li-Rich Mn-based Oxide cathode (LRMO) material was coated with potassium Prussian blue (KPB). In this case, the coating served as a protective layer against the cathode corrosion under the electrolyte influence. Besides, the KPF layer functions as a host for Li$^+$ transport and accumulation, providing enhanced ion conductivity of the material. Therefore, KBP-coated Li-rich cathode delivered an initial discharge capacity of 281.7 mAh g^{-1} with the capacity retention of 85.69%, which is higher compared to 79.52% typical for uncoated LRMO cathode. The cycling life and the rate performance have also been greatly improved due to the surface modification. The capacity value was 176.8 mAh g^{-1} after 100 cycles at the current density of 0.5 C, compared to the capacity of only 135.3 mAh g^{-1} for an unmodified LMRO cathode.

Morphology engineering and developing LLOs nanostructures with unique properties is another strategy used these days with the aim to improve the electrochemical performance of Li-reach materials (Ma et al. 2016a; Xie et al. 2019). Thus, reducing the size of structural elements to the nano level can shorten the diffusion pathway for lithium ions, and in such a way to enhance the capacity rate even at the high charge-discharge rates. A combination of microsized and nanosized elements allows to decrease the side reactions on the electrode surface and to improve simultaneously cycling stability. Ma et al. 2016a described how Li-rich layered oxide materials were sensitized in the form of hierarchical nano/micro rods (Fig. 4.5(a)). The electrochemical measurements of the obtained composite reveal that it delivers high discharge capacities of 300 mAh g^{-1} and 159.5 mAh g^{-1} measured at 1 C and 10 C rates, respectively. The capacity retention was determined to be 96.0% after 80 cycles at 1 C that evidences the excellent cycling stability and rate capability. The remarkably improved cyclability and the rate capability of the composite cathode material are determined by the combination of the nano/micro hierarchical microrod structure and the LLOs-modified surface, which reduce the structural degradation during the charge-discharge cycles. The electrochemical performance of the hierarchical nano/microrod as a LIB cathode material is presented in Fig. 4.5. With an aim to improve the electronic conductivity and Li-ion diffusivity, doping of LLOs compounds is used. Na, Mg are often employed as dopants at the Li site; cations dopants such as Cu, Zn, Cr, Mg, Ti, Al, etc. are used at the transition metal (TM) site; and O substitution by F (Chen et al. 2016; Ding et al. 2017; Wei et al. 2017a; Sorboni et al. 2019; Sun et al. 2019; Zhao et al. 2019a; Çetin et al. 2020b; Hao et al. 2021).

The above described the core-shell strategy of NMC cathode material can be also applied with the aim to improve the electrochemical performance of Li-reach compounds. Ma et al. 2019 developed a core-shell composite cathode material with Ni-reach core (NMC) and Li- and Mn-reach (LMR) as the shell. The fabricated cathode demonstrates superior overall electrochemical performance in comparison with the pristine NMC and Li-reach cathodes. It delivers the initial reversible capacity of 226.6 mAh g^{-1} at the rate of 0.1 C and 120.2 mAh g^{-1} at 10.0 C. The capacity retention was 86.9% of its initial capacity value after 100 cycles at 4.6 V and 1.0 C. Placke et al. 2021 used Li/Mn reach compound (LMR) as the core and NMC as the shell. The authors have investigated the electrochemical performance of the cathode depending on the shell thickness, 10 vol.% (CS10), 25 vol.% (CS25), 50 vol.% (CS50) and demonstrated improved long-term performance of the core-shell materials as compared to the individual core and individual shell material. They managed to achieve the capacity retentions of ≈ 80% after 209 cycles with initial discharge capacities of 244 mAh g^{-1} (CS10), 267 mAh g^{-1} (CS25), and 259 mAh g^{-1} (CS50), resulting in specific energies of 868 Wh kg^{-1}, 957 Wh kg^{-1}, and 943 Wh kg^{-1}, respectively. Though significant progress has already been made during the last 30 years in improving LLOs structural stability and electrochemical performance, still many issues must be addressed before this material moves to wide commercial applications in the batteries for electric vehicles. The main problem to be solved is the voltage drop during prolonged charging-discharging of the electrode (Hao et al. 2021, Çetin et al. 2020b).

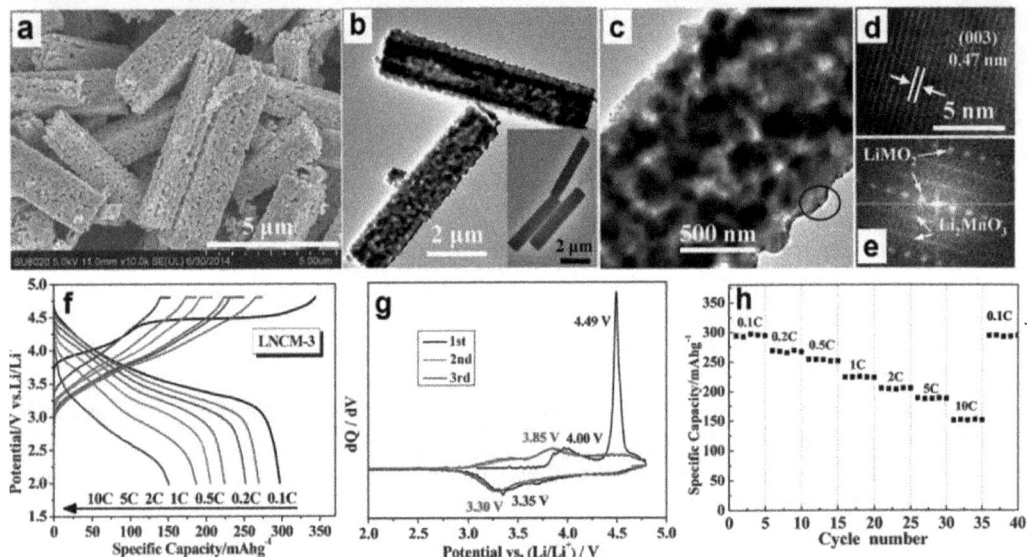

Figure 4.5 (a) FESEM image of hierarchical nano/microrod $0.5Li_2MnO_3 \cdot 0.5$ $LiNi_{1/3}Co_{1/3}Mn_{1/3}O_2$; (b) (c) TEM and HRTEM images of the LLO structure (inset of b: TEM image of precursor; (d) HRTEM image of the circled region in (c); (e) FFT of the HRTEM image in (d). Electrochemical performance of the hierarchical nano/microrod as a LIB cathode material: (f) charge-discharge profiles at various rates in the voltage range of 2.0–4.8 V; (g) differential capacity dQ/dV (Q – capacity; V – voltage of the cells) plots at the rate of 0.1 C, (h) rate capability at different rates (1 C = 250 mAh g^{-1}). (Reprinted with permission from Wiley Online Library. *Source*: Ma et al. 2016a. A General and Mild Approach to Controllable Preparation of Manganese-Based Micro- and Nanostructured Bars for High-Performance Lithium-Ion Batteries. *Angewandte Chemie* 55(2016): 3667–3671).

2.2 Cathode Materials with Spinel Crystal Structure

The next crystal structure belonging to the family of lithium-insertion compounds is the spinel structure, and the most widely used positive electrode material in this family is $LiMn_2O_4$ (LMO) (Jiang et al. 2006; Liu et al. 2006; Chen et al. 2019; Nourhan and Nageh 2020). LMO operates at the voltage 4.0 V and provides the capacity value 110 mA/g, a little less as compared to LCO. The main advantages of this material are safety, cheapness, lower toxicity as compared to LCO and environmental friendliness (Xia et al. 2012; Lee et al. 2014; Marincaş et al. 2020; Patel et al. 2021). When one describes the electrode materials for Li-ion battery the key importance are electronic and ionic conductivities of the material. The electrode material with the spinel structure has 3D ionic and electronic conductivities (Fig. 4.2), nevertheless, the electronic conductivity of

LMO is significantly lower than those of LCO, and it is an important drawback of the material (Patel et al. 2021). Another problem associated with the LMO cathode is Mn dissolution in the electrolyte on the cycling at high operating temperatures, especially above 60°C, results in the capacity loss. Therefore, the main obstacles for the large-scale commercialization of LMO are capacity fade during the charge/discharge process and poor electronic and ionic conductivities (Xia et al. 2012; Lee et al. 2014; Marincaş et al. 2020; Patel et al. 2021). Patel et al. 2021 reported that in order to improve the electron conductivity, $AlPO_4$ coated $LiMn_2O_4$ has been developed. The 3 wt.% $AlPO_4$ coated LMO shows the electronic conductivity of 9.493×10^{-4} Ω^{-1} cm^{-1} at room temperature, which is considerably higher than the conductivity of pure $LiMn_2O_4$ that is around $4.304 \times 10^{-4} \Omega^{-1}$ cm^{-1}. Therefore, LMO-$AlPO_4$ cathode material with advanced properties is an attractive candidate for Li-ion battery applications and it can be a good alternative for the traditional cathode materials (Placke et al. 2021).

To prevent manganese dissolution, a protective metal oxide coating can be deposited on the electrode surface, most often metal oxide layer, SiO_2, SnO_2, MgO, etc. (Gnanaraj et al. 2003; Wang et al. 2006; Arumugam and Kalaignan 2008). Wang et al. 2020d described that an amorphous Al_2O_3/ZrO_2 composite layer with a thickness of 6 nm was coated on the surface of LMO. The coating has been shown to enhance the cathode surface stability and reduced polarization. The composite $Al_2O_3/ZrO_2/LMO$ cathode has shown improved electrochemical characteristics, such as better rate properties and cycle reversibility, especially at elevated temperatures. The modified cathode also demonstrates an enhanced reversible intercalation/deintercalation ability. As it was reported, the material demonstrated the capacity retention of 88.1% at 55°C after 100 cycles at a 1 C rate, which is higher as compared to 61.3% typical for bare LMO. These results confirm, that the protective coating has improved the chemical stability of the LMO cathode surface and reduced the side reaction between the cathode and the electrolyte and of Li^+. Therefore, from one side deposition of the protective layer is an effective strategy that provides separation of the active spinel material from the electrolyte, however, from another side, it is hard to deposit the layer uniformly and continuously, and as a result the cathode material still contacts with the electrolyte partially. Moreover, the coating can increase the overall resistance, leading to a decrease in the capacity value of LMO cathodes.

High electrochemical stability of $LiMn_2O_4$ (LMO) cathode materials can be also achieved by doping with metals, for example, Al, Co, Ag, Ni, Nb, Sc, Fe, etc. (Wei et al. 2006; Zhou et al. 2008; Yi et al. 2013; Wang et al. 2014; Liu et al. 2015; Bhuvaneswari et al. 2019a). Double-doping has shown even better results in terms of improvement of LMO structural and thermal stability (Fang et al. 2015; Susanto et al. 2015). The bi-metal modified LMO samples demonstrate an improved capacity, cyclability, and high rate capability. Synthesized by Yu et al. 2019 Ni and Mg dual-doped LMO composite delivered the first-cycle discharge specific capacity of 112.3 mAh g^{-1} at 1 C and 91.2 mAh g^{-1} at a high current rate of 20 C. Excellent capacity retention of 77% was obtained after 1000 cycles. The sample shows the excellent capacity of 97.6 mAh g^{-1} even at elevated temperature 55°C at the high current rate of 10 C. Therefore, the double metal doping of LMO cathode offers a new opportunity for developing high-performance cathode material for lithium-ion batteries. Although substitution of metals into the crystalline structure of the material is favorable in terms of improving the crystal structure, however, this strategy would surrender the capacity value because the metals dopants do not participate in the electrochemical reaction.

An improvement of electrochemical reversibility and stability of the spinel LMO cathode material at elevated temperatures can also be achieved by the development of core-shell morphologies. Wen et al. 2015 developed a new composite cathode with a bulk $LiMn_2O_4$, as a core that is responsible for delivering high capacity. The spherical LMO core was completely encapsulated by a spinel structure concentration-gradient shell that increased gradually Ni and Co contents. A smooth concentration gradient of the transition metals (Ni, Co and Mn) enabled unimpeded diffusion of Li ions and provided a protective layer suppressing Mn dissolution in the electrolyte. The composition of the outmost layer of the spherical core-shell structure was

$LiNi_{0.45}Mn_{1.45}Co_{0.1}O_4$. The composite cathode has shown to poses excellent cycling stability at elevated temperature, delivering a discharge capacity of 118 mAh g^{-1} between 3.0 and 4.4 V at a rate of 1 C (148 mA g^{-1}) at 55°C with 96% capacity retention after 200 cycles. At the same time, the core-shell cathode has also shown a good rate capability, delivering a high discharge capacity of over 110 mAh g^{-1} even at a rate of 5 C. Therefore, the core-shell structure spinel cathode material with transitional metals concentration gradient is a rather promising cathode material for advanced lithium-ion batteries.

Cathode morphology engineering is another strategy used to improve the rate performance of LMO cathode materials by shortening the diffusion paths for the lithium ions (Xu et al. 2020b). Different micro- and nanostructures have been reported so far, among them hollow micro- and nanospheres (Kumar et al. 2019), nanosheets (Sun et al. 2012), nanofibers (Qian et al. 2014), nanotubes (Chen et al. 2018a), core-shell structures (Deng et al. 2015) and so on. From one side nanostructuring intensifies electrochemical reactions with lithium ions, but at the same time, side reactions are also intensified as a result of high surface area. A combination of micro- and nanostructures in the LMO electrode material has shown promising results, especially hollow hierarchical micro/nanostructures have shown to be very effective in improving the cathode electrochemical performance. In such hierarchical micro/nano-sized architectures the nano-sized active structural elements are responsible for intensified charge transfer and the micro-sized secondary assemblies reinforce the overall structural stability of the composite (Ma et al. 2016b; Xu et al. 2017; Liang et al. 2020). Hollow micro/nano architectures are mainly synthesized through hard or sacrificial templates techniques, that usually utilize numerous complex preparation and formation steps. Therefore, the simple template-free approaches to produce hierarchical spinel hollow cathode materials still remain a great challenge these days.

Surface orientation engineering has proved to greatly influence Mn dissociation during the cycling process. Yang et al. 2017; Xiao et al. 2019 showed that the {111} LMO plane has the denser manganese atoms arrangement as compared to {110} and {100}, planes. Therefore, it interacts with the electrolyte to a smaller extent, and as a result reduces Mn dissolution from the surface, enhancing in such a way the cycling stability. Xu et al. 2020b designed spinel $LiMn_2O_4$ cathode for lithium-ion batteries with micro-nano structured morphology and {111} preferable surface orientation that is capable of functioning at extreme conditions. Using solvothermal method followed with lithiation reaction, the authors have designed LMO nanoparticles with different morphology and surface orientation: micro-spherical (LMO-MS), micro-tubular (LMO-MT), and hybrid sphere-interconnected-tube micro-structured (LMO-MST) (Fig. 4.6(a), (b), (c)).

LMO-MS demonstrates (111) facets, LMO-MT has the high index lattice (311) and LMO-MST exhibits the densest (111) facets at the micro-spherical surface and a new (111) plane appearance on the surface of micro-tubes. It has been proved in the work that materials with different surface lattice orientations have different electrochemical performances (Fig. 4.6(d–g)).

The composite material with LMO-MST morphology has demonstrated an outstanding cycling performance with the capacity retention after 1000 cycles of 84.3% at 10 C. The electrode has superior rate capability up to 10 C (124.2 mAh g^{-1} at 10 C). The electrochemical performances of the cathode at elevated (55°C) and decreased (−5°C) temperatures have shown that under such extreme conditions it still maintains optimal cycling stability. The results of the work by Xu et al. 2020 have proved that the morphology and facets orientation control can significantly improve the electrochemical performance of the spinel cathode materials. Therefore, this promising approach can promote the development and application of spinel electrodes in Li-ion batteries.

A very promising composition for next-generation high-performance lithium-ion batteries is LNMO ($LiNi_{0.5}Mn_{1.5}O_4$), it is one of the most favoring Co free cathode materials, due to the high energy density of 650 Wh kg^{-1}, high thermal stability, high ionic conductivity (Ma et al. 2016; Xu et al. 2017; Liang et al. 2020). When using LNMO as a cathode material, the battery operates higher as compared to LCO, LNO, and LMO voltages (~4.7 V vs. Li) and has a large specific capacity value of 146.6 mAh g^{-1}. However, before this material is commercialized and used in mass production more investigations are required. The biggest limit preventing LMNO from the practical application is a rapid capacity decay during cycling, especially at high voltages and as a result shorten battery lifetime. This material can exist in two phases ordered and disordered ones that have different electrochemical properties, for example, one has higher capacity but shorter life cycle, another has better cyclability but lower capacity (Liang et al. 2020). The approaches to improve LNMO performance are quite similar to LMO modification strategies, and they can be categorized into three main categories: morphology engineering, doping, surface modification (Ma et al. 2016b; Xu et al. 2017; Liang et al. 2020). Various LMNO nanostructures with different morphologies have been reported for LNMO composition, among them nanowires, nanorods, nanoplatelets, etc. (Whittingham 2008; Yang et al. 2014; Xu et al. 2015; Sun et al. 2018; Zhao et al. 2018). The research results have evidenced that the most promising morphology have one-dimensional nanostructures such as nanowires and nanorods, due to short lithium-ion diffusion length and large surface area providing sufficient electrode-electrolyte interaction (Liu et al. 2013; Yang et al. 2014; Xu et al. 2015; Wei et al. 2017b). Similarly, as for LMO cathode, nanoengineering of the surface has shown to intensify side reaction on the electrode-electrolyte interface. That is why these days researchers have moved from synthesizing nanostructured LMNO to microsized morphologies or to integrating both micro and nanoelements. Microplates, core-shell structures, microcubes, octahedra and truncated octahedra have been reported recently (Whittingham 2008; Liu et al. 2013; Yang et al. 2014; Xu et al. 2015; Wei et al. 2017b; Sun et al. 2018; Zhao et al. 2018).

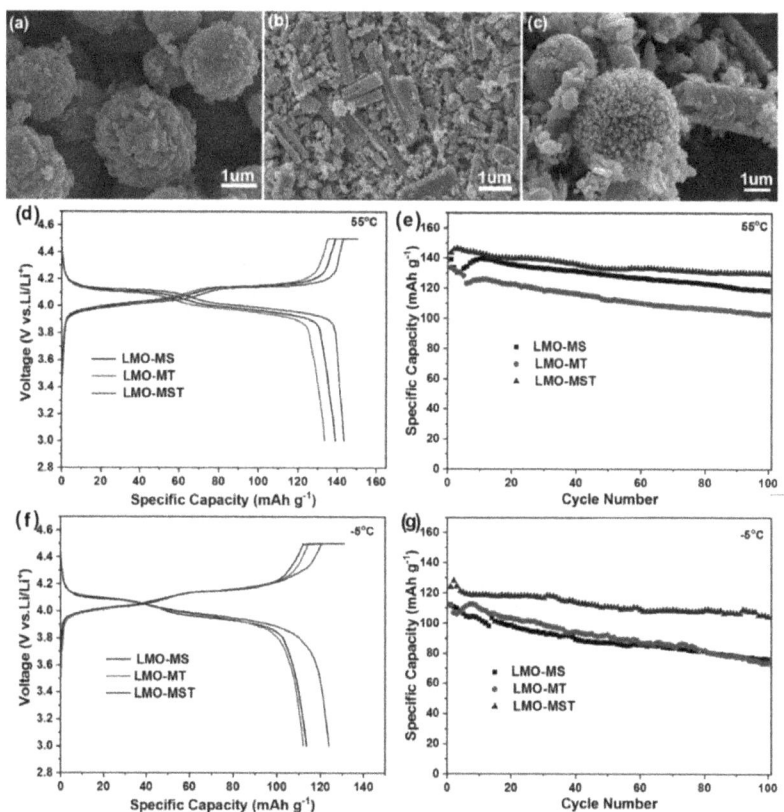

Figure 4.6 SEM images of LMO materials with different morphologies: (a) LMO-MS, (b) LMO-MT, (c) LMO-MST. The first charge/discharge curves and cycle stabilities of the LMO samples at 0.5 C under conditions: (d, e) 55 °C, (f, g) –5 °C. (Reprinted with permission from Copyright Elsevier. *Source*: Xu et al. 2020b. The improved performance of spinel $LiMn_2O_4$ cathode with micro-nano structured sphere-interconnected d-tube morphology and surface orientation at extreme conditions for lithium-ion batteries. *Electrochimica Acta* 358(2020): 136901.

The second modification approach is doping of LNMO, most often metals are used as dopants, among them Na, Mg, Ti, V, Cr, Fe, Zr, Sc, etc. (Liu et al. 2014a; Liu et al. 2017; Wang et al. 2017; Feng et al. 2018; Bhuvaneswari et al. 2019b; Zong et al. 2020a; Kocak et al. 2021), also non-metal dopants are reported, mainly Si and P (Luo et al. 2017; Zong et al. 2020b). The third modification approach is adopted with the purpose to stabilize electrode-electrolyte interphase and decreasing the side reactions by means of LNMO surface engineering. In this case, a protective coating layer is deposited onto the LNMO electrode. There are three main types of protective coatings: electronic-conductive coating, ionic-conductive coating and acid-protective coating. Carbon layers are mainly used to compensate for poor electronic conductivity of spinel LNMO. Due to adjustable conductivity values and interaction with the active materials, different carbon allotropes, such as carbon nanotubes, graphene, graphene oxides have been intensively used to improve the properties of LNMO electrode and as a result the battery performance (Monaco et al. 2015; Hwang et al. 2016; Ku et al. 2019; Gao et al. 2020). Conductive polymer coatings have recently attracted great attention in LIB research because they improve the electronic conductivity of the electrode material, without affecting the capacity value, unlike the previously described carbon coatings (Ma et al. 2016b; Xu et al. 2017; Liang et al. 2020). The improvement of the ionic conductivity of the electrode material also has a significant effect on the performance of high-voltage LIB, that is why a coating with heightened ionic conductivity is becoming more and more popular. Among a wide variety of different ionic conductive layers, the following compounds Li_2SiO_3, Li_4SiO_4, Li_3PO_4 and $LaFeO_3$ that are excellent Li-ion conductors, have been intensively used as the coating to improve the ionic conductivity of the cathode materials (Xu et al. 2016; Yubuchi et al. 2016; Deng et al. 2017; Mou et al. 2018). Different metal oxide coatings have been employed as an acid-protective layer on the LNMO surface, which include SiO_2, TiO_2, Al_2O_3, etc. (Fan et al. 2007; Tao et al. 2017; Chang et al. 2019). The implementation of the above-described strategies of LNMO cathode modification, such as doping, cathode surface coating can help to increase the cycling stability for the application of this material in high voltage LIBs. However, there are still some challenges the researchers should solve before this material goes into mass production and find its practical applications in high-power devices. First of all, coating the cathode surface is difficult to implement for large-scale battery applications. Moreover, the applied protective coating decreases the discharge capacity of LNMO cathodes. Secondly, the relationships between LNMO crystal structure, morphology, defect, etc., and cycling performance should be further investigated.

Another spinel structure worth mentioning is LCMO ($LiCo_{0.5}Mn_{1.5}O_4$), which is the first reported material that has overcome the 5 V voltage barrier (Kuwata et al. 2014). It has the highest potential among all high-voltage spinel structures and has a 30% higher energy density as compared to conventional LCO and LMO cathode materials. Moreover, LCMO experiences very small volume changes during the insertion/extraction of lithium ions, which is a significant advantage for battery application because volume changes often cause cracking that results in the shortening of the battery lifetime. However, it delivers a low specific capacity of 95 mAh g^{-1} at 5 V, the highest possible theoretical capacity of LCMO is 145 mAh g^{-1}. The main drawbacks of this high voltage material are a significant capacity fading during cycling, low cycling stabilities, and cell degradation at high voltages due to the electrolyte decomposition (Kuwata et al. 2014; Reeves-McLaren et al. 2018; Windmüller et al. 2018; Ishigaki et al. 2019; Liu et al. 2021). Use of a solid electrolyte can overcome the problem of cell degradation caused by the use of a conventional liquid electrolyte, which decomposes at voltages more than 4.7 V. The cell with a solid electrolyte can operate stably up to voltages of 5.5 V (Windmüller et al. 2018). However, new additional requirements appear to the cathode material itself in the case of the solid electrolyte usage; the active material must be tightly bonded and densified in order to provide diffusion pathways for electrons and lithium ions (Placke et al. 2017; Windmüller et al. 2018).

There are also difficulties associated with the synthesis of single-phase LCMO because this material loses a significant amount of oxygen at temperatures higher than 600°C. Such oxygen loss during the LCMO formation procedure causes the capacity to fade at 5 V and the appearance of an unexpected peak at around 4 V. Therefore, the efforts of researchers are currently concentrated on improving the capacity and the cycling properties of LCMO together with developing new effective synthesis roots. With an aim to decrease oxygen deficiency, Ishigaki et al. 2019 investigated the effect of LCMO post-deposition annealing in an oxygen atmosphere at different temperatures and for different periods of time. Then, the post-annealed samples were used as the cathode in all-solid-state Li-ion cells and their electrochemical performance investigated. The results have shown that the best discharge capacity in the 5 V region was obtained for the thin films that have been annealed at 600°C for 6 hours. The improvement in the charge-discharge characteristics has been reached through the controlling of the oxygen deficiency by the post-deposition annealing in the oxygen atmosphere. Fluorination is another effective strategy used to reduce the phase decomposition of LCMO during synthesis. Liu et al. 2021 have shown that fluorination improves the structural stability during the charge-discharge process. Besides this, the fluorinated $LiCoMnO_4$ shows a better capacity of 128.1 mAh g^{-1} at 0.1°C and improved rate performance. The fluorinated cathode presents the discharge capacity of 112.7 mAh g^{-1} at 10°C, while the capacity of 38 mAh g^{-1} is observed for the pristine $LiCoMnO_4$ at the same rate. By comparison, the fluorinated cathode material shows high potential for high energy density batteries.

Summarizing the above-described spinel composite materials, it is worth noting, that the spinel electrodes operate at higher voltages as compared to the layered, but the specific capacity of the spinel cathode is less than 150 mAh g^{-1}. As mentioned earlier, for the layered cathode materials the high specific capacity values of 180 mAh g^{-1} (LNC) (Belharouak et al. 2003; Shang et al. 2016) and 200 mAh g^{-1} (NMC) (Lyu et al. 2020; Stephan 2020; Wood et al. 2020) have been reported. The most attractive cathode materials in terms of the high specific capacity value are the layered Li-rich composites, with reported (Manthiram et al. 2016; Pan et al. 2018; Çetin et al. 2020a; Zheng et al. 2020) capacity value of 400 mAh g^{-1}. However, the voltage drop and the structural changes during the charge-discharge cycles, the irreversible initial capacity loss, limit large-scale applications of the lithium-rich layered compounds. Recently researchers decided to combine the spinel and the layered crystal structures into one composite material (Kim et al. 2013; Yi et al. 2017). In such a way a new class of complex composite cathode materials has been developed that simultaneously uses the advantages of both layered and spinel crystal structure and in this case, each phase could compensate for the drawbacks of one other. Such integrated layered-spinel composite cathode utilizes three-dimensional diffusion channels of the spinel structure, which provides a high rate performance and high operating voltage. At the same time, the layered component in the compound is responsible for the high specific capacity value. Such integrated composites have proved themselves as structurally stable cathodes with a high rate performance and high cycle life that can operate at the high voltages with a low voltage fading (Kim et al. 2013; Li et al. 2017; Yi et al. 2017; Liu et al. 2019; Xu et al. 2020).

The results of investigations have shown, that the cubic close-packed oxygen arrays in both the layered and spinel crystal structures are structurally compatible (Lee et al. 2013). This fact makes it possible to integrate the layered Li-rich and the spinel oxides into one composite, which might show both a high capacity, high voltage and an excellent rate capability. The works of (Wu et al. 2013; Feng et al. 2015; Wang et al. 2015b; Zhao et al. 2016) demonstrated that such integration of spinel materials into layered lithium-rich compounds results in an improvement of the electrochemical performance of the lithium-ion battery. In their work, Pang et al. 2019 developed LCO/Al-doped LNMO composite cathode material for high-voltage lithium-ion battery application that has improved the cycling performance and thermal stability. LCO/Al-doped LNMO exhibits improved cycle stability between 3.00 V and 4.55 V. The capacity of the composite can be up to 161.9 mAh g^{-1} with much better cyclability. The cycle-life improvement of lithium-ion batteries with the composite cathode is explained by a delay of the structure collapse and a reduction of cobalt dissolution. Moreover, the application of LCO/Al-doped LNMO in Li-ion cells leads to an increase in the onset temperature of primary exothermic peaks, and a decrease in the maximum rising temperature of thermal shock from 500 C to 160 C.

Kim et al. 2013 have developed Layered-Layered-Spinel (LLS) composite material $Li_{1.1}Mn_{0.97}Ni_{0.265}Cr_{0.1}Co_{0.065}O_3$ for application in high-performance lithium-ion batteries. They investigated the composite cathode electrochemical properties depending on the amount of spinel phase (x = 0, 0.25, 0.5, 0.75, and 1). It has been established that the spinel phase addition significantly influences the electrochemical performance of the composite samples. The LLS composite with 0.5 spinel phase content has been shown to exhibit the best electrochemical performance among the others compositions. The cathode delivers a high reversible capacity of 152 mAh g^{-1} at the rate of 10 C. Moreover, after 300 cycles the composite cathode demonstrates an outstanding capacity retention value of 94.4% with an average coulombic efficiency of 98.3% during cycling at 0.1 C. The excellent electrochemical properties such as the high rate capability and the superior cycle life are explained by the well-ordered mixed structure and doped chromium ions. Therefore, the new integrated layered layered-spinel composite reported in (Kim et al. 2013) has demonstrated to be a very promising candidate to be used as high-performance cathode material in advanced lithium-ion batteries. Therefore, the described above approach of developing new integrating composites that utilizing both layered and spinel crystalline structures have found to be very promising for high energy density applications, particularly for electric vehicles (Kim et al. 2013; Lee at al. 2013; Wu et al. 2013; Feng et al. 2015; Wang et al. 2015; Zhao et al. 2016; Li et al. 2017; Yi et al. 2017; Liu et al. 2019; Pang et al. 2019; Xu et al. 2020c). Several Li-ion batteries manufacturers have already launched new chemistries into production that use a cathode with a combination of layer-spinel materials such as NMC and LMO or NCA and LMO, or LCO and LMO.

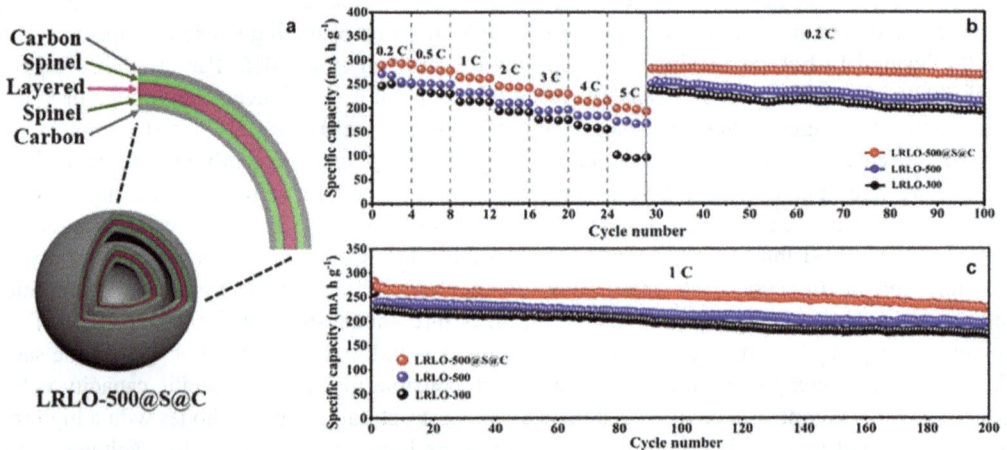

Figure 4.7 (a) Schematic illustration of LRLO-500@S@C structure; (b) rate capability at various rates from 0.2 C to 5 C; (c) cycling performances at 1 C rate (Reprinted with permission from Copyright Elsevier. *Source:* Ma et al. 2019. Double-shell Li-rich layered oxide hollow microspheres with sandwich-like carbon@ spinel@layered@spinel@carbon shells as a high-rate lithium-ion battery cathode. *Nano Energy* 59(2019): 184–196).

Ma et al. 2019 developed a composite cathode material with double-shell morphology for high-rate lithium-ion battery applications. In this case, Li-Rich Layered Oxide (LRLO) hollow microspheres served as a core, and sandwich-like structure carbon/spinel/layered/spinel/ carbon has used a shell (Fig. 4.7). The LRLO core structure can be controlled by varying the calcination temperature of the precursor, two LRLO samples have been sensitized at temperatures 300°C and 500°C, assigned as LRLO-300 and LRLO-500, respectively. LRLO-500 was used for the formation of the double-shell sandwich-like structure, assigned as LRLO-500@S@C. As it was reported in Ma et al. (2019b) the fabricated LRLO-500@S@C composite cathode provided a high initial capacity of 312.5 mAh g^{-1} with a large initial Coulombic efficiency of 89.7%. Even after prolonged cycling (200 cycles) at 1.0 C and 5.0 C a large and stable discharge capacities of 228.3 mAh g^{-1} and 196.1 mAh g^{-1} have been obtained, respectively. The electrochemical characteristics of LRLO-500@S@C in comparison with two bare LRLO samples are presented in Fig. 4.7. The authors explain the outstanding electrochemical performances of the cathode material by multiscale coordinated design based on hierarchical double-shell hollow construction. Moreover, the special heterostructured shells and the introduced oxygen vacancies, which benefit to shorten Li-ion diffusion paths, strengthen structural stability, and reduce side reactions.

2.3   Cathode Materials with Olivine Crystal Structure

Lithium-ion phosphate with the olivine structure LiFePO$_4$ (LFP) has the theoretical capacity value of 170 mAh/g, which is lower than the capacity of the layered electrode materials and higher than that of materials with the spinel structure. The cell voltage value is the lowest among all commercially used cathode materials and is in the range from 3.4 to 4.1 V, and as a result, it has the lowest energy density of 586 Whk g^{-1} (Zhang 2011; Zhang et al. 2012). One of the main drawbacks of the electrode with an olivine structure is the fact that the material has only 1D channels for electrons and ions motion (Fig. 4.2). Therefore, the LFP electrode has low values of both the lithium ionic diffusivity ($\sim 10^{-14}$ cm^2 s^{-1}) and the electronic conductivity ($\sim 10^{-9}$ S cm^{-1}), leading to low power densities of the cell and as a result a poor rate capability (Julien et al. 2014; Chen et al. 2019; Zhao et al. 2021). On the other hand, the material abundance in nature provides a cheap price of LFP based Li-ion cell making it an attractive cathode chemistry due to a possibility of low-cost batteries production. Other important advantages of this crystal structure are a superior cycle life, which is much higher as compared to LCO and LMO, a high intrinsic safety up to temperature 400°C, non-toxicity, and environmental friendliness (Julien et al. 2014; Chen et al. 2019; Zhao et al. 2021).

With an aim to intensify diffusion of lithium-ion, nanostructuring, or in other words decreasing the size of the particles comprising the electrode to is used these days. This strategy decreases diffusion paths due to shortened Li-ion insertion/extraction pathways, reducing in such a way the overall diffusion time and increasing the electrode power (Bi et al. 2013; Gu et al. 2007). The low electronic conductivity can be compensated by using composite materials with some content of carbon or other conductive components (Gu et al. 2007; Bi et al. 2013; Demirocak et al. 2017; Liu et al. 2018). Therefore, carbon coating and decreasing the particle size to nanometers are two important approaches to improving the electrochemical performance of LiFePO$_4$. As was mentioned for layered and spinel cathodes, surface nano-engineering have intensified side reaction and reduced the cathode cycle ability. In the case of olivine LFP cathode, nanostructuring has been found to be an effective strategy that improves the olivine cathode electrochemical performance. Different LiFePO$_4$ morphologies have been reported in literature, among them nanoparticles (Konarova and Taniguchi 2009), nanorods (Wang et al. 2016), nanoflowers (Rangappa et al. 2010), nanobelts (Shao et al. 2014), nanofibers (Toprakci et al. 2012), nanowires (Hosono et al. 2010), nanoplates (Wang et al. 2012), nanospheres (Cheng et al. 2013), etc. It is worth mentioning, that different LiFePO$_4$ cathode morphologies not only reduce lithium ions diffusion length but also enhance the performances of cathode material. Among the wide range of different LFP nanostructures, nanorods morphology has shown to be beneficial to improve the electrochemical performance at high rates (Wang et al. 2016).

It is known that the capacity performance is mainly affected by the distance of electron transport and ion diffusion into the cathode material, which is dependent on the radius value. That is why the nanorods radius is a very important parameter responsible for the better rate. Nanowires "net-like" structure is also a promising morphology because it possesses a better permeation behavior and a better contact of the active material with the electrolyte. LiFePO$_4$ cathode grown in the form of nanowires demonstrate a specific capacity value of 169 mAh g^{-1} (Hosono et al. 2010), which is really close to the theoretically calculated one. Therefore, 1-D nanostructured morphologies, such as nanorods and nanowires, enlarge the electrode-electrolyte interface, and as a result, can yield an enhanced rate performance and cycling capability. As was mentioned above, carbon coating or additives have been shown to be the most effective and cheap way to improve the electronic conductivity of LFP electrodes. Different conductive carbon materials are used these days as carbon additives (Gong et al. 2016, Eftekhari 2017), including graphene (Zhang et al. 2021), graphene oxide (GO) (Chien et al. 2020a), graphene nanosheet (GNS) (Gong et al. 2016; Eftekhari 2017), carbon nanofibers (CNF) (Adepoju et al. 2020), and carbon nanotubes (CNT) (Gong et al. 2016; Eftekhari 2017; Chen et al. 2018b). Table 4.1 summarizes different modification strategies of LFP cathode electrochemical properties.

Table 4.1 Modification of LFP electrochemical properties with carbon

Composite	Modification strategy	Capacity	Cyclic stability
LFP-G (Zhang et al. 2021)	Graphene addition	160 mAh g^{-1} at 0.2 C and 144.2 mAh g^{-1} at 20 C	no obvious capacity fading after 40 cycles
LFP/C/1%rHTGO (Chien et al. 2020a)	Hydrothermal porous graphene oxide addition	160.5, 151.8, 138.8, and 130.3 mAh g^{-1}, at 0.1 C, 1 C, 5 C, and 10 C	capacity retention 99.6% after 300 cycles at 0.1 C and 91.8% after 300 cycles at 10 C
CNF-LFP (Adepoju et al. 2020)	Carbon nanofibers addition	150 mAh g^{-1} at 0.1C	capacity retention of 98.4% at 5 C after 200 cycles
LFP-CNTs (Chen et al. 2018b)	Carbon nanotubes addition	160.3 mAh g^{-1} at 0.3 C	capacity fading only 0.4% after 50 cycles
LFP-G-CNT (Chen et al. 2018b)	Graphene and carbon nanotubes additives	168.4 mAh g^{-1} at 0.1 C and 103.7 mAh g^{-1} at 40 C	is capable of 82% capacity retention after 3000 cycles
LFP-C-G (Guan et al. 2019)	Activated carbon and graphene addition	167.3 mAh g^{-1} at 0.1 C 66 mAh g^{-1} at an extremely high rate of 100 C	the capacity retention of 82% after 3000 cycles
LFP-C core-shell (Bao et al. 2017)	Formation of core shell nanorods with carbon	173.8 mAh g^{-1}, 114 mAh g^{-1} at 0.1 C and 10 C	capacity retention 96.5% after 500 cycles at 0.1 C
LFP-C (Toprakci et al. 2011)	Carbon coating	125, 141 and 136 mAh g^{-1} at 0.1 C for calcination temperatures 600, 700 and 800 °C	no obvious capacity fading after 100 cycles
LFP/FeS/C (Hongtong et al. 2019)	Doping with FeS and carbon addition	155 mAh g^{-1} at 0.2 C	no obvious capacity fading after 50 cycles

Carbon nanofibers have recently been introduced as a conductive additive to enhance the LFP electronic conductivity because of their excellent electrical conductivity and a high aspect ratio. In their work (Adepoju et al. 2020) developed CNF-LFP composite cathode and it was shown, that the cathode delivers the discharge capacity of ~150 mAh g^{-1} at 0.1 C that is higher as compared to the pristine LFP cathode. The electrode has an excellent rate performance at >5 C capability, the high cycling stability with the capacity retention of 98.4% at 5 C after 200 cycles. The LFP/ CNF composite material has demonstrated an excellent rate of performance at 10 C and 20 C when compared to the pure LFP cathode, which typically fades out at 20 C. The composite still operates even at extremely high rates of 30 C, 40 C, and 50 C. The enhanced performance at capability rates higher than 5 C is attributed to the creation of long-range conductive networks and bridges by adding CNFs into the LFP structure. Therefore, this cathode chemistry with the enhanced battery performance indicates promise for high-power applications.

Many research groups have devoted their attention to developing LFP-CNTs composites with improved properties. Developed in (Chen et al. 2018b) LFP-CNTs composite has an initial discharge capacity of 160.3 mAh g^{-1} at 0.3 C, and the capacity fading was only 0.4% after 50 cycles. Synthesized LFP-CNTs composite cathode material with 7 wt.% CNTs showed an initial discharge capacity of 152.7 mAh g^{-1} at 0.18 C discharge rate with a capacity retention value of 97.77% after 100 cycles. Graphene has shown excellent potential as an additive to LFP cathode material for high-power lithium-ion batteries. It has attracted great attention because of its increased surface area and excellent electrical conductivity. Developed by (Zhang et al. 2021) LFP/graphene composite has an initial charge-discharge capacity of 141 mAh g^{-1} at 0.1 C and retains 72% of the initial capacity at 10 C. However, the rate-performance enhancement of LFP/CNTs and LFP/graphene composite materials is still limited and further investigations are required. With this purpose in their work (Chen et al. 2018b) developed nanocrystalline LiFePO$_4$/graphene/carbon nanotubes (LFP-G-CNT) composite with a significantly enhanced electrochemical ability. The electrochemical measurements prove that the three-component LFP-G-CNT composite performs better than two-component LFP-G and LFP-CNT composites. The electrode has a high initial discharge capacity of 168.4 mAh g^{-1} at 0.1 C and 103.7 mAh g^{-1} at 40 C and excellent cycling stability. Guan et al. 2019 described? LFP/activated carbon/graphite composite (assigned as LFP/AC/G) with hierarchical porous architecture that was developed for superior high-rate lithium-ion storage. This cathode composition delivers a remarkable high capacity of 66 mAh g^{-1} at the extremely high rate of 100 C, and excellent life-cycle stability. It has been shown in the work that the electrode is capable of 82% capacity retention after 3000 cycles. Due to its outstanding properties, this chemistry has a high potential for applications in high-power Li-ion batteries. The electrochemical properties of the LFP/AC/G composite electrode are given in Fig. 4.8.

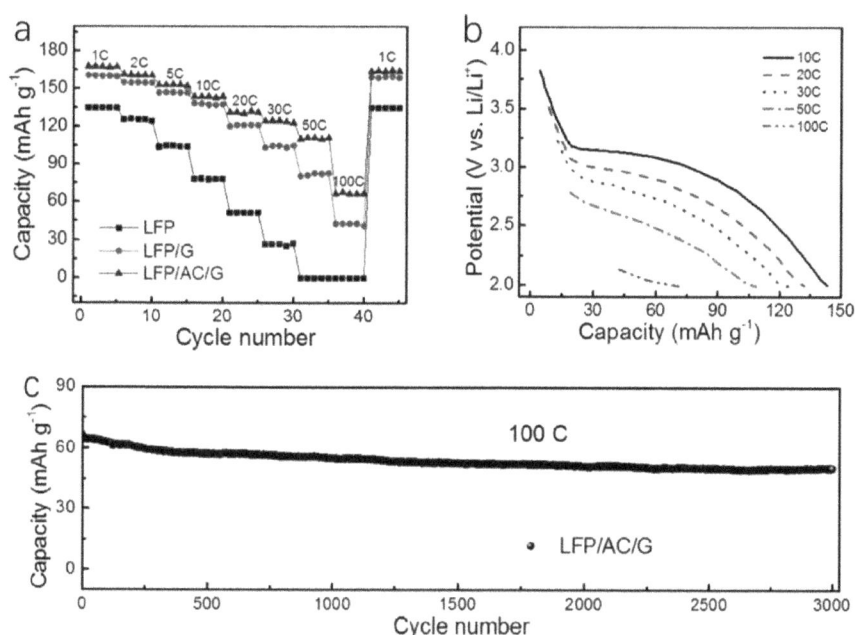

Figure 4.8 (a) Rate performances of LFP/AC/G composite in comparison with bare LFP and LFP/G composite at rates from 1 C to 100 C; (b) charge-discharge characteristics of LFP/AC/G, LFP/G, and LFP samples; (c) long-term cycle performance of LFP/AC/G composite at rate 100 C. (Reprinted with permission from Copyright Elsevier. *Source:* Guan et al. 2019. LiFePO$_4$/activated carbon/graphene composite with capacitive-battery characteristics for superior high-rate lithium-ion storage. *Electrochimica Acta* 294 (2019): 148–155).

In another study (Chien et al. 2020a) described how the electrochemical performance of two-component LFP/C composites was improved by the use of Graphene Oxides (GOs) with various morphologies as conductive additives to LFP/C. Three different GOs morphologies have been investigated in the work, including pristine GO, three-dimensional GO and hydrothermal porous GO (HTGO). It was shown, that among different morphologies, the addition of 1 wt.% of HTGO to LFP/C leads to the best electrochemical performance. Such a composite (denoted as LFP/C/1%rHTGO) shows discharge capacities 160.5, 151.8, 138.8, and 130.3 mAh g^{-1}, at 0.1 C, 1 C, 5 C, and 10 C, respectively. The capacity retention was measured to be 99.6% after 300 cycles at 0.1 C and 91.8% after 300 cycles at 10 C.

A coating with carbon of 1D LFP nanostructures, among them nanofibers, nanorods, and nanowires has advanced the development of new types of composite in the form of 1D LFP/Carbon core-shell nanostructures (Zhu et al. 2011; Toprakci et al. 2011; Bao et al. 2017; Hongtong et al. 2019). In their work (Bao et al. 2017) described how mono-dispersed LiFePO$_4$-C core-shell [001] nanorods have been successfully developed for high power Li-ion battery cathode applications. In this case, LFP monodisperse nanorods serve as a core and carbon coating as a shell. The size reduction along [010] direction together with the perfect LFP olivine lattice structure and smooth carbon coating on the top of the mono-dispersed LiFePO$_4$ allowed to create the core-shell nanorods structure that delivers an excellent high rate capacity value with outstanding cycle stability. The discharge capacity of the core-shell nanorods structure was established to be 173.8 mAh g^{-1} at 0.1 C that is higher than the theoretical value. It can be explained by 1D morphology and by carbon films, which have a higher theoretical capacity than LFP. At the rate of 10 C, the reversible specific capacity of the composite cathode was found to be 114 mAh g^{-1}. After 500 cycles, the capacity retention was found to be 96.5%. The authors of the work (Hongtong et al. 2019) combined three different modification strategies, namely control of morphology and size of a 1D material; doping; and coating with carbon. They developed FeS doped LFP/carbon nanofibers composite with enhanced electrochemical properties. In the composite, FeS doped LiFePO$_4$ nanofibers of core have been coated with a carbon layer, that served as a shell. FeS phase was evenly distributed within the core region of the LiFePO$_4$ fibers, the dopant is known to be electrochemically inactive within this operating voltage of Li-ion cell and acts as a conductive additive to the LFP cathode. Cation doping with Mg, Na and Al$_3$ ions has improved the electrochemical performance of the undoped LiFePO$_4$/FeS/C composite fiber even more. The coulombic efficiency of the composite was found to be higher than 98%. The discharge capacity at a 0.2 C reached 155 mAh/g value for Al3+ doped LFP/FeS/Carbon. LFP-C core-shell nanocomposites with advanced electrochemical properties have been synthesized in the form of nanoparticles (Huang et al. 2016), nanopillows (Xu et al. 2014), etc. Nevertheless, 1D nanorods show better properties in terms of reverse capacity value, rate capability, and capacity retention.

LFP has shown great potential when used as an additive to cathode material with layered or spinel structure (Wie et al. 2020). In their works (Liu et al. 2014b; Wu et al. 2016) demonstrated that LFP additive could effectively reduce the heat release during charge-discharge processes and improve the reversible capacities of NMC electrodes. To improve the electrochemical performance of the NMC-LFP composite even further, carbon conductive additives can be used. For example, (Wang et al. 2007) described how Ni-rich NMC-LFP-graphite composite has been developed. The cell shows outstanding life-cycle properties, after 500 cycles, it still delivers a discharge capacity of 165.3 mAh g^{-1}, corresponding to a capacity retention of 91.65% at 1 C, in the voltage range from 3.0 to 4.2 V. The achieved value is much higher than that of 70.65% for pristine NCM after 500 cycles at 1 C. As already mentioned, LCO electrodes struggle from structural changes and capacity fade at high operating voltages, Wang et al. (2007) showed that composite LCO-LFP electrode demonstrates significantly improved electrochemical performance at high potentials and high temperatures. The life-cycle properties, electrochemical performance and safety issues of NCA electrode chemistry can also be improved by electrode modification with LFP (Wie et al. 2020). In this composite NCA is responsible for high voltage and high energy densities and LFP improves the material stability and cyclability. Developed by (Wie et al. 2020) NCA-LFP composite electrodes with 20 wt.% LFP have enhanced electrochemical performance and safety.

The addition of LFP allowed decreasing the internal stress value by about 62.9% during cycling at 1 C as compared to the pristine NCA. The electrode is also more thermally stable, in this case, the exothermic peak moved to the high-temperature zone by about 16.8 °C. The discharge capacity and capacity retention ratio were improved by 19 and 17% at 1 C, respectively.

Another strategy to improve the electrode properties is a partial substitution of iron with manganese and finding an optimal proportion in order to increase the cell voltage and not have a simultaneous drop in the ionic conductivity. The latest research has shown that besides using manganese as an additive to the electrode composition, cobalt, nickel, and vanadium can also be used. For example, Liu et al. (2020) confirmed that the addition of an appropriate amount of doped Ni to LFP/C composite could significantly improve the capacity value and the cycling performance. The specific capacity of the composite after sufficient activation can reach 175.8 mAh g^{-1} at 0.2 C that is higher than the LFP theoretical capacity value of 170 mAh g^{-1}. Even at high rate capabilities, such as 10 C, the capacity retention ratio was 93.9% after 200 cycles.

Many other phosphates chemistries with olivine structures are currently under intense scientific investigation for potential application as cathodes in LIBs. Among them lithium cobalt phosphate $LiCoPO_4$ (LCP) Zhang et al. (2018b), Lithium Nickel Phosphate $LiNiPO_4$ (LNP) (Babu et al. 2016; Huma Nasir et al. 2020), Lithium Vanadium Phosphate $Li_3V_2(PO_4)_3$ (LVP) (Rui et al. 2014; Liu et al. 2016), lithium manganese phosphate ($LiMnPO_4$) (Bakenov and Taniguchi 2011). Moreover, mixed olivine $Li(MM')PO_4$ phases have been developed recently, where (M, M' = Fe, Mn, Co, Ni, Mg, Zn, Al) are used. The most widespread mixed chemistries are lithium nickel cobalt phosphate ($LiNiCoPO_4$), lithium iron manganese phosphate ($LiFeMnPO_4$), lithium manganese iron cobalt phosphate ($LiMnFeCoPO_4$), etc. (Zhang et al. 2012; Ma et al. 2019). These alternative olivine materials operate at higher than traditional LFP voltages, around 4.8 V nominally, and have higher theoretical values of the energy density of around 670 Wh/kg, but they suffer from low ionic and electronic conductivities and undergo structural changes during cycling. Many laborataries are currently working on the improvement of the electrochemical performance of the transition metal phosphor-olivines. Different strategies have been employed, between the transition metals substitution; silicon doping; morphology engineering; coating with conducting medium like carbon or organic polymers. However, they still exhibit poor electrochemical performance characteristics and need some major improvements before using in mass production (Bakenov and Taniguchi 2011; Rui et al. 2014; Babu et al. 2016; Liu et al. 2016; Zhang et al. 2018; Huma Nasir et al. 2020).

Recently, researchers developing methods to prepare the two-components composite cathode materials where LFP is combined with another type of olivine phosphate, for example, LFP-LCP (Jang et al. 2013), LFP-LMP (Zaghib et al. 2012; Tan et al. 2017), LFP-LVP (He et al. 2016; Cheng et al. 2017) composites. LVP cathode materials have a higher operating voltage of 4.55 V and as a result higher energy density (896 Wh/kg). Moreover, the addition of LVP into LFP intensifies the electron conductivity and Li-ion diffusivity as compared to the initial material. Thus, the development of a new synthesis method to prepare the LFP-LVP composite cathode materials allows to combine the advantages of both LFP and LVP (He et al. 2016; Cheng et al. 2017; Chien et al. 2020b). These two-component olivine composites have even better electrochemical properties when different carbon conductive additives are introduced into their composition. In their work (Chien et al. 2020b) described that LFP/LVP/carbon composite cathode materials at a molar ratio of LFP/LVP 9:1 were developed and the electrochemical performance of the cathode was investigated depending on the amount of conductive carbon and type of carbon additive. As carbon additives, Graphene Oxide (GO), graphene nanosheet (GNS) and carbon nanotube CNT) have been tested. The results of the investigations have shown that the best discharge capacities were obtained for LFP/LVP/C/GNS + CNT composite. In this case, the cathode had the following electrochemical characteristics: the discharge capacity of 160.1 mAh/g at 0.1 C/0.1 C; 129.3 mAh/g at 0.2 C/5 C; and 117.8 mAh/g at 0.2 C/10 C. The electrode showed outstanding life-cycle capabilities, after 100 cycles the capacity retention was 97.60% and after 500 cycles it had a value of 85.20% at 1 C/1 C. The results obtained confirm that this cathode chemistry has excellent

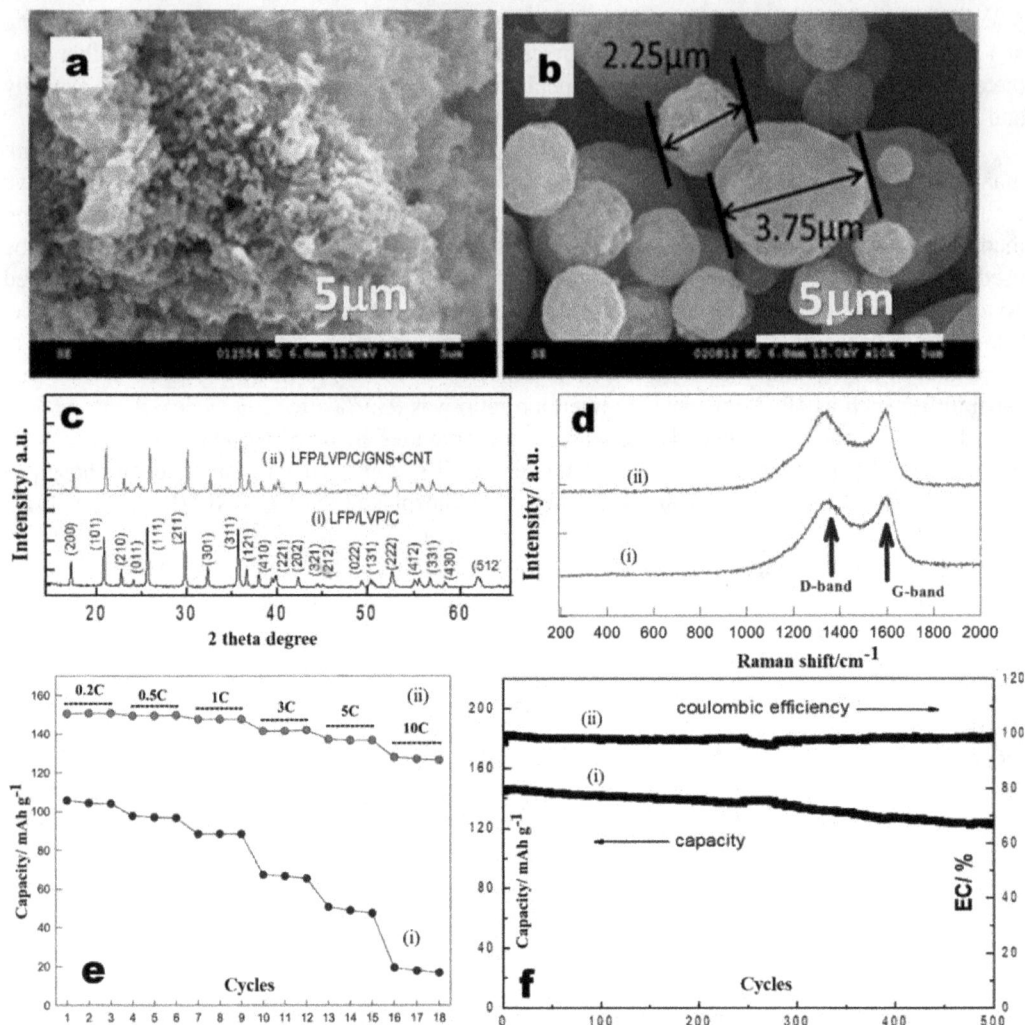

Figure 4.9 SEM images of the (a) LFP/LVP/C and (b) LFP/LVP/C/GNS + CNT composite cathode materials. (c) XRD patterns of LFP/LVP/C and LFP/LVP/C/GNS + CNT composites; (c) corresponding micro-Raman spectra; (e) Cycling performances of the prepared LFP/LVP/C and LFP/LVP/C/GNS + CNT composite cathode materials; (f) the long-term cycling stability performance of LFP/LVP/C/GNS + CNT composite cathode at 1 C/1 C for 500 cycles. (Reprinted with permission from Copyright Elsevier. *Source:* Chien et al. 2020b. Preparation of LiFePO$_4$/Li$_3$V$_2$(PO$_4$)$_3$/C composite cathode materials and their electrochemical performance analysis. *Journal of Alloys and Compounds* 847(2020): 156447).

discharge capacity and long-term cycling stability. Therefore, this material is a very attractive candidate for use in lithium-ion batteries for electric vehicles applications. SEM, XRD, Raman spectra and electrochemical properties of composite LFP/LVP/C without the addition of GNS + CNT in comparison with LFP/LVP/C/GNS + CNT composite are given in Fig. 4.9. As shown in the figure, the addition of GNS + CNT improves discharge capacity and cycling stability of the composite.

Zhao et al. 2019b developed composite cathode 5LiFe$_{0.9}$Mn$_{0.1}$PO$_4$·4Li$_3$V$_2$(PO$_4$)$_3$/C with excellent electrochemical properties for high capacity Li-ion batteries application. During cycling in the potential range of 2.4 and 4.8 V, the cathode delivered the specific capacities of 195, 173 and 158 mAh g^{-1} at the rates 0.1 C, 0.5 C and 2 C. The cathode also shows a good reversible capacity of 160 mAh g^{-1} after 100 cycles with the capacity retention of 88% of the initial value at 0.1 C. Even at a high current density of 2 C, the cathode delivers a good reversible capacity of 140 mAh g^{-1} after 100 cycles. Therefore, the composite cathode has better electrochemical properties as compared to a single active component (LiFePO$_4$, LiMnPO$_4$ or Li$_3$V$_2$(PO$_4$)$_3$). Therefore, the development of composite materials by combining advantages of different materials is a highly attractive approach of novel material synthetics with improved electrochemical performance, environmental safety, thermal stability and life-cycle performance.

3   CONCLUSION

The grating part of commercially available Li-ion battery cells uses transitional metal oxides with layered, spinel and olivine crystal structure as a cathode material. It is caused by the fact that these materials have the structure with available diffusion channels for Li-ions motion and reversible insertion/extraction. Analysis of literature has shown, that the use of nanocomposite materials as cathode materials in Li-ion batteries can improve specific capacity value, cycle life, safety, rate capability of the existing traditional cathode materials. Numerous research works have shown, that one component cathode material alone is not capable of satisfying continuously growing demands for high-energy-density Li-ion batteries applications. Summarizing different material modification techniques one can conclude that the main approaches used to improve the electrochemical performance of the cathode are:

- nanoengineering or in other words tailoring the electrode material morphology by decreasing the size of structural elements to nano level or formation cathode with the structure that combines micro- and nanoelements in one electrode material;
- partial substitution of one metal in the metal oxide structure with another transition metal, or even substitution with more than one metal type, in this way researchers, are trying to find a perfect cathode chemistry with advanced electrochemical properties;
- doping with foreign elements or compounds;
- to decrease the interaction of the active cathode material with the electrolyte, different protective coatings are deposited onto the cathode surface;
- in order to increase the conductivity of the electrode material carbon additives or carbon coating are used, including different forms of carbon such as graphene, graphene oxide, carbon nanotubes, etc.;
- development of gradient core-shell nanostructured materials that use a concentration gradient chemistry with the aim to improve stability and cycling performance of the electrode;
- formation of composite material that combines several cathode chemistries with the same crystal structure, for example, LCO-NMC, LCO-NCA, LFP-LVP, etc.;
- development of integrated composite materials that combine different crystal structures in one cathode, for example, layered-spinel, layered-spinel-layered, spinel-olivine, olivine-layered, etc.

The latest reported cathode materials combine several modification strategies in one structure; such an approach allows obtaining an outstanding electrochemical property when using this cathode in full Li-ion cell. One can expect the application of these advanced cathode materials in the commercial Li-ion cells with high energy density in the near future.

ACKNOWLEDGMENT

This work was supported by the Ministry of Science and Education of Ukraine within the research project 0119U100763.

References

Adepoju, A.A. and Q.L. Williams. 2020. High C-rate performance of $LiFePO_4$/carbon nanofibers composite cathode for Li-ion batteries. Current Applied Physics 20: 1–4.

Arumugam, D. and G.P. Kalaignan. 2008. Synthesis and electrochemical characterizations of Nano-SiO_2-coated $LiMn_2O_4$ cathode materials for rechargeable lithium batteries. Journal of Electroanalytical Chemistry 624: 197–204.

Babu, K.V., L.S. Devi, V. Veeraiah and K. Anand. 2016. Structural and dielectric studies of $LiNiPO_4$ and $LiNi_{0.5}Co_{0.5}PO_4$ cathode materials for lithium-ion batteries. Journal of Asian Ceramic Societies 4: 269–276.

Bakenov, Z. and I. Taniguchi. 2011. $LiMnPO_4$ olivine as a cathode for lithium batteries. The Open Materials Science Journal 5: 222–227.

Bao, L., G. Xu, X. Sun, H. Zeng, R. Zhao, X. Yang, et al. 2017. Mono-dispersed $LiFePO_4$@C core-shell [001] nanorods for a high power Li-ion battery cathode. Journal of Alloys and Compounds 708: 685–693.

Belharouak, I., H. Tsukamoto and K. Amine. 2003. $LiNi_{0.5}Co_{0.5}O_2$ as a long-lived positive active material for lithium-ion batteries. Journal of Power Sources 119–121: 175–177.

Bhuvaneswari, S., U.V. Varadaraju, R. Gopalan and R. Prakash. 2019a. Structural stability and superior electrochemical performance of Sc-doped $LiMn_2O_4$ spinel as cathode for lithium ion batteries. Electrochimica Acta 301: 342–351.

Bhuvaneswari, S., U.V. Varadaraju, R.Gopalan and R. Prakash. 2019b. Sc-doping induced cation-disorder in $LiNi_{0.5}Mn_{1.5}O_4$ spinel leading to improved electrochemical performance as cathode in lithium ion batteries. Electrochimica Acta 327: 135008.

Bi, Z., X. Zhang, W. He, D. Min and W. Zhang. 2013. Recent advances in $LiFePO_4$ nanoparticles with different morphology for high-performance lithium-ion batteries. RSC Advances 3: 19744–19751.

Cao, F.-F, H. Ye and Y.-G. Guo. 2019. Nanostrucutres and nanomaterials for lithium-ion batteries. pp. 89–158. In: Y.-G. Guo (ed.). Nanostrucutres and Nanomaterials for Batteries. Springer: Singapore.

Cao, H., F. Du, J. Adkins, Q. Zhou, H. Dai, P. Sun, et al. 2020. Al-doping induced superior lithium ion storage capability of $LiNiO_2$ spheres. Ceramics International 46: 20050–20060.

Çetin, B., Z. Camtakan and N. Yuca. 2020a. Effect of Mn, Ni, Co transition metal ratios in lithium rich metal oxide cathodes on lithium ion battery performance. Materials Today: Proceedings 33: 2490–2494.

Çetin, B., Z. Camtakan and N. Yuca. 2020b. Synthesis and characterization of li-rich cathode material for lithium ion batteries. Materials Letters 273: 127927.

Chang, Q., A. Wei, W. Li, X. Bai, L. Zhang, R. He, et al. 2019. Structural and electrochemical characteristics of Al_2O_3-modified $LiNi_{0.5}Mn_{1.5}O_4$ cathode materials for lithium-ion batteries. Ceramics International 45: 5100–5110.

Chen, H., Q. Hu, Z. Huang, Z. He, Z. Wang, H. Guo, et al. 2016. Synthesis and electrochemical study of Zr-doped $Li[Li_{0.2}Mn_{0.54}Ni_{0.13}Co_{0.13}]O_2$ as cathode material for Li-ion battery. Ceramics International 42: 263–269.

Chen, F., W. Zhang, J. Fu, Z. Yang, X. Fan, W. Zhang, et al. 2018a. Sequential precipitation induced interdiffusion: A general strategy to synthesize microtubular materials for high performance lithium-ion battery electrodes. Journal of Materials Chemistry A 6: 18430–18437.

Chen, Y., H. Zhang, Y. Chen, G. Qin, X. Lei and L.Y. Liu. 2018b. Graphene-carbon nanotubes-modified $LiFePo_4$ cathode materials for high-performance lithium-ion batteries. Materials Science Forum 913: 818–830.

Chen, Z., W. Zhang and Z. Yang. 2019. A review on cathode materials for advanced lithium ion batteries: Microstructure designs and performance regulations. Nanotechnology 31: 012001– 012037.

Cheng, F., S. Wang, A.H. Lu and W.C. Li. 2013. Immobilization of nanosized $LiFePO_4$ spheres by 3D coralloid carbon structure with large pore volume and thin walls for high power lithium-ion batteries. Journal of Power Sources 229: 249–257.

Cheng, W., L. Wang, Z. Sun, Z. Wang, Q. Zhang, D. Lv, et al. 2017. Preparation and characterization of $LiFePO_{4-x}Li_3V_2(PO_4)_3$ composites by two-step solid-state reaction method for lithium-ion batteries. Materials Letters 198: 172–175.

Chien, W.-C., Y.-R. Li, S.-H. Wu, Y.-S. Wu, Z.-H. Wu, Y.-J. James Li, et al. 2020a. Modifying the morphology and structure of graphene oxide provides high-performance $LiFePO_4/C/rGO$ composite cathode materials. Advanced Powder Technology 31: 4541–4551.

Chien, W.-C., J.-S. Jhang, S.-H. Wu, Z.-H. Wu and C.-C. Yang. 2020b. Preparation of $LiFePO_4/Li_3V_2(PO_4)_3/C$ composite cathode materials and their electrochemical performance analysis. Journal of Alloys and Compounds 847: 156447.

Cho, J., T.-G. Kim, C. Kim, J.-G. Lee, Y.-W. Kim, et al. 2005. Comparison of Al_2O_3- and $AlPO_4$-coated $LiCoO_2$ cathode materials for a Li-ion cell. Journal of Power Sources 25: 58–64.

Choi, D., J. Kang and B. Han. 2019. Unexpectedly high energy density of a Li-ion battery by oxygen redox in $LiNiO_2$ cathode: First-principles study. Electrochimica Acta 20: 166–172.

Cui, P., Z. Jia, L. Li and T. He. 2011. Preparation and characteristics of Sb-doped $LiNiO_2$ cathode materials for Li-ion batteries. Journal of Physics and Chemistry of Solids 72: 899–903.

Demirocak, D.E., S.S. Srinivasan and E.K. Stefanakos. 2017. A review on nanocomposite materials for rechargeable Li-ion batteries. Applied Sciences 7: 731–757.

Deng, J., J. Pan, Q. Yao, Z. Wang, H. Zhou and G. Rao. 2015. Porous core-shell $LiMn_2O_4$ microellipsoids as high-performance cathode materials for Li-ion batteries. Journal of Power Sources 278: 370–374.

Deng, Y., J. Mou, H. Wu, N. Jiang, Q. Zheng, K.H. Lam, et al. 2017. A superior Li_2SiO_3-composited $LiNi_{0.5}Mn_{1.5}O_4$ cathode for high-voltage and high-performance lithium-ion batteries. Electrochimica Acta 235: 19–31.

Ding, Z.P., M.Q. Xu, J.T. Liu, Q. Huang, L.B. Chen, P. Wang, et al. 2017. Understanding the enhanced kinetics of gradient-chemical-doped lithium-rich cathode material. ACS Applied Materials & Interfaces 9: 20519–20526.

Ding, C.-C., L. Jia, X. Wang, Y. Li, Z.-Q. Wang and L.-N. Wu. 2021. An investigation of local structures for $3d_3$, $3d_5$ and $3d_7$ ions doing in the layered $LiCoO_2$ cathode materials. Journal of Physics and Chemistry of Solids 150: 109800.

Divakaran, A.M., M. Minakshi, P.A. Bahri, S. Paul, P. Kumari, A.M. Divakaran, et al. 2021. Rational design on materials for developing next generation lithium-ion secondary battery. Progress in Solid State Chemistry 62: 100298.

Eftekhari, A. 2017. $LiFePO_4/C$ nanocomposites for lithium-ion batteries. Journal of Power Sources 343: 395–411.

Fan, Y., J. Wang, Z. Tang, W. He and J. Zhang. 2007. Effects of the nanostructured SiO_2 coating on the performance of $LiNi_{0.5}Mn_{1.5}O_4$ cathode materials for high-voltage Li-ion batteries. Electrochimica Acta 52: 3870–3875.

Fang, D.-L., J.-C. Li, X. Liu, P.-F. Huang, T.-R. Xu, M.-C. Qian, et al. 2015. Synthesis of a Co–Ni doped $LiMn_2O_4$ spinel cathode material for high-power Li-ion batteries by a sol–gel mediated solid-state route. Journal of Alloys and Compounds 640: 82–89.

Feng, X., Z. Yang, D. Tang, Q. Kong, L. Gu, Y. Wang, et al. 2015. Performance improvement of Li-rich layer-structured $Li_{1.2}Mn_{0.54}Ni_{0.13}Co_{0.13}O_2$ by integration with spinel $LiNi_{0.5}Mn_{1.5}O_4$. Physical Chemistry Chemical Physics 17: 1257–1264.

Feng, S., X. Kong, H. Sun, B. Wang, T. Luo, G. Liu, et al. 2018. Effect of Zr doping on $LiNi_{0.5}Mn_{1.5}O_4$ with ordered or disordered structures. Journal of Alloys and Compounds 749: 1009–1018.

Fu, C., G. Li, D. Luo, Q. Li, J. Fan and L. Li. 2014. Nickel-rich layered microspheres cathodes: lithium/nickel disordering and electrochemical performance. ACS Applied Materials & Interfaces 6: 15822–15831.

Gao, C., H. Liu, S. Bi, S. Fan, X. Meng, Q. Li, et al. 2020. Insight into the effect of graphene coating on cycling stability of $LiNi_{0.5}Mn_{1.5}O_4$: Integration of structure-stability and surface-stability. Journal of Materiomics 6: 712–722.

Gnanaraj, J.S., V.G. Pol, A. Gedanken and D. Aurbach. 2003. Improving the high-temperature performance of $LiMn_2O_4$ spinel electrodes by coating the active mass with MgO via a sonochemical method. Electrochemistry Communications 5: 940–945.

Gong, C., Z. Xue, S. Wen, Y. Ye and X. Xie. 2016. Advanced carbon materials/olivine $LiFePO_4$ composites cathode for lithium ion batteries. Journal of Power Sources 318: 93–112.

Gu, H.-B., D.-K. Jun, G.-C. Park, B. Jin and E.M. Jin. 2007. Nanosized LiFePO$_4$ cathode materials for lithium ion batteries. Journal of Nanoscience and Nanotechnology 7: 3980–4.

Guan, Y., J. Shen, X. Wei, Q. Zhu, X. Zheng, S. Zhou, et al. 2019. LiFePO$_4$/activated carbon/graphene composite with capacitive-battery characteristics for superior high-rate lithium-ion storage. Electrochimica Acta 294: 148–155.

Hao, G., Q. Lai and H. Zhang. 2021. Nanostructured Mn-based oxides as high-performance cathodes for next generation Li-ion batteries. Journal of Energy Chemistry 59: 547–571.

He, W., C. Wei, X. Zhang, Y. Wang, Q. Liu, J. Shen, et al. 2016. Li$_3$V$_2$(PO$_4$)$_3$/LiFePO$_4$ composite hollow microspheres for wide voltage lithium-ion batteries. Electrochimica Acta 219: 682–692.

Hongtong, R., P. Thanwisai, R. Yensano, J. Nash and S. Srilomsak. 2019. Core-shell electrospun and doped LiFePO$_4$/FeS/C composite fibers for Li-ion batteries. Journal of Alloys and Compounds 804: 339–347.

Hosono, E., Y.G. Wang, N. Kida, M. Enomoto, N. Kojima, M. Okubo, et al. 2010. Synthesis of triaxial LiFePO$_4$ nanowire with a VGCF core column and a carbon shell through the electrospinning method. ACS Applied Materials & Interfaces 2: 212–218.

Hua, W., B. Schwarz, R. Azmi, M. Müller, M.S. Dewi Darma, M. Knapp, et al. 2020. Lithium-ion (de) intercalation mechanism in core-shell layered Li(Ni,Co, Mn)O$_2$ cathode materials. Nano Energy 78: 105231.

Huang, Z., P. Luo and D. Wang. 2017. Preparation and characterization of core-shell structured LiFePO$_4$/C composite using a novel carbon source for lithium-ion battery cathode. Journal of Physics and Chemistry of Solids 102: 115–120.

Hudaya, C., J.H. Park, J.K. Lee and W. Choi. 2014. SnO$_2$-coated LiCoO$_2$ cathode material for high-voltage applications in lithium-ion batteries. Solid State Ionics 256: 89–92.

Huma Nasir, M., N.K. Janjua and J. Santoki. 2020. Electrochemical performance of carbon modified LiNiPO$_4$ as Li-ion battery cathode: A combined experimental and theoretical study. Journal of Electrochemical Society 167: 130526.

Hwang, B.J., C.Y. Chen, M.Y. Cheng, R. Santhanam and K. Ragavendran. 2010. Mechanism study of enhanced electrochemical performance of ZrO$_2$-coated LiCoO$_2$ in high voltage region. Journal of Power Sources 195: 4255–4265.

Hwang, T., J. KeeLee, J. Mun and W. Choi. 2016. Surface-modified carbon nanotube coating on high-voltage LiNi$_{0.5}$Mn$_{1.5}$O$_4$ cathodes for lithium ion batteries. Journal of Power Sources 322: 40–48.

Ishigaki, N., N. Kuwata, A. Dorai, T. Nakamura, K. Amezawa and J. Kawamura. 2019. Effect of post-deposition annealing in oxygen atmosphere on LiCoMnO$_4$ thin films for 5 V lithium batteries. Thin Solid Films 686: 137433.

Itou, Y. and Y. Ukyo. 2005. Performance of LiNiCoO$_2$ materials for advanced lithium-ion batteries. Journal of Power Sources 146: 39–44.

Jang, I.C., H.H. Lim, S.B. Lee, K. Karthikeyan, V. Aravindan, K.S. Kang, et al. 2013. Preparation of LiCoPO$_4$ and LiFePO$_4$ coated LiCoPO$_4$ materials with improved battery performance. Journal of Alloys and Compounds 479: 21–324.

Jian, Z., W. Wang, M. Wang, Y. Wang, N.A. Yeung, M. Liu, et al. 2018. Al$_2$O$_3$ coated LiCoO$_2$ as cathode for high-capacity and long-cycling Li-ion batteries. Chinese Chemical Letters 29: 1768–1772.

Jiang, C., E. Hosono and H. Zhou. 2006 Nanomaterials for lithium-ion batteries. Nanotoday 1: 28–33.

Johnston, W.D., R.R. Heikes and D. Sestrich. 1958. The preparation, crystallography, and magnetic properties of the Li$_x$Co$_{(1-x)}$O system. Journal of Physics and Chemistry of Solids 7: 1–13.

Julien, C.M., A. Mauger, K. Zaghib and H. Groult. 2014. Comparative issues of cathode materials for Li-ion batteries. Inorganics 2: 132–154.

Kalluri, S., M. Yoon, M. Jo, S. Park, S. Myeong, J. Kim, et al. 2017. Surface engineering strategies of layered LiCoO$_2$ cathode material to realize high-energy and high-voltage Li-ion cells. Advanced Energy Materials 7: 1601507.

Kim, B., J.-G. Lee, M. Choi, J. Cho and B. Park. 2004. Correlation between local strain and cycle-life performance of AlPO$_4$-coated LiCoO$_2$ cathodes. Journal of Power Sources 16: 190–192.

Kim, D., G. Sandi, J.R. Croy, K.G. Gallagher, S.-H. Kang, E. Lee, et al. 2013. Composite 'Layered-Layered-Spinel' cathode structures for lithium-ion batteries. Journal of the Electrochemical Society 160: A31.

Kocak, T., L. Wu, J. Wang, U. Savaci, S. Turan and X. Zhang. 2021. The effect of vanadium doping on the cycling performance of $LiNi_{0.5}Mn_{1.5}O_4$ spinel cathode for high voltage lithium-ion batteries. Journal of Electroanalytical Chemistry 881: 114926.

Konarova, M. and I. Taniguchi. 2009. Physical and electrochemical properties of $LiFePO_4$ nanoparticles synthesized by a combination of spray pyrolysis with wet ball-milling. Journal of Power Sources 194: 1029–1035.

Ku, D.J., J.H. Lee, S.J. Lee, M. Koo and B.J. Lee. 2019. Effects of carbon coating on $LiNi_{0.5}Mn_{1.5}O_4$ cathode material for lithium ion batteries using an atmospheric. Surface and Coatings Technology 376: 25–30.

Kumar, N., J.R. Rodriguez, V.G. Pol and A. Sen. 2019. Facile synthesis of 2D graphene oxide sheet enveloping ultrafine 1D $LiMn_2O_4$ as interconnected framework to enhance cathodic property for Li-ion battery. Applied Surface Science 463: 132–140.

Kuwata, N., S. Kudo, Y. Matsuda and J. Kawamura. 2014. Fabrication of thin-film lithium batteries with 5-V-class LiCoMnO4 cathodes. Solid State Ionics 262: 165–169.

Lee, E.S., A. Huq and A. Manthiram. 2013. Understanding the effect of synthesis temperature on the structural and electrochemical characteristics of layered-spinel composite cathodes for lithium-ion batteries. Journal of Power Sources 240: 193–203.

Lee, M.-J., S. Lee, P. Oh, Y. Kim and J. Cho. 2014. High Performance $LiMn_2O_4$ cathode materials grown with epitaxial layered nanostructure for Li-ion batteries. Nano Letters 14: 993-999.

Li, D., H. Zhang, C. Wang, D. Song, X. Shi and L. Zhang. 2017. New structurally integrated layered-spinel lithium-cobalt-manganese-oxide composite cathode materials for lithium-ion batteries. Journal of Alloys and Compounds 696: 276–289.

Liang, G., V.K. Peterson, K.W. See, Z. Guo and W.K. Pang. 2020. Developing high-voltage spinel $LiNi_{0.5}Mn_{1.5}O_4$ cathodes for high-energy density lithium-ion batteries: Current achievements and future prospects. Journal of Materials Chemistry A 8: 15373–15398.

Linden, D. and T.B. Reddy. 2002. Handbook of Batteries. McGraw-Hill, New York: USA.

Liu, H.K., G.X. Wang, Z. Guo, J. Wang, K. Konstantinov. 2006. Nanomaterials for lithium-ion rechargeable batteries. Journal of Nanoscience and Nanotechnology 6: 1–15.

Liu, J., W. Liu, S. Ji, Y. Zhou, P. Hodgson and Y. Li. 2013. Electrospun spinel $LiNi_{0.5}Mn_{1.5}O_4$ hierarchical nanofibers as 5 V cathode materials for lithium-ion batteries. ChemPlusChem 78: 636–641.

Liu, M.-H., H.-T. Huang, C.-M. Lin, J.-M. Chen and S.-C. Liao. 2014a. Mg gradient-doped $LiNi_{0.5}Mn_{1.5}O_4$ as the cathode material for Li-ion batteries. Electrochimica Acta 120: 133–139.

Liu, L., X. Yan, Y. Wang, D. Zhang, F. Du, C. Wang, et al. 2014b. Studies of the electrochemical properties and thermal stability of $LiNi_{1/3}Co_{1/3}Mn_{1/3}O_2/LiFePO_4$ composite cathodes for lithium ion batteries. Ionics 20: 1087–1093.

Liu, H., R. Tian, Y. Jiang, X. Tan, J. Chen, L. Zhang, et al. 2015. On the drastically improved performance of Fe-doped $LiMn_2O_4$ nanoparticles prepared by a facile solution-gelation route. Electrochimica Acta 180: 138–146.

Liu, C., R. Massé, X. Nan and G. Cao. 2016. A promising cathode for Li-ion batteries: $Li_3V_2(PO_4)_3$. Energy Storage Materials 4: 15–58.

Liu, G., J. Zhang, X. Zhang, Y. Du, K. Zhang, G. Li, et al. 2017. Study on oxygen deficiency in spinel $LiNi_{0.5}Mn_{1.5}O_4$ and its Fe and Cr-doped compounds. Journal of Alloys and Compounds 725: 580–586.

Liu, J., S. Guo, C. Hu, H. Lyu, X. Yan and Z. Guo. 2018. Advanced nanocomposite electrodes for lithium-ion batteries. pp. 7–32. In: Z. Guo, Y. Chen and N.L. Lu (eds). Multifunctional Nanocomposites for Energy and Environmental Applications. Wiley-VCH Verlag GmbH & Co. KGaA, Weinheim: Germany.

Liu, J., J. Wang, Y. Ni, Y. Zhang, J. Luo, F. Cheng, et al. 2019. Spinel/Lithium-rich manganese oxide hybrid nanofibers as cathode materials for rechargeable lithium-ion batteries. Small Methods 3: 1900350.

Liu, Y., Y.-J. Gu, G.-Y. Luo, Z.-L. Chen, F.-Z. Wu, X.-Y. Dai, et al. 2020. Ni-doped $LiFePO_4/C$ as high-performance cathode composites for Li-ion batteries. Ceramics International 46: 14857–14863.

Liu, S., H. He and C. Chang. 2021. Understanding the improvement of fluorination in 5.3V $LiCoMnO_4$ spinel. Journal of Alloys and Compounds 860: 158468.

Luo, Y., H. Li, T. Lu, Y. Zhang, S.S. Mao, Z. Liu, et al. 2017. Fluorine gradient-doped $LiNi_{0.5}Mn_{1.5}O_4$ spinel with improved high voltage stability for Li-ion batteries. Electrochimica Acta 238: 237–245.

Lyu, Y., X. Wu, Z. Feng, T. Cheng, Y. Liu, M. Wang, et al. 2020a. An overview on the advances of $LiCoO_2$ cathodes for lithium-ion batteries. Advanced Energy Materials 11: 2000982–2001011.

Lyu, P., Y. Huo, Z. Qu and Z. Rao. 2020b. Investigation on the thermal behavior of Ni-rich NMC lithium-ion battery for energy storage. Applied Thermal Engineering 166: 114749.

Ma, G., S. Li, W. Zhang, Z. Yang, S. Liu, X. Fan, et al. 2016a. A general and mild approach to controllable preparation of manganese-based micro- and nanostructured bars for high performance lithium-ion batteries. Angewandte Chemie 55: 3667–3671.

Ma, J., P. Hu, G. Cui and L. Chen. 2016b. Surface and interface issues in spinel $LiNi_{0.5}Mn_{1.5}O_4$: Insights into a potential cathode material for high energy density lithium ion batteries. Chemistry of Materials 28: 3578–3606.

Ma, F., Y. Wu, G. Wei, S. Qiu, J. Qu and T. Qi. 2019a. Comparative study of simple and concentration gradient shell coatings with $Li_{1.2}Ni_{0.13}Mn_{0.54}Co_{0.13}O_2$ on $LiNi_{0.8}Mn_{0.1}Co_{0.1}O_2$ cathodes for lithium-ion batteries. Solid State Ionics 341: 115034.

Ma, Y., P. Liu, Q. Xie, G. Zhang, H. Zheng, Y. Cai, et al. 2019b. Double-shell Li-rich layered oxide hollow microspheres with sandwich-like carbon@spinel@layered@spinel@carbon shells as high-rate lithium ion battery cathode. Nano Energy 59: 184–196.

Manthiram, A., J.C. Knight, S.-T. Myung, S.-M. OH and Y.-K. Sun. 2016. Nickel-rich and lithium-rich layered oxide cathodes: Progress and perspectives. Advanced Energy Materials 6: 1501010.

Marincaş, A.-H., F. Goga, S.-A. Dorneanu and P. Ilea. 2020. Review on synthesis methods to obtain $LiMn_2O_4$-based cathode materials for Li-ion batteries. Journal of Solid State Electrochemistry 24: 473–497.

Monaco, S., F.D. Giorgio, L.D. Col, M. Riché, C. Arbizzani and M. Mastragostino. 2015. Electrochemical performance of $LiNi_{0.5}Mn_{1.5}O_4$ composite electrodes featuring carbons and reduced graphene oxide. Journal of Power Sources 278: 733–740.

Mou, J., Y. Deng, L. He, Q. Zheng, N. Jiang and D. Lin. 2018. Critical roles of semi-conductive $LaFeO_3$ coating in enhancing cycling stability and rate capability of 5 V $LiNi_{0.5}Mn_{1.5}O_4$ cathode materials. Electrochimica Acta 260 (2018) 101–111.

Muto, S., K. Tatsumi, Y. Kojima, H. Oka, H. Kondo, K. Horibuchi, et al. 2012. Effect of Mg-doping on the degradation of $LiNiO_2$-based cathode materials by combined spectroscopic methods. Journal of Power Sources 205: 449–455.

Nie, Y., W. Xiao, C. Miao, J. Wang, Y. Tan, M. Xu, et al. 2021. Improving the structural stability of Ni-rich $LiNi_{0.81}Co_{0.15}Al_{0.04}O_2$ cathode materials with optimal content of trivalent Al ions doping for lithium ions batteries. Ceramics International 47: 9717–9726.

Nourhan, M. and A.K. Nageh. 2020. Recent advances in the design of cathode materials for Li-ion batteries. RSC Advances 10: 21662–21685.

Ozoemena, K.I. and S. Chen. 2016. Nanomaterials in Advanced Batteries and Super-capacitors. Springer International Publishing, Cham: Switzerland.

Pan, H., S. Zhang, J. Chen, M. Gao, Y. Liu, T. Zhu, et al. 2018. Li- and Mn-rich layered oxide cathode materials for lithium-ion batteries: A review from fundamentals to research progress and applications. Molecular Systems Design and Engineering 3: 748–780.

Pang, P., Z. Wang, X. Tan, Y. Deng, J. Nan, Z. Xing, et al. 2019. $LiCoO_2$@$LiNi_{0.45}Al_{0.05}Mn_{0.5}O_2$ as high-voltage lithium-ion battery cathode materials with improved cycling performance and thermal stability. Electrochimica Acta 327: 135018.

Patel, S., R.K. Singh and R. Kumar. 2021. Structural and electronic property of $AlPO_4$ modified $LiMn_2O_4$ material for Li-ion battery application. Materials today: Proceedings 42: 776–780.

Phattharasupakun, N., J. Wutthiprom, S. Duangdangchote, S. Sarawutanukul, C. Tomon, F. Duriyasart, et al. 2021. Core-shell Ni-rich NMC-nanocarbon cathode from scalable solvent-free mechanofusion for high-performance 18650 Li-ion batteries. Energy Storage Materials 36: 485–495.

Pistoia, G. 2014. Lithium-Ion Batteries: Advances and Applications. Elsevier, Amsterdam: Netherlands.

Placke, T., R. Kloepsch, S. Dühnen and M. Winter. 2017. Lithium ion, lithium metal, and alternative rechargeable battery technologies: the odyssey for high energy density. Journal of Solid State Electrochemistry 21: 1939–1964.

Placke, T., M. Winter and R. Schmuch. 2021. Finding the sweet spot: Li/Mn-rich cathode materials with fine-tuned core-shell particle design for high-energy lithium ion batteries. Electrochimica Acta 366: 137413.

Primer, A. 2019. Li-ion Battery Chemistries. Elsevier, Amsterdam: Netherlands.

Qian, M., J. Huang, S. Han and X. Cai. 2014. Preparation and electrochemical performance of the interconnected $LiMn_2O_4$ fibers. Electrochimica Acta 120: 16–22.

Rangappa, D., K. Sone, T. Kudo and I. Honma. 2010. Directed growth of nanoarchitectured $LiFePO_4$ electrode by solvothermal synthesis and their cathode properties. Journal of Power Sources 195: 6167–6171.

Reeves-McLaren, N., M. Hong, H. Alqurashi, L. Xue, J. Sharp, A.J. Rennie, et al. 2018. The spinel $LiCoMnO_4$: 5V cathode and conversion anode. Energy Procedia 151: 158–162.

Roziera, P. and J.R. Tarascon. 2015. Review-Li-rich layered oxide cathodes for next generation Li-ion batteries: chances and challenges. Journal of Electrochemical Society 162: A2490–A2499.

Rui, X., Q. Yan, M. Skyllas-Kazacos and T.M. Lim. 2014. $Li_3V_2(PO_4)_3$ cathode materials for lithium-ion batteries: A review. Journal of Power Sources 258: 19–38.

Ryu, H.-H., N.-Y. Park, J. Hyun Seo, Y.-S. Yu and M. Sharma. 2020. A highly stabilized Ni-rich NCA cathode for high-energy lithium-ion batteries. Materials Today 36: 73–82.

Shang, L., H. Li, H. Lai, D. Li, Q. Wu, L. Yang, et al. 2016. Hierarchical $LiNixCoyO_2$ mesostructures as high-performance cathode materials for lithium ion batteries. Journal of Power Sources 326: 279–284.

Shao, D., J. Wang, X. Dong, W. Yu, G. Liu, F. Zhang, et al. 2014. Preparation and electrochemical performances of $LiFePO_4$/C composite nanobelts via facile electrospinning. Journal of Materials Science: Materials in Electronics 25: 1040–1046.

ShinKo, H., J. HanKim, J. Wang and J. Dae Lee. 2017. Co/Ti co-substituted layered $LiNiO_2$ prepared using a concentration gradient method as an effective cathode material for Li-ion batteries. Journal of Power Sources 372: 107–115.

Sivajee Ganesh, K., B. Purusottam Reddy, P. Jeevan Kumar and O.M. Hussain. 2018. Influence of Zr dopant on microstructural and electrochemical properties of $LiCoO_2$ thin film cathodes by RF sputtering. Journal of Electroanalytical Chemistry 828: 71–79.

Sorboni, Y.G., H. Arabi and A. Kompany. 2019. Effect of Cu doping on the structural and electrochemical properties of lithium-rich $Li_{1.25}Mn_{0.50}Ni_{0.125}Co_{0.125}O_2$ nanopowders as a cathode material. Ceramics International 45: 2139–2145.

Stephan, A.K. 2020. A pathway to understand NMC cathodes. Joule 4: 1632–1633.

Sun, Y.-K., J.-M. Han, S.-T. Myung, S.-W. Lee and K. Amine. 2006. Significant improvement of high voltage cycling behavior AlF_3-coated $LiCoO_2$ cathode. Electrochemistry Communications 8: 821–826.

Sun, W., F. Cao, Y. Liu, X. Zhao X. Liu and J. Yuan. 2012. Nanoporous $LiMn_2O_4$ nanosheets with exposed {111} facets as cathodes for highly reversible lithium-ion batteries. Journal of Materials Chemistry 22: 20952–20957.

Sun, W., Y. Li, K. Xie, S. Luo, G. Bai, X. Tan, et al. 2018. Constructing hierarchical urchin-like $LiNi_{0.5}Mn_{1.5}O_4$ hollow spheres with exposed {111} facets as advanced cathode material for lithium-ion batteries. Nano Energy 54: 175–183.

Sun, Y.Y., Q. Wu and L. Zhao. 2019. A new doping element to improve the electrochemical performance of $Li_{1.2}Mn_{0.54}Ni_{0.13}Co_{0.13}O_2$ materials for Li-ion batteries. Ceramics International 45: 1339–1347.

Susanto, D., H. Kim, J.-Y. Kim, S.H. Lim, J.W. Yang, S.A. Choi, et al. 2015. Effect of (Mg, Al) double doping on the thermal decomposition of $LiMn_2O_4$ cathodes investigated by time-resolved X-ray diffraction. Current Applied Physics 15: S27–S31.

Taha, T.A. and M.M. El-Molla. 2020. Green simple preparation of $LiNiO_2$ nanopowder for lithium ion battery. Journal of Materials Research and Technology 9: 7955–7960.

Tan, Q., B. Yan, Y. Xu, Y. Chen and J. Yang. 2017. Preparation and electrochemical performance of carbon-coated $LiFePO_4$/$LiMnPO_4$-positive material for a Li-ion battery. Particuology 30: 144–150.

Tao, S., F. Kong, C. Wu, X. Su, T. Xiang, S. Chen, et al. 2017. Nanoscale TiO_2 membrane coating spinel $LiNi_{0.5}Mn_{1.5}O_4$ cathode material for advanced lithium-ion batteries. Journal of Alloys and Compounds 705: 413–419.

Toprakci, O., L. Ji, Z. Lin, H.A.K. Toprakci and X. Zhang. 2011. Fabrication and electrochemical characteristics of electrospun LiFePO$_4$/carbon composite fibers for lithium-ion batteries. Journal of Power Sources 196: 7692–7699.

Toprakci, O., H.A.K. Toprakci, L.W. Ji, G.J. Xu, Z. Lin and X. Zhang. 2012. Carbon nanotube-loaded electrospun LiFePO$_4$/carbon composite nanofibers as stable and binder-free cathodes for rechargeable lithium-ion batteries. ACS Applied Materials & Interfaces 4: 1273–1280.

Tuccillo, M., O. Palumbo, M. Pavone, A.B. Muñoz-García, A. Paolone, S. Brutti, et al. 2020. Analysis of the phase stability of LiMO$_2$ layered oxides (M = Co, Mn, Ni). Crystals 10: 526–550.

Uddin, M.-J., P. Kumar and A. Sung-Jin Cho. 2017. Nanostructured cathode materials synthesis for lithium-ion batteries. Materials Today: Energy 5: 138–157.

Wang, Z., L. Liu, L. Chen and X. Huang. 2002. Structural and electrochemical characterizations of surface-modified LiCoO$_2$ cathode materials for Li-ion batteries. Solid State Ionics 148: 335–342.

Wang, L., J. Zhao, S. Guo, X. He and C. Jiang. 2006. Investigation of SnO$_2$-modified LiMn$_2$O$_4$ composite as cathode material for lithium-ion batteries. International Journal of Electrochemical Science 5: 1113–1126.

Wang, H., W.-D. Zhang, L.-Y. Zhu and M.-C. Chen. 2007. Effect of LiFePO$_4$ coating on electrochemical performance of LiCoO$_2$ at high temperature. Solid State Ionics 178: 131–136.

Wang, L., X.M. He, W.T. Sun, J.L. Wang, Y.D. Li and S. Fan. 2012. Crystal orientation tuning of LiFePO$_4$ nanoplates for high rate lithium battery cathode materials. Nano Letters 12: 5632–5636.

Wang, Z., J. Du, Z. Li and Y. Wu. 2014. Sol-gel synthesis of Co-doped LiMn$_2$O$_4$ with improved high-rate properties for high-temperature lithium batteries. Ceramics International 40: 3527–3531.

Wang, C.-C., Y.-C. Lin and P.-H. Chou. 2015a. Mitigation of layer to spinel conversion of a lithium-rich layered oxide cathode by substitution of Al in a lithium ion battery. RSC Advances 5: 68919–68928.

Wang, D., R. Yu, X. Wang, L. Ge and X. Yang. 2015b. Dependence of structure and temperature for lithium-rich layered-spinel microspheres cathode material of lithium ion batteries. Scientific Reports 5: 8403.

Wang, Y., B. Zhu, Y. Wang and F. Wang. 2016. Solvothermal synthesis of LiFePO$_4$ nanorods as high-performance cathode materials for lithium ion batteries. Ceramics International 42: 10297–10303.

Wang, J.-f., D. Chen, W.W. Wang and G.-c. Liang. 2017. Effects of Na$^+$ doping on crystalline structure and electrochemical performances of LiNi$_{0.5}$Mn$_{1.5}$O$_4$ cathode material. Transactions of Nonferrous Metals Society of China 27: 2239–2248.

Wang, L., Z. Wu, J. Zou, P. Gao, X. Niu, H. Li, et al. 2019a. Li-free cathode materials for high energy density lithium batteries. Joule 3: 2086–2102.

Wang, F., Y. Jiang, S.L. Lin, W. Wang, C. Hu, Y. Wei, et al. 2019b. High-voltage performance of LiCoO$_2$ cathode studied by single particle microelectrodes-influence of surface modification with TiO$_2$. Electrochimica Acta 295: 1017–1026.

Wang, Y., Q. Zhang, Z.-C. Xue, L. Yang and J. Wang. 2020a. An in situ formed surface coating layer enabling LiCoO$_2$ with stable 4.6V high-voltage cycle performances. Advanced Energy Materials 10: 2001413.

Wang, Y., T. Cheng, Z.E. Yu, Y. Lyu, B and Guo. 2020b. Study on the effect of Ni and Mn doping on the structural evolution of LiCoO$_2$ under 4.6V high-voltage cycling. Journal of Alloys and Compounds 842: 155827.

Wang, C.-C., J.-W. Lin, Y.-H. Yu, K.-H. Lai, S.-M. Lee, K.-F. Chiu et al. 2020c. Nanolaminated ZnO-TiO$_2$ coated lithium-rich layered oxide cathodes by atomic layer deposition for enhanced electrochemical performance. Journal of Alloys and Compounds 842: 155845.

Wang, S., C. Luo, Y. Feng, G. Fan, L. Feng, M. Ren, et al. 2020d. Electrochemical properties and microstructures of LiMn$_2$O$_4$ cathodes coated with aluminum zirconium coupling agents. Ceramics International 46: 13003–13013.

Wei, Y., K.W. Nam, K.B. Kim and G. Chen. 2006. Spectroscopic studies of the structural properties of Ni substituted spinel LiMn$_2$O$_4$. Solid State Ionics 177: 29–35.

Wei, X., P. Yang, H. Li, S. Wang, Y. Xing, X. Liu and S. Zhang. 2017a. Synthesis and properties of mesoporous Zn-doped Li$_{1.2}$Mn$_{0.54}$Co$_{0.13}$Ni$_{0.13}$O$_2$ as cathode materials by a MOFs-assisted solvothermal method. RSC Advances 7: 35055–35059.

Wei, Q., F. Xiong, S. Tan, L. Huang, E.H. Lan, B. Dunn, et al. 2017b. Energy storage: Porous one-dimensional nanomaterials: Design, fabrication and applications in electrochemical energy storage. Advanced Materials 29: 1602300.

Wei, Y., C. Zhou, D. Zhao and G. Wang. 2020. Enhanced electrochemical performance and safety of $LiNi_{0.8}Co_{0.15}Al_{0.05}O_2$ by $LiFePO_4$ modification. Chemical Physics Letters 751: 137480.

Wen, W., S. Chen, Y. Fu, X. Wang and H. Shu. 2015. A coreeshell structure spinel cathode material with a concentration gradient shell for high performance lithium-ion batteries. Journal of Power Sources 274: 219–228.

Whittingham, M.S. 2008. Inorganic nanomaterials for batteries. Dalton Transactions 40: 5424–5431.

Windmüller, A., C. Dellen, S. Lobe, C.-L. Tsai, S. Möller, Y. Jung, et al. 2018. Thermal stability of 5V $LiCoMnO_4$ spinels with LiF additive. Solid State Ionics 320: 378–386.

Wood, M., J. Li, R.E. Ruther, Z. Du, E.C. Self, H.M. Meyer III, et al. 2020. Chemical stability and long-term cell performance of low-cobalt, Ni-Rich cathodes prepared by aqueous processing for high-energy Li-ion batteries. Energy Storage Materials 24: 188–197.

Wu, F., N. Li, Y. Su, H. Shou, L. Bao, W. Yang, et al. 2013. Spinel/layered heterostructured cathode material for high capacityand high-rate Li-ion batteries. Advanced Materials 25: 3722–3726.

Wu, Z., S. Ji, T. Liu, Y. Duan, S. Xiao, Y. Lin, et al. 2016. Aligned Li+ tunnels in core-shell $Li(Ni_xMn_yCo_2)O_2$@$LiFePO_4$ enhances its high voltage cycling stability as Li-ion battery cathode. Nano Letters 16: 6357–6363.

Xia, H., Z. Luo and J. Xie. 2012. Nanostructured $LiMn_2O_4$ and their composites as high-performance cathodes for lithium-ion batteries. Progress in Natural Science: Materials International 22: 572–584.

Xiao, Y., X.-D. Zhang, Y.-F. Zhu, P.-F. Wang, Y.-X. Yin, X. Yang, et al. 2019. Suppressing manganese dissolution via exposing stable {111} facets for high-performance lithium-ion oxide cathode. Advanced Science 6: 1801908.

Xie, Y., S. Chen, W. Yang, H. Zou, Z. Lin and J. Zhou. 2019. Improving the rate capability and decelerating the voltage decay of Li-rich layered oxide cathodes by constructing a surface-modified microrod structure. Journal of Alloys and Compounds 772: 230–239.

Xu, G., F. Li, Z.H. Tao, X. Wei, Y. Liu, X. Li, et al. 2014. Monodispersed $LiFePO_4$@C core-shell nanostructures for a high power Li-ion battery cathode. Journal of Power Sources 246: 696–702.

Xu, R., X. Zhang, R. Chamoun, J. Shui, J.C.M. Li, J. Lu, et al. 2015. Enhanced rate performance of $LiNi_{0.5}Mn_{1.5}O_4$ fibers synthesized by electrospinning. Nano Energy 15: 616–624.

Xu, Y.-H., S.-X. Zhao, Y.-F. Deng, H. Deng and C.-W. Nan. 2016. Improved electrochemical performance of 5V spinel $LiNi_{0.5}Mn_{1.5}O_4$ microspheres by F-doping and Li_4SiO_4 coating. Journal of Materiomics 2: 265–272.

Xu, X., S. Deng, H. Wang, J. Liu and H. Yan. 2017. Research progress in improving the cycling stability of high-voltage $LiNi_{0.5}Mn_{1.5}O_4$ cathode in lithium-ion battery. Nano-Micro Letters 9: 22.

Xu, Z., L. Ci, Y. Yuan, X. Nie, J. Li, J. Cheng, et al. 2020a. Potassium prussian blue-coated Li-rich cathode with enhanced lithium ion storage property. Nano Energy 75: 104942.

Xu, C., J. Li, X. Feng, J. Zhao, C. Tang, B. Ji, et al. 2020b. The improved performance of spinel $LiMn_2O_4$ cathode with micro-nanostructured sphere-interconnecte d-tube morphology and surface orientation at extreme conditions for lithium-ion batteries. Electrochimica Acta 358: 136901.

Xu, L., Z. Sun, Y. Zhu, Y. Han, M. Wu, Y. Ma, et al. 2020c. A Li-rich layered-spinel cathode material for high capacity and high rate lithium-ion batteries fabricated via a gas-solid reaction. Science China Materials 63: 2435–2442.

Yang, Z., Q. Qiao and W. Yang. 2011. Improvement of structural and electrochemical properties of commercial $LiCoO_2$ by coating with LaF_3. Electrochimica Acta 56: 4791–4796.

Yang, S., J. Chen, Y. Liu and B. Yi. 2014. Preparing $LiNi_{0.5}Mn_{1.5}O_4$ nanoplates with superior properties in lithium-ion batteries using bimetal-organic coordination-polymers as precursors. Journal of Materials Chemistry A 2: 9322–9330.

Yang, C., Y. Deng, M Gao, X. Yang, X. Qin and G. Chen. 2017. High-rate and long-life performance of a truncated spinel cathode material with off-stoichiometric composition at elevated temperature. Electrochimica Acta 225: 198–206.

Yang, S.-Q., P.-B. Wang, H.-X. Wei, L.-B. Tang, X.-H. Zhang, Z.-J. He, et al. 2019a. $Li_4V_2Mn(PO_4)_4$-stablized $Li[Li_{0.2}Mn_{0.54}Ni_{0.13}Co_{0.13}]O_2$ cathode materials for lithium ion batteries. Nano Energy 63: 103889.

Yang, Z., J. Zhong, J. Li, Y. Liu, B. Niu and F. Kang. 2019b. Li-rich layered oxide coated by nanoscale MoOx film with oxygen vacancies and lower oxidation state as a high-performance cathode material. Ceramics International 45: 439–448.

Yazami, R. 2014. Nanomaterials for Lithium-Ion Batteries: Fundamentals and Applications. CRC Press Taylor and Francis Group, New York: USA.

Yi, T.-F., L.-C. Yin, Y.-Q. Ma, H.-Y. Shen, Y.-R. Zhu and R.-S. Zhu. 2013. Lithium-ion insertion kinetics of Nb-doped $LiMn_2O_4$ positive-electrode material. Ceramics International 39: 4673–4678.

Yi, L., Z. Liu, R. Yu, C. Zhao, H. Peng, M. Liu, et al. 2017. Li-rich layered/spinel heterostructured special morphology cathode material with high rate capability for Li-ion batteries. ACS Sustainable Chemical Engergy 5: 11005–11015.

Yoshio, M., R.J. Brodd and A. Kozawa. 2009. Lithium-Ion Batteries: Science and Technology. Springer Science +Business Media, Cham: Switzerland.

Youp Song, M., H. Uk Kim and H. Ryoung Park. 2014. Electrochemical properties of $LiNiO_2$ cathode after TiO_2 or ZnO addition. Ceramics International 40: 4219–4224.

Yu, Y., M. Xiang, J. Guo, C. Su, X. Liu, H. Bai, et al. 2019. Enhancing high-rate and elevated-temperature properties of Ni-Mg co-doped $LiMn_2O_4$ cathodes for Li-ion batteries. Journal of Colloid and Interface Science 555: 64–71.

Yuan, X., H. Liu and J. Zhang. 2011. Lithium-Ion Batteries: Advanced Materials and Technologies. CRC Press Taylor and Francis Group, New York: USA.

Yubuchi, S., Y. Ito, T. Matsuyama, A. Hayashi and M. Tatsumisago. 2016. 5V class $LiNi_{0.5}Mn_{1.5}O_4$ positive electrode coated with Li_3PO_4 thin film for all-solid-state batteries using sulfide solid electrolyte. Solid State Ionics 285: 79–82.

Zaghib, K., M. Trudeau, A. Guerfi, J. Trottier, A. Mauger, R. Veillette, et al. 2012. New advanced cathode material: $LiMnPO_4$ encapsulated with $LiFePO_4$. Journal of Power Sources 204: 177–181.

Zha, G., W. Hu, S. Agarwal, C. Ouyang, N. Hu and H. Hou. 2021. High performance layered $LiNi_{0.8}Co_{0.07}Fe_{0.03}Mn_{0.1}O_2$ cathode materials for Li-ion battery. Chemical Engineering Journal 409: 128343.

Zhang, W.-J. 2011. Structure and performance of $LiFePO_4$ cathode materials: A review. Journal of Power Sources 196: 2962–2970.

Zhang, Y., Q.-Y. Huo, P.-P. Du, L.-Z. Wang, A.-G. Zhang, Y.-h. Song, et al. 2012. Advances in new cathode material $LiFePO_4$ for lithium-ion batteries. Synthetic Metals 162: 1315–1326.

Zhang, X.-D., J.-L. Shi, J.-Y. Liang, Y.-X. Yin and J.-N. Zhang. 2018a. Suppressing surface lattice oxygen release of Li-rich cathode materials via heterostructured spinel $Li_4Mn_5O_{12}$ coating. Advanced Materials 30: 1801751.

Zhang, M., N. Garcia-Araez and A.L. Hector. 2018b. Understanding and development of olivine $LiCoPO_4$ cathode materials for lithium-ion batteries. J. Mater. Chem. A 6: 14483–14517.

Zhang, B., Y. Xu, J. Wang, X. Ma, W. Hou and X. Xue. 2021. Electrochemical performance of $LiFePO_4$/graphene composites at low temperature affected by preparation technology. Electrochimica Acta 368: 137575.

Zhao, J., H. Wang, Z. Xie, S. Ellis, X. Kuai, J. Guo, et al. 2016. Tailorable electrochemical performance of spinel cathode materials via *in-situ* integrating a layered Li_2MnO_3 phase for lithium-ion batteries. Journal of Power Sources 333: 43–52.

Zhao, H., F. Li, X. Shu, J. Liu, T. Wu, Z. Wang, et al. 2018. Environment-friendly synthesis of high-voltage $LiNi_{0.5}Mn_{1.5}O_4$ nanorods with excellent electrochemical properties. Ceramics International 44: 20575–20580.

Zhao, T.L., R.X. Ji, H.B. Yang, Y.X. Zhang, X.G. Sun, Y. Li, et al. 2019a. Distinctive electrochemical performance of novel Fe-based Li-rich cathode material prepared by molten salt method for lithium-ion batteries. Journal of Energy Chemistry 33: 37–45.

Zhao, W., X. Xiong, Y. Yao, B. Liang, Y. Fan, S. Lu, et al. 2019b. $5LiFe_{0.9}Mn_{0.1}PO_4 \cdot 4Li_3V_2(PO_4)_3$/C composites as high capacity cathode materials for lithium-ion batteries. Applications of Surface Science 483: 1166–1173.

Zhao, S., Z. Guo, K. Yan, K. Wan, F. He, B. Sun, et al. 2021. Towards high-energy-density lithium-ion batteries: Strategies for developing high-capacity lithium-rich cathode materials. Energy Storage Materials 34: 716–734.

Zhenfei, C., M. Yangzhou, H. Xuanning, Y. Xiaohui, Y. Zexinc, Z. Shihong, et al. 2020. High electrochemical stability Al-doped spinel $LiMn_2O_4$ cathode material for Li-ion batteries. Journal of Energy Storage 27: 101036.

Zheng, J., G. Teng, C. Xin, Z. Zhuo, J. Liu, Q. Li, et al. 2017. Role of superexchange interaction on tuning of Ni/Li Disordering in Layered $Li(NixMnyCoz)O_2$. Journal of Physical Chemistry Letters 8: 5537–5542.

Zheng, J., Y. Ye, T. Liu, Y. Xiao, C. Wang, F. Wang, et al. 2019. Ni/Li disordering in layered transition metal oxide: Electrochemical impact, origin, and control. Accounts of Chemical Research 52: 2201–2209.

Zheng, H., X. Han, W. Guo, L. Lin, Q. Xie, P. Liu, et al. 2020. Recent developments and challenges of Li-rich Mn-based cathode materials for highenergy lithium ion batteries. Materials Today: Energy 18: 100518.

Zhong, Z., L. Chen, C. Zhu, W. Ren, L. Kong, Y. Wan, et al. 2020. Nano $LiFePO_4$ coated Ni rich composite as cathode for lithium ion batteries with high thermal ability and excellent cycling performance. Journal of Power Sources 464: 228235.

Zhou, W.-J., B.-L. He and H.-L. Li. 2008. Synthesis, structure and electrochemistry of Ag-modified $LiMn_2O_4$ cathode materials for lithium-ion batteries. Materials Research Bulletin 43: 2285–2294.

Zhu, C., Y. Yu, L. Gu, K. Weichert and J. Maier. 2011. Electrospinning of highly electroactive carbon-coated single-crystalline $LiFePO_4$ nanowires. Angewandte Chemie 50: 6278–6282.

Zong, B., Y. Lang, S. Yan, Z. Deng, J. Gong, J. Guo, et al. 2020a. Influence of Ti doping on microstructure and electrochemical performance of $LiNi_{0.5}Mn_{1.5}O_4$ cathode material for lithium-ion batteries. Materials Today: Communications 24: 101003.

Zong, B., Z. Deng, S. Yan, Y. Lang, J. Gong, J. Guo, et al. 2020b. Effects of Si doping on structural and electrochemical performance of $LiNi_{0.5}Mn_{1.5}O_4$ cathode materials for lithium-ion batteries. Powder Technology 364: 725–737.

Chapter 5

Graphene Nanocomposites for Pressure Sensors Applications

Victor K. Samoei, Surendra Maharjan*
and Ahalapitiya H. Jayatissa*

The University of Toledo, 2801 W. Bancroft St, Toledo, OH - 43606, USA
vsamoei@rockets.utoledo.edu; smaharj7@rockets.utoledo.edu; ahalapitiya.jayatissa@utoledo.edu

1   INTRODUCTION

Graphene is a single two-dimensional layer of carbon atoms bonded in a hexagonal lattice structure. It is a single-atom-thick sheet of sp^2 hybridized carbon atoms arranged in a honeycomb lattice structure (Katsnelson 2007). It has received increasing attention due to its unique physicochemical properties such as high surface area, excellent conductivity, high mechanical strength and ease of functionalization and mass production (Adhikari and Majumdar 2004; An et al. 2017). Other properties of graphene include good chemical stability, excellent thermal conductivity, high young's modulus, a high value of white light transmittance and exceptionally fast mobility of charge carriers. These fascinating properties have attracted extensive research interest in recent years with ever-increasing scientific and technological applications (Angus 2013; Asthana et al. 2014; Ji et al. 2016).

Graphene Oxide (GO) sheets have gained a lot of interest as a nanofiller for polymer nanocomposites due to their excellent compatibility with organic polymers. Composite fibers can be prepared mainly in two processes: mixing and spinning (Ji et al. 2016). Graphene can be added in three different ways throughout the mixing process: (i) solvent mixing, (ii) melt processing, and (iii) *in-situ* polymerization. Some researchers also tried to coat graphene onto the fiber surface after spinning. Graphene/polymer fibers can be prepared using a variety of spinning methods: wet-spinning, melt and electrical spinning.

For various commercial applications, graphene must be modified to customize its solubility in an organic medium. Individual graphene integration in polymer matrices to form advanced multifunctional composites is one of the most promising routes because polymer composites typically have a high specific modulus and specific strength, and a wide range of applications in the aerospace, automobile and defense industries (Itapu and Jayatissa 2018). Furthermore, traditional processing technologies may efficiently process and build polymer composites into precisely formed components while preserving the structure and properties of graphene. Graphene has a

---

*Corresponding Author

higher surface-to-volume ratio than carbon nanotubes (CNTs), a promising filler for composites before graphene was isolated but the inner nanotube surface is inaccessible to polymer molecules (Mittal et al. 2015).

Exfoliation of graphite, reduction of Graphene Oxide (GO), Chemical Vapor Deposition (CVD), solvothermal, epitaxial growth, molecular beam epitaxy and electrically assisted synthesis are among the most widely used methods for producing graphene on a laboratory scale (Dhand et al. 2013; Ji et al. 2016). Among all the strategies to produce graphene, chemical vapor deposition on transition metal substrates has emerged as the most promising method for producing graphene since it is affordable and yields large-area graphene (Zhang et al. 2013).

2 PROPERTIES OF GRAPHENE NANOPARTICLES

The graphene honeycomb lattice is made up of two carbon atom sub-lattices that are connected by σ bonds. Each carbon atom in the lattice has a π orbital that contributes to a delocalized network of electrons as shown in Fig. 5.1 (Zhu et al. 2010). The quality of graphene has been characterized with the Raman spectroscopy (Sidorov et al. 2012; Gautam et al. 2017). Bilayer graphene, which consists of two stacked monolayers and where the quasiparticles are massive chiral fermions, has a quadratic low-energy band structure that generates very different scattering properties from those of the monolayer. These properties have made the bilayer a subject of interest (Abergel et al. 2010).

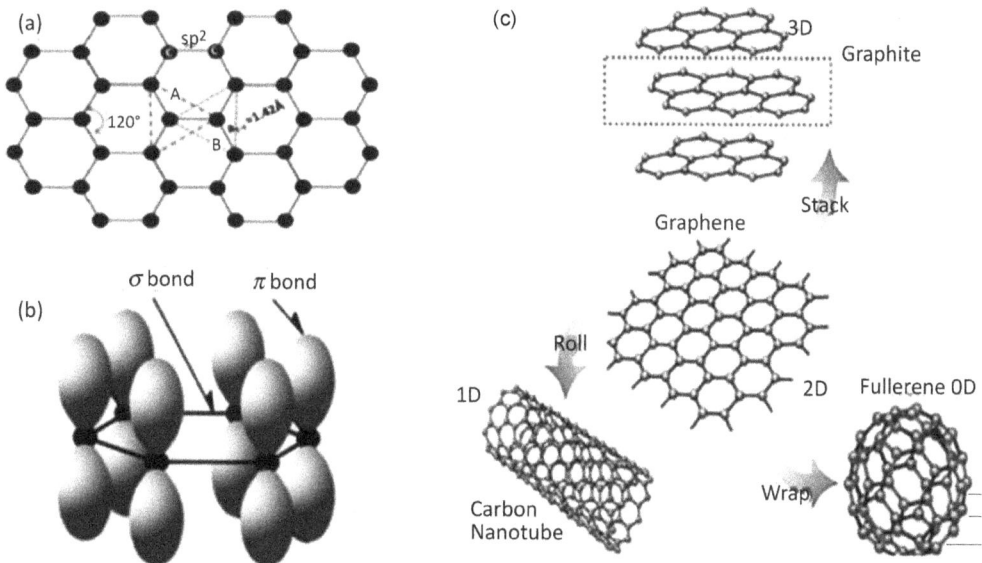

Figure 5.1 (a) Schematic structure of honeycomb lattice structure. (b) Schematic of sp^2 hybrid c–c bond structure containing in-plane ζ bond and perpendicular π bond, and (c) Graphene stack to a 3D graphite, a wrap to a 2D and a 1D carbon nanotube.

Many graphene characteristics measured in experiments have exceeded those obtained in any other material, with some reaching theoretically predicted limits such as room-temperature electron mobility of $2.5 \times 10^5 \, cm^2 V^{-1} s^{-1}$ (Novoselov et al. 2012), Young's modulus of 1 TPa, intrinsic strength of 130 GPa, the optical absorption of exactly $\pi\alpha \approx 2.3\%$ (in the infrared limit, where α is the fine structure constant), complete impermeability to any gases, ability to tolerate extremely high densities of electric current.

Many superior properties of graphene justify its nickname as a 'miracle material'. However, some of these properties have only been attained in the highest-quality graphene samples (mechanically exfoliated graphene) and graphene placed on specific substrates like hexagonal boron nitride. Although these technologies are continually developing, no equivalent properties have been seen on graphene created using other ways. When mass-produced graphene exhibits the same remarkable performance as the best samples obtained in research facilities, graphene will be of much greater importance for industrial applications. The thermal conductivity is approximately 5000 $Wm^{-1}K^{-1}$ for a suspended monolayer graphene 'flake', produced (Zhu et al. 2010; Guo et al. 2017). Graphene is an isotropic conductor; it conducts heat in all directions. The thermal conductivity increases logarithmically due to the stable bonding pattern as well as being a 2D material. As graphene is more resistant to tear than steel and is also lightweight and flexible, its conductivity could have some attractive real-world applications (Warner et al. 2013; Lotsch 2015). Nature provides us with many other 2D crystals, such as boron nitride and molybdenum disulfide. Being structurally related to graphene but having their distinctive properties, they offer the possibility of fine-tuning material and device characteristics to suit a particular technology better or to be used in combination with graphene.

3   PREPARATION METHODS OF GRAPHENE

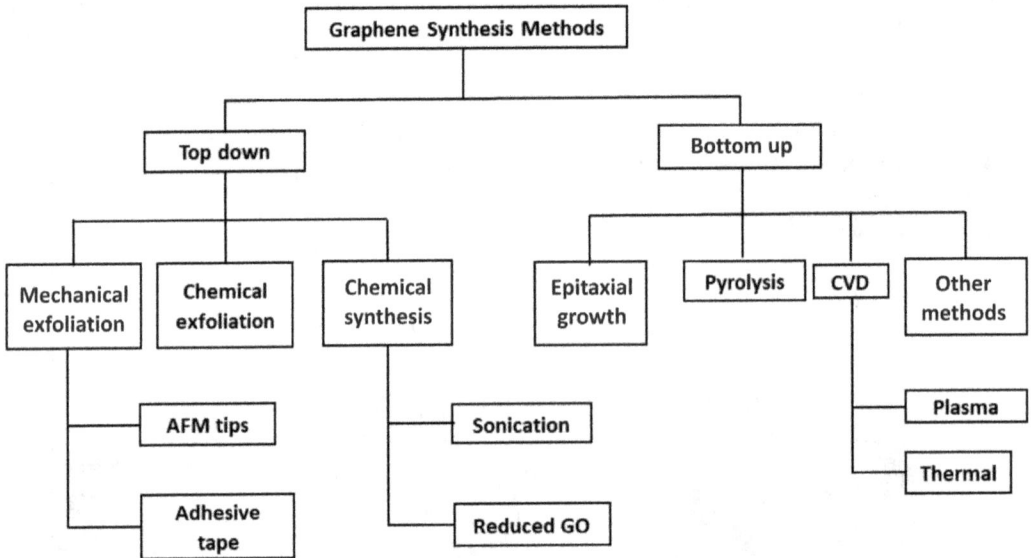

Figure 5.2   A process flow chart of graphene synthesis.

Intensive research efforts have been made over the last few years on the mass production of graphene to be introduced into practical applications. Each production method attributes different characteristics to the final material and has different possibilities for upscaling (Papageorgiou et al. 2017). Table 5.1 gives a comparison of the thermal properties of graphene films and the fabrication process. The timely innovation of a simple approach proposed by Novoselov's group in 2004 was highlighted after a long and coherent series of unsuccessful attempts to build a single-layer graphene sheet. The group repeatedly peeled a graphite crystal using an adhesive tape until a specific limit and later transferred the thinned-out graphite onto an extremely thin (<300 nm) appropriately colored, oxidized silicon wafer (Soldano et al. 2010). This remarkable discovery led to the onset of mass-scale production of graphene and its utilization in various polymeric, electronic industries. In the past four decades, various unsuccessful attempts have been made to achieve large-scale production of pure, defect-free graphene sheets (Soldano et al. 2010).

Table 5.1 The dependence of thermal properties on the fabrication methods of graphene films (Fu et al. 2020)

Assembled structures	Raw materials	Exfoliation/ Dispersion methods	Film formation methods	Heat treatment (°C)	Thermal conductivity (Wm⁻¹K⁻¹)
Graphene film	GO	Not mentioned	Self-assembling	1000	60
Graphene film	GO	Ultrasonication	Vacuum filtration	1200	1043.5
Graphene film	GO	Ultrasonication	Self-assembling	2850	1100
Graphene film	Graphene	Ball milling	Vacuum filtration	2850	1434
Graphene film	Thermal exfoliation	Ultrasonication	Electro-spray deposition	2850	1434
Graphene film	Graphite oxide	Not mentioned	Self-assembling	3000	1950
Graphene film	GO	Shear mixing	Self-assembling	2850	3214

3.1　Hydrothermal Synthesis

The hydrothermal synthesis method involves different techniques to crystallize the nanoparticles substances. The synthesis method usually takes place at a high vapor pressure level and a high-temperature aqueous solution, therefore the terms 'Hydro' and 'Thermal'. The heterogeneous reactions for synthesizing inorganic materials in aqueous media are also done above ambient temperature and pressure. The morphology of the materials can be controlled, either in low-pressure or high-pressure conditions depending on the vapor pressure of the main composition in the reaction. The hydrothermal synthesis can generate nanomaterials that are not stable at elevated temperatures and the nanomaterials produced by this method with high vapor pressures have a minimum loss of materials. These are considered as a significant advantage over others. Graphene is commonly synthesized using heated rGO sheets as precursors, which are then treated with oxidizing chemicals to introduce epoxy groups on the carbon lattice at cleavage sites (Gan et al. 2020).

3.2　Chemical Vapor Deposition

Among all the strategies to produce graphene, Chemical Vapor Deposition (CVD) on transition metal substrates has become the most promising approach, which is inexpensive and produces large-area graphene with precision and lasting stability (Zhang et al. 2013). The technique is currently costly due to high energy usage. However, once the transfer process is optimized this method may indeed be cost-effective. Several issues need to be resolved before graphene CVD technology can become widely used (Novoselov et al. 2012). Graphene growth on thin films of metals needs to be achieved, simultaneously gaining control of the domain (grain) size, ripples, doping level and the number of layers. Controlling the number of graphene layers and their relative crystallographic orientation is crucial since it will enable various applications that would need double, triple or even thicker graphene layers. Simultaneously, the transfer process should be improved and optimized with the objectives of minimizing the damage to graphene and recovering the sacrificial metal (Huang 2018).

　　The complication of the transfer process might be comparable to the growth of graphene itself. However, several applications rely on the conformal growth of graphene on the surface of the metal and do not require graphene transfer at all. Graphene's high thermal and electrical conductivities, as well as its outstanding barrier qualities, allow it to enhance copper interconnects in integrated circuits. As graphene is inert, it is a great gas barrier and forms a conformal layer on metal surfaces with the most intricate topographies, providing corrosion protection.

　　The development of graphene growth on arbitrary surfaces and/or at low temperatures (for example, utilizing plasma-enhanced CVD or other methods) with a minimum number of defects would be game-changing breakthroughs. The former would eliminate the time-consuming and costly transfer step, allowing for greater integration of this 2D crystal with other materials (like Si or GaAs). The latter would improve compatibility with modern microelectronic technologies and allow significant energy saving (Muñoz et al. 2017).

CVD graphene has been investigated for gas sensor applications (Gautam and Jayatissa 2011; Gautam and Jayatissa 2012(a–d)). Widely used gas sensing mechanisms are based on the change of electrical conductivity in graphene films in the presence of reducing or oxidizing gases. Field effect transistor-based gas sensors were also investigated (Gautam and Jayatissa, 2012(d)). In order to enhance the sensor performance, a few modifications of graphene surfaces were used: (a) coating of graphene layer with catalyst materials such as Pt (Gautam and Jayatissa 2012(a)), Coating of gold nanoparticles (Gautam and Jayatissa 2012(b)), and coating with surfaces of graphene with other metals (Gautam and Jayatissa 2012(c)).

3.3 Sol-gel Method

In this method, a solution is transformed into a gel and then deposited on a surface. The processes performed in low concentrated polymer-solvent solutions is another attractive route to develop hybrid membranes because it allows an *in-situ* dispersion of metal-based nanoparticles within the polymeric matrix, achieving a suitable interfacial morphology between the continuous and the dispersed phase (Angus 2013).

3.4 Other Growth Methods

Although there are several other growth methods, it is unlikely that they will become commercially viable in the next decade. However, some of these methods have specific advantages that should be investigated further. Using a chemistry-driven bottom-up strategy, surface-assisted coupling of molecular monomer precursors into linear polyphenylenes and subsequent cyclodehydrogenation is an innovative way to build high-quality graphene nanoribbons and even more complex structures (such T- and Y-shaped connections). Molecular beam epitaxy has been used to grow chemically pure graphene, but it is unlikely to be used on a large scale because of its much higher cost than CVD methods. Laser ablation is a potentially interesting growth technique allowing the deposition of graphene nanoplatelets on arbitrary surfaces (Novoselov et al. 2012). Because this approach is in direct competition with the spray-coating of chemically exfoliated graphene, it is unlikely to be widely adopted. Following the realization of the potential importance of graphene as a replacement for semiconductor materials and Indium Tin Oxide (ITO), alternative ways for graphene production have been explored. Other approaches have been sought to produce the same hydrophilicity in graphite without the time and material requirements of Hummer's method. Very recently, glow discharge treatment has been proven to introduce oxygen species into the lattices

Table 5.2 Advantages and disadvantages for techniques currently used to produce graphene (Soldano et al. 2010)

Categories	Advantages	Disadvantages
Mechanical exfoliation	Low-cost and easy No special equipment needed, SiO$_2$ thickness is tuned for better contrast	Serendipitous Uneven films Labor intensive (not suitable for large-scale production)
Epitaxial growth	Most even films (of any method) Large scale area	Difficult control for morphology and adsorption energy High-temperature process
Graphene oxide	Straight-forward up-scaling Versatile handling of the suspension Rapid process	Fragile stability of the colloidal dispersion Reduction to graphene is only partial

of all forms of graphitic materials for example buckyballs, CNTs, graphene, carbon nanofibers and graphite. The structure of the resulting graphene/graphite oxides are very similar to Hummers GO and can be treated thermally to selectively reduce epoxides. Unlike the Hummers method, plasma functionalization requires no strong acids, can proceed at room temperature and can be completed very quickly, often in a matter of seconds or minutes (Cooper et al. 2012).

4 GRAPHENE VERSUS TRADITIONAL MATERIALS

Graphene has a nominal bandgap of zero, whereas other semiconductors have a finite bandgap. Normally, differently doped materials should be used to study electron and hole transportation through a semiconductor. However, in graphene, the nature of a charge carrier changes from an electron to a hole or vice versa at the Dirac point, and the Fermi level is always within the conduction or valence band, but in typical semiconductors, when pinned by impurity states, the Fermi level frequently falls within the bandgap. Dispersion in graphene is chiral. This has something to do with some extremely different material phenomena, such as Klein tunneling.

The dispersion relation of graphene is linear, whereas semiconductors have quadratic dispersion. Many of the impressive physical and electronic properties of graphene can be consequences of this fact. Graphene is much thinner than a traditional 2D Electron Gas (2DEG). The effective thickness of a standard 2DEG in a quantum well or heterostructure is typically approximately 5–50 nm. This is due to the constraints on construction and the fact that the confined electron wave functions have an evanescent tail that stretches into the barriers. Graphene on the other hand is only a single layer of carbon atoms, generally regarded to have a thickness of about 3 Å (twice the carbon-carbon bond length). Electrons conducting through graphene are constrained in the z-axis to a much greater extent than those that conduct through a traditional 2DEG. Graphene has been found to have a finite minimum conductivity, even in the case of vanishing charge carriers (Cooper et al. 2012).

5 APPLICATIONS OF GRAPHENE

The exceptional electrical properties of graphene have attracted applications for future electronics such as ballistic transistors, field emitters, components of integrated circuits, transparent conducting electrodes and sensors. Engineering, electronics, medicinal, energy, industrial and domestic design are just a few of the possibilities for graphene. Shen et al. (2012) extensively reviewed the biomedical applications of graphene including drug delivery, gene delivery, cancer therapy, biosensing and bioimaging, GO-based antibacterial materials and scaffolds for tissue/cell culturing (Choi et al. 2010). Similarly, Huang (2018) and Choi et al. explained various phenomena associated with graphene and graphene-based materials and their applications in the field of memory devices for electronics, ranging from electrochemical sensors to instrumentation (Dhand et al. 2013).

5.1 Electronic Devices

With high electrical conductivity, high carrier mobility and moderately high optical transmittance in the visible range of the spectrum, graphene-based polymer composites have been used as the electrodes for solar cells, organic solar cells, liquid crystal devices, Organic Light-Emitting Diodes (OLEDs) and field emission devices. Transparent conducting films are used in many electronic devices like solar cells, touch screens, flat panel displays, etc. CVD-grown graphene and Chemically Modified Graphene (CMG) have been used for the development of such films through different approaches. Graphene films were also investigated for electronic device applications such as solar cells (Gautam et al. 2017; Shi and Jayatissa 2018). Energy Storage Li-ion battery (LIB) is considered as one of the most promising storage systems because of its high absolute potential against the standard hydrogen cell (3.04 V) and its low atomic weight (6.94 gmol), which leads to the large energy density with a theoretical value up to 400 $Whkg^{-1}$. Supercapacitor or Ultra-capacitor is another type of electrochemical energy storage device that provides high power density (10 $kWkg^{-1}$), short charge and discharge time, and long cycling life as compared to the battery devices. Graphene derivatives and conducting polymers are combined and used as the hybrid type of supercapacitor, the combination of EDLC and pseudo-capacitors. Due to the change of conductance as a function of surface adsorption, large specific area and low Johnson noise, graphene has proved to be a promising candidate to detect a variety of molecules such as gases, chemicals and biomolecules. Conductive polymer nanocomposites usually exhibit a positive temperature coefficient, but a recent study described the opposite behavior in PVDF. This makes it suitable for the application in temperature sensors. The fabrication of graphene-based composites has been widely investigated regarding the application in mechanical devices (Wang and Jayatissa 2015a, b).

5.2   Medical Application

Graphene-polymer nanocomposites have been extensively studied for biomedical applications such as drug and gene delivery, tissue engineering and prosthetic bones. Graphene has been investigated for drug loading and delivery because strong interaction exists between hydrophobic drugs and aromatic regions of graphene sheets (Mohan et al. 2018; Silva et al. 2018). Graphene and graphene-oxides also have tunable surface properties, which enhance interaction with polar liquids such as bio-fluids. In addition, these materials exhibit biocompatibility in cell attachment and cell growth. However, the long-term effect of graphene inclusion in the human body needs more research (Justino et al. 2017).

5.3   Automotive Industry

Graphene-based composites have been investigated with a great interest for use as electrode material in electrochemical supercapacitors in electrical/hybrid automobiles owing to their unique combination of properties such as high lightweight, compatibility with other materials, good electrical conductivity and controlled pore size distribution. This material has recently been used in supercapacitor devices to replace conventionally used carbon (Jishnu et al. 2020). Typically, graphitic carbon-based materials are randomly oriented with respect to the current collectors in a conventionally stacked geometry in supercapacitors leading to lower utilization of the electrochemical surface area of graphene layers and consequently limits the extent of the EDL (Electrical Double Layer) formed at the interface.

5.4   Application in Superhydrophobic Coatings

Superhydrophobic materials based on graphene have been developed. Graphene has attracted the attention of many researchers in recent years due to many outstanding features which make them suitable for a passive layer formation that protects metals from oxidation and corrosion (Berry 2013; Asthana et al. 2014; Tian et al. 2015; Rinaldi et al. 2016). A multilayer of graphene is formed by stacking several sheets on top of another up to the point where the structure becomes graphite which is usually about 30 layers (Cooper et al. 2012; Li et al. 2017; Papageorgiou et al. 2017; Kang et al. 2019). The unique physicochemical properties of graphene have led to a wide range of applications in sensors, solar cells and many other diverse fields (Shao et al. 2010; Justino et al. 2017). Jishnu et al. 2020 have developed a superhydrophobic graphene-based material with self-cleaning and anti-corrosion performance. The graphene-based material exhibited a great super hydrophobicity among other characteristics such as anti-sticking, self-cleaning, anti-corrosion and low friction. Table 5.1 given in the preparation methods of graphene provides a summary of the synthesis methods and characteristics of graphene-based superhydrophobic materials provided by Jishnu et al. (2020). Other coatings exist based on the demand for polymer-based coatings for various applications is high. For example, in fluorine carbon coatings, surfaces are converted from super-hydrophilic to superhydrophobic to maintain the Cassie–Baxter state stability by reducing the surface free energy to a quarter compared with intrinsic silicon (Kim et al. 2011).

5.5 Sensor Applications (Gas and Biosensors)

The operational principle of graphene-based gas or bioelectronic sensors is based on the change of graphene's electrical conductivity due to the adsorption of molecules on the graphene surface. The change in conductivity can be attributed to the change in carrier concentration of graphene due to the absorbed gas molecules acting as donors or acceptors (Yuan and Shi 2013; Varghese et al. 2015; Wang et al. 2016; Singh et al. 2017). Furthermore, some interesting properties of graphene aid to increase its sensitivity up to single atom or molecular level detection. First, graphene is a two-dimensional (2D) material and its whole volume i.e., all carbon atoms are exposed to the analyte of interest (Ho et al. 2013). Second, graphene is highly conductive with low Johnson noise, therefore, a little change in carrier concentration can cause a notable variation of electrical conductivity. Third, graphene has very few crystal defects ensuring a low level of noise caused by thermal switching. Finally, four probe measurements can be made on a single-crystal graphene device with ohmic electrical contacts having low resistance (Ma et al. 2010).

Graphene possesses good sensing properties towards NO_2, NH_3, H_2O, and CO. Graphene sensing properties were fully recoverable after exposure to the analyte of interest, by vacuum annealing at 150°C or by illumination to UV for a short time. Furthermore, it was also demonstrated that the chemical doping of graphene by both holes and electrons, in high concentration, did not affect the mobility of graphene (Hosseingholipourasl et al. 2020). Due to the change of conductance as a function of surface adsorption, large specific area and low Johnson noise, graphene has proved to be a promising candidate to detect a variety of molecules such as gases, pressure chemicals and biomolecules. Conductive polymer nanocomposites usually exhibit a positive temperature coefficient, but a recent study described the opposite behavior (negative temperature coefficient) in PVDF. Graphene's properties are particularly sensitive to the environment because it is a two-dimensional fabric with no mass. As a result, it is natural to consider graphene for sensor applications ranging from magnetic field measurements to DNA sequencing, and from tracking the velocity of surrounding fluids to strain gauges. The latter is probably the most competitive application. Graphene is the only crystal that can be stretched by 20%, thus enhancing the working range of such sensors significantly (Singh et al. 2018).

Currently, graphene gas detectors, although extremely sensitive, have only a minor competitive edge over the existing device. Low selectivity and water toxicity limit their applicability, although such detectors can be made cheaply so that they could be used in some niche applications. Functionalization might improve the selectivity of graphene sensors, but because it is rather an expensive method, it is probably most suitable for bio-sensing (Yu et al. 2019).

The major advantage of graphene sensors is their multi-functionality. Multidimensional measurements, such as strain, gas environment, pressure and magnetic field, can all be done with a single device. In this sense, graphene offers unique opportunities. With the development of increasingly interactive consumer electronic devices, such sensors will certainly find their way into many products (Ha et al. 2018).

The anomalously high energy splitting between the zero-energy and first Landau levels in graphene makes it an ideal material for developing universal resistance standards based on the quantum Hall effect, and such devices are already in use by various metrological facilities (Smith et al. 2019).

6 PRESSURE SENSORS

Flexible pressure sensors are attracting great interest from researchers and are widely applied in various new electronic equipment because of their distinct characteristics with high flexibility, high sensitivity and lightweight. Examples include Electronic skin (E-skin) and wearable flexible sensing devices (Liu et al. 2019; Li et al. 2019; Zhang et al. 2019; Chen and Yan 2020).

Herrmann et al. (2007) are the first to demonstrate the functionalized gold NP-based sensitive strain gauges for sensor applications. Their study mainly focuses on the tunneling model and sensitivity of the functionalized NP films, which depends on various factors, including the size of the nanoparticle, interparticle distance and the conductance of the binding molecules. The drawbacks of this system are the sensitivity and detection range of weight that is not varied based on suitable applications. Secondly, the properties of NP-based sensors, mainly depending on the substrate and the linker molecules (Vaka et al. 2020).

The flaws in this system could be impregnated of AuNWs onto tissue paper and sandwiched between PDMS instead of using only AuNW and high resistance of 2.5 ± 0.4 MΩcm^{-1} which is not suitable for a sensor. To overcome these problems, a new device with a higher sensitivity factor and well-organized structure has been designed. It demonstrates how functionalized AuNPs can be placed between two graphene layers to form a new pressure sensor. The advantage of using graphene hybrid is its sensing mechanism, excellent electronic and electrochemical properties which helps to improve the sensitivity. When compared with earlier reports, this graphene hybrid device exhibits better sensitivity, which is desired for the surface to be uniform and have a good transfer of electrons for checking the sensitivity of a device.

Thin-film sensors have several advantages over traditional sensors. Thin film technology permits a sensor to be thin and compact while still preserving exceptional precision and long-term stability. Many types of substrates, including ceramic materials and high-grade specialty steels, can be deposited on and fused to thin films. According to studies, these sensors offer a faster and more precise reaction time, and their smaller size allows for more intimate placement and less external circuitry (Kimura 2019).

7 APPLICATION OF PRESSURE SENSORS AND THE CURRENT MARKET

Pressure sensors are vital components of wearable electronics for human-machine interaction, health monitoring, disease prevention, and so on. State-of-the-art pressure sensors often require the use of encapsulating materials such as graphene composites and others like polydimethylsiloxane (PDMS) to provide mechanical durability, which significantly deteriorates the detection limit, permeability and wearing comfort of the pressure sensors. Pressure sensor applications are categorized into different groups such as the human-machine interface, electronic skin, and machine monitoring (Liu et al. 2021).

7.1 Human-Machine Interface

Human-machine interface refers to devices and software installed on machines that assist human-machine communication. Flexible pressure sensors play an important role in translating mechanical input from humans to electrical signals for controlling the machine or for providing feedback. Hang et al. (2020) demonstrated a highly stretchable and self-healing sensor with reduced graphene oxide. The sensor can detect various human motions and can transmit the data to the smartphone combined with a readout and wireless system (Hang et al. 2020). For example, a smart glove with integrated pressure sensors was able to control the movement of the robot arm to play music, and epidermal acoustic sensors and a speech recognition algorithm could translate the commands from a human to control a virtual character in computer games. Flexible touch panels have been developed with transparent pressure sensor arrays to fulfill the requirements of flexible displays (Tee et al. 2014; Jung et al. 2019). Kim et al. (2011) reported a surface-capacitive touch system as a stretchable ionic touch panel, using touching fingers as a ground electrode to determine the contact position. A wearable touch panel was demonstrated to write words, play music and play chess. In addition, pressure sensors can also be used in security applications, such as signature collectors for the detection of signing characteristics (Li et al. 2018; 2020; Zhang et al. 2019).

7.2   Electronic Skin

Electronic skin (e-skin) is designed to mimic the comprehensive nature of human skin, it refers to a network of flexible electronics with various sensing functionalities that emulate the functions of human skin, but are not limited to the functions of human skin (Jeon et al. 2019). For practical applications, it can help disabled people to restore their sensing ability lost in accidents, e.g., prosthetics or build the biomimetic cognition of surrounding stimulus on the surface of artificial robotics. Pressure sensors are an indispensable part of electronic skin, and typically require large-area coverage with small sensing pixels. Various prototypes of pressure sensor arrays installed on the hand, tongue or robotics have been developed in recent years. Graphene is a promising material in the development of these electronics skin sensors. Yu Quing Liu fabricated graphene-based electronic skin with laser reduction of graphene oxide (GO) and Laser-Induced Graphene (LIG) on polyimide (PI) (An et al. 2017; You et al. 2020).

Tian et al. developed a graphene-based resistive pressure sensor with a sensitivity of 0.96 kPa^{-1} in the range from 0 to 50 kPa (Tian et al. 2015). This pressure sensor is made of two LRGO films, which are perpendicular to each other. When force was applied to the LRGO film, the contact area increases and the electrical pathways become more. Therefore, the current increases with a fixed voltage. It can detect pressing, bending, twisting forces (Tian et al. 2015).

The combination of different functional electronic components is often required to realize electronic skin. Commonly used functional components include sensors, flexible energy supply and transmitting and processing components. The last one is essential for stimulus transportation, especially at the device/neuron interface. Kim et al. (2011) reported an artificial afferent nerve consisting of piezoresistive pressure sensors, ring oscillators and synaptic transistors to mimic the function of biological mechanoreceptors, nerve fibers and biological synapses. The sensor not only correctly identified braille characters but also actuated real muscle tissues when connected to the nerves of a discoid cockroach (Zhang et al. 2020).

7.3   Health Monitoring

Flexible pressure sensors have been broadly applied in healthcare services. They can be used in voice recognition, heart rate and respiration rate monitoring, head position adjustment, plantar pressure mapping, wound monitoring. Various piezoresistive and capacitive pressure sensors have demonstrated the capability of precise pulse measurement which contains abundant physiological and pathological information. Luo et al. reported that epidermal pulses measured by a flexible piezoresistive sensor can be used for beat-to-beat Blood Pressure (BP) tracking through the Pulse Transit Time (PTT) method with the aid of electrocardiogram (ECG) electrodes (Luo et al. 2016).

Several issues exist when flexible pressure sensors are used in cardiovascular monitoring. For example, when measuring epidermal pulses on a wrist, recording quality signals requires the sensor to be precisely located above the radial artery. Otherwise, the measured signals can be distorted, preventing useful physiological information from extraction. Although the alignment problem may be addressed by using densely packed pressure sensor arrays, the required power will be multiplied by the number of sensors. To address this problem, Fan et al. (2018) proposed a wearable liquid capsule sensor platform embedded with a piezoresistive pressure sensor. In this sensor design, the pulsation signals can be precisely collected and transmitted through flexible elastomeric membranes of the capsule to the liquid-embedded pressure sensor, which substantially relaxes the typical stringent alignment requirements between the sensor and the artery to 8.5 mm. Another important problem is motion artifact, which compromises the accuracy of pressure measurement. Although algorithms based on data processing to eliminate motion artifacts and recover signals have been proposed, little progress on motion-tolerant pressure sensors has been reported. For certain healthcare applications, e.g., intraocular pressure measurements, sensors need to be implanted inside the body and continuously operate over a long period of time; accordingly, the implanted devices need to have good biocompatibility and biodegradability to avoid irritability and to bypass the need for an extraction surgery when they are no longer needed.

The quality and reproducibility of constituent materials play a critical role in the commercial viability of pressure sensors (Eaton and Smith 1997). Pressure sensors with a wide detection range and high compressibility are highly desirable in future portable devices. Recently, various pressure-sensing mechanisms including transistor sensing, capacitive sensing, piezoelectric sensing, triboelectric sensing and resistive sensing have been used to construct pressure sensors with excellent performance. Resistive sensors in particular draw extensive attention due to their simple device structures and low energy consumption (Chen et al. 2018).

8 THE ADVANCEMENT OF PRESSURE SENSORS

With the advancement in wearable electronics technology, many efforts have been made to develop highly sensitive flexible pressure sensors using non-toxic and novel materials, which can measure and identify biomolecules, toxic gases, pressure, temperature with less complexity in design. These sensors are used as a transducer in day-to-day used materials such as smartwatches, phones, textiles and medical equipment (Gu et al. 2019). The ease of fabrication and simplicity in designing the system are the primary reasons behind its widespread attention in recent times. Flexible pressure sensors consist of a mechano-electric transducer coated on a flexible substrate. When the substrate and the film are deflected due to external pressure, responses can be collected from these transducers, which are proportional to the magnitude of the pressure. The response signal can be calibrated for the magnitude of the pressure.

The discovery of graphene and its advanced electronic properties have resulted in many applications sensors (Nag et al. 2018). The high flexibility of graphene in the transverse direction along with its piezo resistivity has made it the ideal material for flexible pressure sensor applications. However, the complexity in the mass production of graphene layers has restricted the use of graphene in commercial applications (Buzaglo et al. 2017). To address this problem, investigators started focusing on the use of conductive polymeric materials such as PVDF, PMMA and PU as a matrix with dispersed graphene and other nanoscale electronic materials as active materials. Most of these conductive polymers have mechano-electric properties. For instance, PVDF is a piezoelectric material and PMMA is piezoresistive, which also contributed to the response. Experiments with doped polymeric matrixes were also conducted to increase the response of the sensor (transducer) due to an increase in mechano-electric charge carriers (Adhikari and Majumdar 2004). Presently a number of investigations have been undertaken to improve the response of flexible pressure sensors by using a combination of primary charge carriers and matrixes (Tian et al. 2015; Tao et al. 2017).

It has been found that the sensitivity of the flexible pressure sensor depends on the microstructural dispersion of nanoparticles and the compatibility of the binder and the nanoparticles. The binder/ particle dispersion should allow for the development of a greater number of conduction routes with a slight change in sensing pressure. Tian et al. (2015) conducted a resistive pressure sensor for a wide pressure range (0~50 KPa) by creating a form-like structure based on Laser-Scribed Graphene (LSG). It was found that the sensitivity of the film was high at the low-pressure range (<5 KPa), and as the pressure increased (> 5 KPa) the sensitivity dropped. The relative increase in conduction paths decreases with an increase in pressure thus, reducing the sensitivity (Tao et al. 2017).

9 CHALLENGES AND TRENDS OF WEARABLE SENSORS

Medical and health data systems based on the internet are expanding quickly. Precision medicine is an emerging medical field that focuses on making an accurate diagnosis, and wearable sensors will provide a possibility for the development of this field. Doctors can build efficient health care programs on time based on data analysis of clinical research and patients' routine monitoring. To obtain these data accurately, continuous and repeated measurements will be conducted by wearable devices, which are implemented for sports and sleep monitoring; however, they fall short of all indicators for precision medicine. Combining precision medicine with wearable sensors and big data and applying this to the construction of medical platforms could allow one to effectively analyze large measurement data (Xu et al. 2018).

Initially, motion sensors such as acceleration sensors were invented as wearable sensors for recording movement. Then pressure sensors, photoelectric sensors and temperature sensors were developed and combined into wearable devices for blood sugar, blood pressure and heart rate monitoring. These sensors are mainly utilized to measure physical quantities, but there are some difficulties to monitor chemical quantities, such as the measurement of particles content (Xu and Yuan 2021). Furthermore, market adoption was influenced by available working hours, therefore minimizing power usage should be considered for future growth. Because of large amounts of data involving personal habits and privacy stored in wearable sensors, it is necessary to establish a platform to supervise data security for healthy and stable development of wearable sensors. Apple Inc. has developed a software framework based on a smartphone app and wearable devices for solving data reliability, security and confidentiality problems, among others, which promoted wearable devices with higher effectiveness in health management and medical treatment (Zang et al. 2015).

Wearable devices like smartwatches and wristbands, which are currently seeing tremendous growth, are indications of wearable product adoption. In the future, the research on implantable wearable devices and thinner skin sensors will become a breakthrough point, and wearable sensors are moving toward higher performance and smaller sizes.

10 CONCLUSION

In this chapter, the properties and applications of graphene-nanocomposites have been reviewed. Preparation methods for a high-quality graphene nanocomposite in a cost-effective manner and on the desired scale are important for many applications. The different techniques that have been employed to prepare graphene such as hydrothermal synthesis, sol-gel and chemical vapor deposition are discussed briefly. The application of graphene nanocomposites in pressure sensors is also described.

■ References

Abergel, D.S.L., V. Apalkov, J. Berashevich, K. Ziegler and T. Chakraborty. 2010. Properties of graphene: A theoretical perspective. Advances in Physics. 59: 261–482.

Adhikari, B. and S. Majumdar. 2004. Polymers in sensor applications. Progress in Polymer Science (Oxford) 29: 699–766.

An, J., T.S.D. Le, Y. Huang, Z. Zhan, Y. Li, L. Zheng, et al. 2017. All-graphene-based highly flexible noncontact electronic skin. ACS Applied Materials and Interfaces 9: 44593–44601.

Angus, M.H. 2013. Recent developments in deposition techniques for optical thin films and coatings. pp. 3–23. In: H. Angus Macleod (ed.). Optical Thin Films and Coatings: From Materials to Applications Woodhead Publishing, Tucson, AZ: United States.

Asthana, A., T. Maitra, R. Büchel, M.K. Tiwari and D. Poulikakos. 2014. Multifunctional superhydrophobic polymer/carbon nanocomposites: Graphene, carbon nanotubes, or carbon black? ACS Applied Materials and Interfaces 6: 8859–8867.

Berry, V. 2013. Impermeability of graphene and its applications. Carbon. 62: 1–10.

Buzaglo, M., I.P. Bar, M. Varenik, L. Shunak, S. Pevzner and O. Regev. 2017. Graphite-to-graphene: Total conversion. Advanced Materials 29: 1603528.

Chen, X., C. Liu, S. Liu, B. Lyu and D. Li. 2018. A high compressibility pressure—sensitive structure based on CB@PU yarn network. Sensors (Switzerland) 18: 4141.

Chen, W. and X. Yan. 2020. Progress in achieving high-performance piezoresistive and capacitive flexible pressure sensors: A review. Journal of Materials Science and Technology 43: 175–188.

Choi, W., I. Lahiri, R. Seelaboyina and Y.S. Kang. 2010. Synthesis of graphene and its applications: A review. Critical Reviews in Solid State and Materials Sciences 35: 52–71.

Cooper, D.R., B. D'Anjou, N. Ghattamaneni, B. Harack, M. Hilke, A. Horth, et al. 2012. Experimental review of graphene. ISRN Condensed Matter Physics 2012: 1–56.

Dhand, V., K.Y. Rhee, H. Ju Kim and D. Ho Jung. 2013. A comprehensive review of graphene nanocomposites: Research status and trends. Journal of Nanomaterials 2013. 158: 1–158.

Eaton, W.P. and J.H. Smith. 1997. Micromachined pressure sensors: Review and recent developments. Smart Materials and Structures 6: 530–539.

Fan, X., Y. Huang, X. Ding, N. Luo, C. Li, N. Zhao and S.C. Chen. 2018. Alignment-free liquid-capsule pressure sensor for cardiovascular monitoring. Advanced Functional Materials 28: 1805045.

Fu, Y., J. Hansson, Y. Liu, S. Chen, A. Zehri, M.K. Samani, et al. 2020. Graphene related materials for thermal management. 2D Materials 7: 012001.

Gan, Y.X., A.H. Jayatissa, Z. Yu, X. Chen and M. Li. 2020. Hydrothermal synthesis of nanomaterials. Journal of Nanomaterials 2020: 3.

Gautam, M. and A.H. Jayatissa. 2011. Gas sensing properties of graphene synthesized by chemical vapor deposition. Materials Science and Engineering: C 31: 1405–1411.

Gautam, M. and A.H. Jayatissa 2012a. Adsorption kinetics of ammonia sensing by graphene films decorated with platinum nanoparticles. Journal of Applied Physics 111: 094317.

Gautam, M. and A.H. Jayatissa 2012b. Ammonia gas sensing behavior of graphene surface decorated with gold nanoparticles. Solid-State Electronics 78: 159–165.

Gautam, M. and A.H. Jayatissa 2012c. Detection of organic vapors by graphene films functionalized with metallic nanoparticles. Journal of Applied Physics 112: 114326.

Gautam, M. and A.H. Jayatissa 2012d. Graphene based field effect transistor for the detection of ammonia. Journal of Applied Physics 112: 064304.

Gautam, M., Z. Shi and A.H. Jayatissa. 2017. Graphene films as transparent electrodes for photovoltaic devices based on cadmium sulfide thin films. Solar Energy Materials and Solar Cells 163: 1–8.

Gu, Y., T. Zhang, H. Chen, F. Wang, Y. Pu, C. Gao, et al. 2019. Mini review on flexible and wearable electronics for monitoring human health information. Nanoscale Research Letters 14: 1–5.

Guo, H., X. Li, B. Li, J. Wang and S. Wang. 2017. Thermal conductivity of graphene/poly(vinylidene fluoride) nanocomposite membrane. Materials and Design 114: 355–363.

Ha, M., S. Lim and H. Ko. 2018. Wearable and flexible sensors for user-interactive health-monitoring devices. Journal of Materials Chemistry B 6: 4043–4064.

Hang, C.Z., X.F. Zhao, S.Y. Xi, Y.H. Shang, K.P. Yuan, F. Yang, et al. 2020. Highly stretchable and self-healing strain sensors for motion detection in wireless human-machine interface. Nano Energy 76: 105064.

Herrmann, J., K.H. Müller, T. Reda, G.R. Baxter, B.D. Raguse, G.J.J.B. De Groot, et al. 2007. Nanoparticle films as sensitive strain gauges. Applied Physics Letters 91: 183105.

Ho, K.I., J.H. Liao, C.H. Huang, C.Y. Su and C.S. Lai. 2013. Electrical probing of multi-ions solution by using graphene-based sensor. Proceedings - Winter Simulation Conference 2013: 37–39.

Hosseingholipourasl, A., S.H.S. Ariffin, Y.D. Al-Otaibi, E. Akbari, F.K.H. Hamid, S.S.R. Koloor, et al. 2020. Analytical approach to study sensing properties of graphene-based gas sensor. Sensors (Switzerland) 20: 1506.

Huang, W. 2018. Graphene oxide nanopapers. pp. 1–26. *In*: W. Huang (ed.). Nanopapers: From Nanochemistry and Nanomanufacturing to Advanced Applications. A volume in Micro and Nano Technologies. Elsevier. 14: 1067–1084.

Itapu, B.M. and A.H. Jayatissa. 2018. A review in graphene/polymer composites. Chemical Science International Journal 23: 1–16.

Jeon, S., S.C. Lim, T.Q. Trung, M. Jung and N.E. Lee. 2019. Flexible multimodal sensors for electronic skin: Principle, materials, device, array architecture, and data acquisition method. Proceedings of the IEEE 107: 2065–2083.

Ji, X., Y. Xu, W. Zhang, L. Cui and J. Liu. 2016. Review of functionalization, structure, and properties of graphene/polymer composite fibers. Composites Part A: Applied Science and Manufacturing 87: 29–45.

Jishnu, A., J., S Jayan, A. Saritha, A.S. Sethulekshmi and G. Venu. 2020. Superhydrophobic graphene-based materials with self-cleaning and anticorrosion performance: An appraisal of neoteric advancement and future perspectives. Colloids and Surfaces, A: Physicochemical and Engineering Aspects. 606: 125395.

Jung, M., S.K. Vishwanath, J. Kim, D.K. Ko, M.J. Park, S.C. Lim, et al. 2019. Transparent and flexible mayan-pyramid-based pressure sensor using facile-transferred indium tin oxide for bimodal sensor applications. Scientific Reports 9: 1–11.

Justino, C.I.L., A.R. Gomes, A.C. Freitas, A.C. Duarte and T.A.P. Rocha-Santos. 2017. Graphene based sensors and biosensors. Trends in Analytical Chemistry 91: 53–66.

Kang, S., T.H. Kang, B.S. Kim, J. Oh, S. Park, I.S. Choi, et al. 2019. 2D reentrant micro-honeycomb structure of graphene-CNT in polyurethane: High stretchability, superior electrical/thermal conductivity, and improved shape memory properties. Composites Part B: Engineering 162: 580–588.

Katsnelson, M.I. 2007. Graphene: Carbon in two dimensions. Materials Today 10: 20–27.

Kim, B.S., S. Shin, S.J. Shin, K.M. Kim and H.H. Cho. 2011. Control of superhydrophilicity/super hydrophobicity using silicon nanowires via electroless etching method and fluorine carbon coatings. Langmuir 27: 10148–10156.

Kimura, M. 2019. Sensor applications of thin-film devices originating in display technologies. Journal of the Society for Information Display 27: 741–756.

Li, F., L. Xie, G. Sun, Q. Kong, F. Su, H. Lei, et al. 2017. Regulating pore structure of carbon aerogels by graphene oxide as 'shape-directing' agent. Microporous and Mesoporous Materials 240: 145–148.

Li, J., R. Bao, J. Tao, Y. Peng and C. Pan. 2018. Recent progress in flexible pressure sensor arrays: From design to applications. Journal of Materials Chemistry C 6: 11878–11892.

Li, X.P., Y. Li, X. Li, D. Song, P. Min, C. Hu, et al. 2019. Highly sensitive, reliable, and flexible piezoresistive pressure sensors featuring polyurethane sponge coated with MXene sheets. Journal of Colloid and Interface Science 842: 54–62.

Li, L., J. Zheng, J. Chen, Z. Luo, Y. Su, W. Tang, X. Gao, et al. 2020. Flexible pressure sensors for biomedical applications: From *ex vivo* to *in vivo*. Advanced Materials Interfaces 7: 2000743.

Liu, K., Z. Zhou, X. Yan, X. Meng, H. Tang, K. Qu, et al. 2019. Polyaniline nanofiber wrapped fabric for high performance flexible pressure sensors. Polymers 11: 1120.

Liu, Z., K. Chen, A. Fernando, Y. Gao, G. Li, L. Jin, et al. 2021. Permeable graphited hemp fabrics-based, wearing-comfortable pressure sensors for monitoring human activities. Chemical Engineering Journal 403: 126191.

Lotsch, B.V. 2015. Vertical 2D heterostructures. Annual Review of Materials Research 45. 85–109.

Luo, N., W. Dai, C. Li, Z. Zhou, L. Lu, C.C.Y. Poon, et al. 2016. Wearable sensors: Flexible piezoresistive sensor patch enabling ultralow power cuffless blood pressure measurement. Advanced Functional Materials 26: 1178–1187.

Ma, F., Z. Zhang, H. Jia, X. Liu, Y. Hao and B. Xu. 2010. Adsorption of cysteine molecule on intrinsic and Pt-doped graphene: A first-principal study. Journal of Molecular Structure: THEOCHEM 955: 134–139.

Mittal, G., V. Dhand, K.V. Rhee, S.J. Park and W.R. Lee. 2015. A review on carbon nanotubes and graphene as fillers in reinforced polymer nanocomposites. Journal of Industrial and Engineering Chemistry 21: 11–25.

Mohan, V.B., K.-tak Lau, D. Hui and D. Bhattacharyya. 2018. Graphene-based materials and their composites: A review on production, applications, and product limitations. Composites Part B: Engineering 142: 200–220.

Muñoz, R., C. Munuera, J.I. Martínez, J. Azpeitia, C. Gómez-Aleixandre and M. García-Hernández. 2017. Low temperature metal free growth of graphene on insulating substrates by plasma assisted chemical vapor deposition. 2D Materials 4: 015009.

Nag, A., A. Mitra and S.C. Mukhopadhyay. 2018. Graphene and its sensor-based applications: A review. Sensors and Actuators, A: Physical 270: 177–194.

Novoselov, K.S., V.I. Fal'Ko, L. Colombo, P.R. Gellert, M.G. Schwab and K. Kim. 2012. A roadmap for graphene. Nature 490: 192–200.

Papageorgiou, D.G., I.A. Kinloch and R.J. Young. 2017. Mechanical properties of graphene and graphene-based nanocomposites. Progress in Materials Science 90: 75–127.

Rinaldi, A., A. Tamburrano, M. Fortunato and M.S. Sarto. 2016. A flexible and highly sensitive pressure sensor based on a PDMS foam coated with graphene nanoplatelets. Sensors (Switzerland) 16: 2148.

Shao, Y., J. Wang, H. Wu, J. Liu, I.A. Aksay and Y. Lin. 2010. Graphene based electrochemical sensors and biosensors: A review. Electroanalysis 22: 1027–1036.

Shen, H., L. Zhang, M. Liu and Z. Zhang. 2012. Biomedical applications of graphene. Theranostics 2(3): 283–294.

Shi, Z. and A.H. Jayatissa. 2018. The impact of graphene on the fabrication of thin film solar cells: Current status and future prospects. Materials 11: 36.

Sidorov, A.N., G.W. Sławiński, A.H. Jayatissa, F.P. Zamborini and G.U. Sumanasekera. 2012. A surface-enhanced Raman spectroscopy study of thin graphene sheets functionalized with gold and silver nanostructures by seed-mediated growth. Carbon 50: 699–705.

Silva, M., N.M. Alves and M.C. Paiva. 2018. Graphene-polymer nanocomposites for biomedical applications. Polymers for Advanced Technologies 29: 687–700.

Singh, E., M. Meyyappan and H.S. Nalwa. 2017. Flexible graphene-based wearable gas and chemical sensors. ACS Applied Materials and Interfaces 9: 34544–34586.

Singh, J., A. Rathi, M. Rawat and M. Gupta. 2018. Graphene: From synthesis to engineering to biosensor applications. Frontiers of Materials Science 12. 1–20.

Smith, A.T., A.M. LaChance, S. Zeng, B. Liu and L. Sun. 2019. Synthesis, properties, and applications of graphene oxide/reduced graphene oxide and their nanocomposites. Nano Materials Science 1. 31–47.

Soldano, C., A. Mahmood and E. Dujardin. 2010. Production, properties, and potential of graphene. Carbon 48: 2127–2150.

Tao, L.Q., K.N. Zhang, H. Tian, Y. Liu, D.Y. Wang, Y.Q. Chen, et al. 2017. Graphene-paper pressure sensor for detecting human motions. ACS Nano 11: 8790–8795.

Tee, B.C.K., A. Chortos, R.R. Dunn, G. Schwartz, E. Eason and Z. Bao. 2014. Tunable flexible pressure sensors using microstructure elastomer geometries for intuitive electronics. Advanced Functional Materials 24: 5427–5434.

Tian, H., Y. Shu, X.F. Wang, M.A. Mohammad, Z. Bie, Q.Y. Xie, et al. 2015. A graphene-based resistive pressure sensor with record-high sensitivity in a wide pressure range. Scientific Reports 5: 1–6.

Vaka, M., M.Z. Bian and N.D. Nam. 2020. Highly sensitive pressure sensor based on graphene hybrids. Arabian Journal of Chemistry 13: 1917–1923.

Varghese, S.S., S. Lonkar, K.K. Singh, S. Swaminathan and A. Abdala. 2015. Recent advances in graphene-based gas sensors. Sensors and Actuators, B: Chemical 218. 160–183.

Wang, W. and A.H. Jayatissa. 2015a. Comparison study of graphene based conductive nanocomposites using poly (methyl methacrylate) and polypyrrole as matrix materials. Journal of Materials Science: Materials in Electronics 26: 7780–7783.

Wang, W. and A.H. Jayatissa. 2015b. Computational and experimental study of electrical conductivity of graphene/poly (methyl methacrylate) nanocomposite using Monte Carlo method and percolation theory. Synthetic Metals 204: 141–147.

Wang, T., D. Huang, Z. Yang, S. Xu, G. He, X. Li, et al. 2016. A review on graphene-based gas/vapor sensors with unique properties and potential applications. Nano-Micro Letters 8: 95–119.

Warner, J.H., F. Schaffel, M. Rummeli and A. Bachmatiuk. 2013. Graphene: Fundamentals and Emergent Applications. Newnes, Oxford; U.K.

Wei, J., T. Vo and F. Inam. 2015. Epoxy/graphene nanocomposites - processing and properties: A review. RSC Advances 5: 73510–73524

Xu, F., X. Li, Y. Shi, L. Li, W. Wang, L. He, et al. 2018. Recent developments for flexible pressure sensors: A review. Micromachines 9: 580.

Xu, J. and K. Yuan. 2021. Wearable muscle movement information measuring device based on acceleration sensor. Measurement 167: 108274.

You, R., Y.Q. Liu, Y.L. Hao, D.D. Han, Y.L. Zhang and Z. You. 2020. Laser fabrication of graphene-based flexible electronics. Advanced Materials 32: 1901981.

Yu, G.H., Q. Han and L.T. Qu. 2019. Graphene fibers: Advancing applications in sensor, energy storage and conversion. Chinese Journal of Polymer Science (English Edition) 37: 535–547.

Yuan, W. and G. Shi. 2013. Graphene-based gas sensors. Journal of Materials Chemistry A 1: 10078–10091.

Zang, Y., F. Zhang, C. Di, and D. Zhu. 2015. Advances of flexible pressure sensors toward artificial intelligence and health care applications. Materials Horizons. 2: 140–156.

Zhang, Y., L. Zhang and C. Zhou. 2013. Review of chemical vapor deposition of graphene and related applications. Accounts of Chemical Research 46: 2329–2339.

Zhang, T., Z. Li, K. Li and X. Yang. 2019. Flexible pressure sensors with wide linearity range and high sensitivity based on selective laser sintering 3D printing. Advanced Materials Technologies 4: 100679.

Zhang, X., Y. Zhuo, Q. Luo, Z. Wu, R. Midya, Z. Wang, et al. 2020. An artificial spiking afferent nerve based on Mott memristors for neurorobotics. Nature Communications 11: 1–9.

Zhu, Y., S. Murali, W. Cai, X. Li, J.W. Suk, J.R. Potts, et al. 2010. Graphene and graphene oxide: Synthesis, properties, and applications. Advanced Materials 22: 3906–3924.

Chapter 6

Application of Nanocomposite Solid Films in Tribology

Bodhi Ravindran Manu and Ahalapitiya H. Jayatissa*

The University of Toledo, 2801 W. Bancroft St, Toledo, OH - 43606, USA

bravind@rockets.utoledo.edu;
ahalapitiya.jayatissa@utoledo.edu

1   INTRODUCTION

A composite having nano-meter size particles dispersed in them is known as a nanocomposite. Nanocomposites are widely used as surface coatings to reduce friction and wear. Their use as a lubricant is more profound in extreme conditions. An extreme condition refers to high variations in temperatures or pressure, radiation, corrosion or very high operating parameters such as speed and load. In such cases, the use of liquid lubrication is inadequate or inconvenient (Vaz et al. 2000). They find applications in medical, automotive, aeronautical and astronomical industries when most available technologies fail to provide the desired lubrication. These nanocomposite coatings are also known as a solid lubricant since they are solid films coated on surfaces of the mating parts to reduce friction and wear. Nanocomposite coatings possess superior wear and mechanical properties credited to the nanoparticles dispersed in them.

Due to the small size of nanoparticles, they have high surface energy and form strong interfacial bonds with the matrix material. This increases the mechanical and tribological properties. Based on the matrix material, nanocomposites can be divided into metallic, ceramic or polymeric (Manu et al. 2021). Aeronautical and astronomical applications rely largely on metal and ceramic-based solid lubricants. In these conditions, the lubricants are subjected to extreme fluctuations in temperatures, pressure and radiation. Ceramics can withstand high temperatures and pressure conditions compared with metals. They are also stable to oxidation and corrosion. They are used in extreme temperatures in the range of 1000 to 1250°C, where most metals oxidize or melt. Metals, on the other hand, are used in the range of 400 to 700°C. Ceramics also possess radiation stability and find its use as lubricants in nuclear reactors. In most applications, metal and ceramic-based nanocomposites consist of 2D materials such as MoS_2 graphite dispersed in

---

them. Their structure is known for reducing the friction coefficient, whereas nanocomposites have superior wear resistance. Nanocomposites with 2D materials dispersed are well-known coatings that possess low friction coefficients with high wear resistance.

Polymers show a very low friction coefficient. However, the viscoelastic nature (dependence of mechanical properties on temperature) makes them difficult to be used in extreme conditions. However, they possess excellent biocompatibility, coupled with low friction enabling them to be the go-to material for bio implants. The high and advanced knowledge of polymers and their chemistry helps in tailoring their properties very close to that of biological tissues (Ramli et al. 2016). It is also possible to manufacture polymer-based implants with complex structures with high dimensional accuracy; which is pivotal for implants life. Polymers such as PMMA, PEEK-based nanocomposites are currently replacing ceramic and metal matrix-based implants due to the high success rates of surgeries. Polymer nanocomposites are also used as coatings on bearings to provide toughness and reduce pitting failure.

The type and size of the fillers greatly determine the tribological and mechanical properties of composites. Their performance is also highly sensitive to environmental conditions and can lead to deterioration. For instance, 2D materials such as MoS_2 based nanocomposites have been extensively used as a lubricant in space applications. Their degradation under humidity challenged scientists and engineers and created problems during the assembly and shelf life of coated components before launch (Manu and Jayatissa 2020). Carbon-based filters such as graphene and CNT, on the other hand, need humidity for effective functioning (Li et al. 2014). This makes the use of nanocomposites for tribo-applications challenging with high dependence on every aspect of working conditions. To increase the working range of these composites, adaptive lubricating system, where two or more lubricants are dispersed in a matrix material; each providing lubrication at a particular working condition.

The nano-sized dispersions in these composites are responsible for their desirable properties. However, the manufacture of nanocomposites with high reliability on particle size dispersion is still challenging. The high surface energy of nanoparticles can lead to amalgamation and increase their size. The formation of a micro or a macro-sized dispersant reduces filler-matrix interfacial strength and can drastically deteriorate its performance. Hence confirming the presence of uniform nano dispersion in the matrix is important for reliable performance. This increases the cost of nanocomposite manufacture, which is required in mass production.

The importance of nanocomposites in tribological applications is described. The desirable qualities that make it superior to conventional liquid lubrication; reasons for their widespread acceptance are covered. The challenges, drawbacks of these coatings are explained from a cost and reliability standpoint. A comprehensive review of the application of metal, ceramic and polymer-based nanocomposites is covered in this chapter.

2 TYPES OF NANOCOMPOSITES FOR TRIBO-APPLICATIONS

2.1 Metal Matrix Composites (MMC)

Metal matrix composites are usually used in applications where the property of metal and ceramics need to be combined. In other words, an increase in wear strength while maintaining metallic properties such as ductility. MMC's are widely used in aerospace applications where weight saving is of high importance. For most MMC, metals such as copper, aluminium, magnesium are usually preferred. On the other hand, reinforcement materials are generally, metal carbides, nitrides and oxides. Sometimes 2D materials such as MoS_2 and graphite are also used for the dual purpose of lubrication and strengthening.

2.2   Ceramic Matrix Composites (CMC)

Metal matrix composites are susceptible to temperature, especially to thermal cycling and can lead to thermal cycling failure. To counteract this problem, ceramic-based nanocomposites are mostly employed under high or low-temperature applications. CMC possesses less ductility than MMC, however, it possesses superior strength and chemical properties. Ceramic matrix composites are usually manufactured using graphite or CNT nanoparticles to produce self-lubricating composites for tribo-applications.

2.3   Polymer Matrix Composites (PMC)

For most tribo-applications, polymers with high cross-linked density are used as the matrix material. With high chemical stability and the advanced knowledge of polymer chemistry, it is possible to manufacture polymers with precise control over particle size and dispersion. Figure 6.1 shows the publications statistics of nanocomposite coatings in tribology over the past decades.

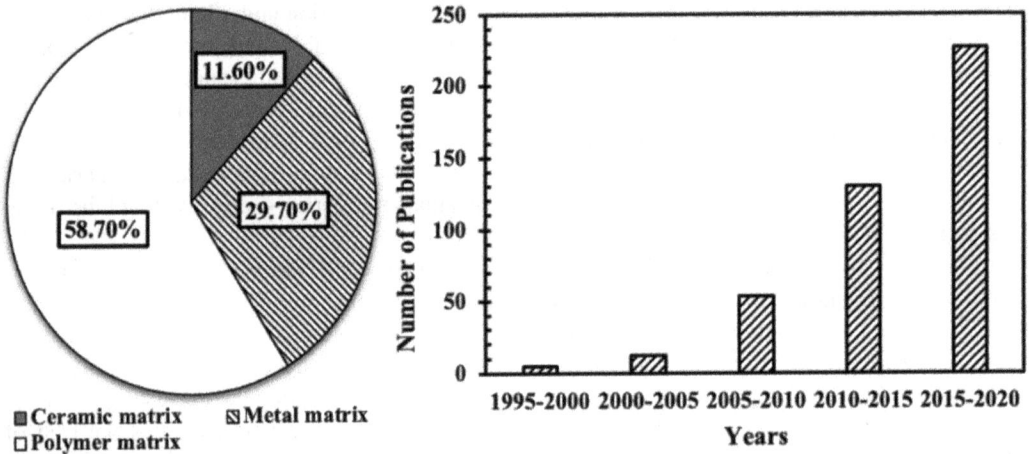

Figure 6.1   Statistics of nanocomposite publications. (a) Relative fraction of publications on ceramic metal and polymer matrix composites from 2000 to 2020; (b) number of publications on tribology of nanocomposites over a 5-year period from 1995 to 2020; web of science keywords used "ceramic matrix", "metal matrix", "polymer matrix" and "nanocomposite tribology".

3   APPLICATIONS OF NANOCOMPOSITES IN BIO-TRIBOLOGY

Bio-tribology is the field of study of bio implants and their friction, wear and compatibility with the human body. Earlier accounts of prosthetic use included wood, metal castings, however, with advancement in science and technology, the ability to manufacture materials that mimic the properties of human body parts, significant interest has been developed on the use of advanced materials such as alloys, nanocomposites to use as a joint replacement and bio-implants material. Metal, ceramic and polymer-based nanocomposites are commonly used in bio-applications. The success of these bio implants depends on many factors, some patient-specific and others general.

The patient-specific aspect of a prosthetic surgery depends on a person's posture and gait. The loadings and allergies to implant material are patient-specific and determine the choice of material and design. Biocompatibility is a blanket term that determines the choice of the material. Since different biological tissues possess different properties, regenerative skills and mechanical structures, the foreign implant should possess high compatibility with these human tissues along with the required tribo-properties for the desired purpose; something hard to achieve and makes the success of these prosthetics very centralized to the patient.

Metallic and ceramic-based nanocomposites were the first class of materials used for artificial implants. Stainless steel, Co–Cr-based composites and ceramics such as Al_2O_3 and zirconia are examples of metal and ceramic-based bio implant materials. However, the biological reaction of metals with human tissues to produce metal ions reduced their interest over time. Polymer nanocomposites showed better biocompatibility and manufacturability. This has resulted in its widespread use as an implant material. UHMWPE and PEEK-based nanocomposites are extensively used for bone and joint replacements. PMMA finds application as a dental implant material. Young's modulus and fracture toughness of different bio implant materials are shown in Fig. 6.2. Next different nanocomposites used as an implant material, their advantages and shortcomings are focused.

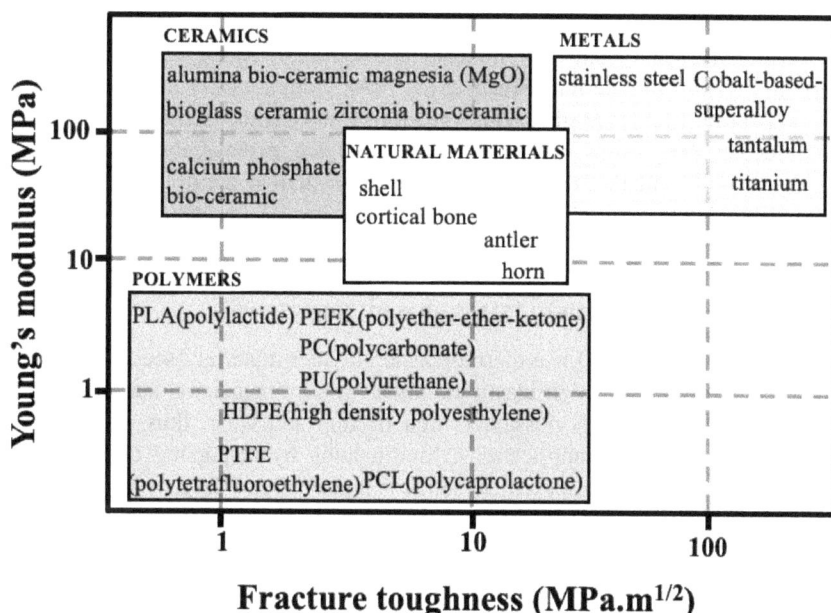

Figure 6.2 Mechanical properties of natural and synthetic bio implant materials. (Reprinted with permission from: Mota 2018. Polymer nanocomposites used as scaffolds for bone tissue regeneration. *Mat. Sci. and Appli.* 9,8(2018): 679–697).

3.1 Metallic and Ceramic-based Biomedical Materials

As stated earlier, metal and ceramic composites were the first class of biomedical material used as implants and prosthetics. Stainless steel was first studied for application in bone fracture readjustments; followed by the use of Co–Cr-based composites for artificial hip and knee joints (Werner et al. 2007). Metals such as beta-titanium and magnesium alloys and their composites were also used for implant applications. However, the dissociation of metal ions in bodily fluids was a serious problem with severe health hazards. Due to this, surface treatment of these metals was of utmost importance; scientists have attempted methods such as coatings (Fu et al. 1998; Golden et al. 2007), laser treatment (Yilbas and Hashmi 2000), plasma modification (Ali et al. 2010) and surface texturing (Hu et al. 2012) to eliminate contact of human tissues with metal ions. Presently, metal composites are not preferred for implants due to ion dissociation and stress shielding effects. Stress shielding occurs because of a higher elastic modulus of metal and ceramic compared with bone materials.

Ceramics are more popular than metals as they have a higher wear resistance with a porous structure like bones. Hydroxyapatite has been intensively explored (Moseke and Gbureck 2010) for its osteo inductivity effect; zirconia is another bio implant ceramic. However, pure zirconia does not have good mechanical properties, hence they are doped with MgO, CaO, Al_2O_3 and other oxides. Significant research on stabilizing, improving the surface hardness and inducing bioactivity on zirconia by different chemical treatments has been conducted (Nath et al. 2008). Other frequently used bio-ceramics include Silicon Nitride (Si_3N_4) (Salgueiredo et al. 2008; Dante and Kajdas 2012) and Silicon Carbide (SiC) (Varela-Feria et al. 2002; Gonzalez et al. 2003).

3.2   Polymer-based Biomedical Materials

Polymer nanocomposites are the most researched and used materials in medical applications. They possess high chemical inertness, biocompatibility; their mechanical properties can be tailored closer to that of bones. The advanced knowledge of polymer chemistry has resulted in a higher interest in polymer nanocomposites for medical applications ranging from implants, scaffolds, drug delivery and tissue engineering (Freed et al. 1994; Soppimath et al. 2001; O'brien 2011). Polymers such as Poly (Methyl Methacrylate) (PMMA), Poly (Ether Ether Ketone) (PEEK), and Ultra-High-Molecular-Weight-Polyethylene (UHMWPE) are extensively used in bio-tribology applications. UHMWPE and PEEK possess properties similar to natural bones and are used in artificial bone and joint surgeries. PMMA on the other hand finds dental applications such as retainers and artificial teeth.

3.2.1   Poly (Methyl Methacrylate) (PMMA)-based Biomaterials

Polymethyl methacrylate (PMMA) is a thermoplastic polymer material used in dental implants. It has good mechanical properties, high elastic modulus, good dimensional stability and biological properties. However, pure PMMA possesses high friction and wear, thus restricting its use as an antifriction material for tribo-applications. Modification by dispersing organic, inorganic or metallic micro/nanoparticles into a polymer matrix is an effective way to improve its mechanical and tribological properties. Fillers such as carbon nanotubes (CNTs), silicon dioxide (SiO_2), Titanium Oxides (TiO_2) are widely used to produce PMMA based nanocomposites.

3.2.2   Poly (Ether Ether Ketone) (PEEK)-based Biomaterials

Polyether ether ketone (PEEK) is a semicrystalline thermoplastic with high mechanical and wear-resistance properties. They also possess good thermal stability. For any composite material, the wear resistance and friction coefficient have a complementary effect. It is impossible to get a low friction coefficient without compromising the wear resistance. Hence, research on nanocomposites is an optimization problem where the effect of nanocomposite type, size, dispersion, composition on friction and wear has to be experimentally determined and optimized for a specified application. Figure 6.3 shows the effect of nanomaterial concentration of SiO_2, Silicon Nitrides (Si_3N_4), and silicon Carbide (SiC) on wear rate and friction of PEEK-based nanocomposites. At a nanoparticle concentration of 7.5 wt.%, the PEEK-based films gave the best performance and are the preferred filler concentration for this nanocomposite.

The orientation of nanoparticles can improve the tribo-performance, even at the same weight percentage of nanoparticles. Highly oriented nanofillers can provide better tribo-properties for PMMA. However, the integration of inorganic and organic nanoparticles in the same matrix can lead to varied results. The interaction between different fillers can cause chemical reactions; reduce the interfacial strength, thus deteriorating its performance. Solid lubricating materials such as PTFE, graphite and WS_2 are widely used as nanofillers in the PMMA matrix.

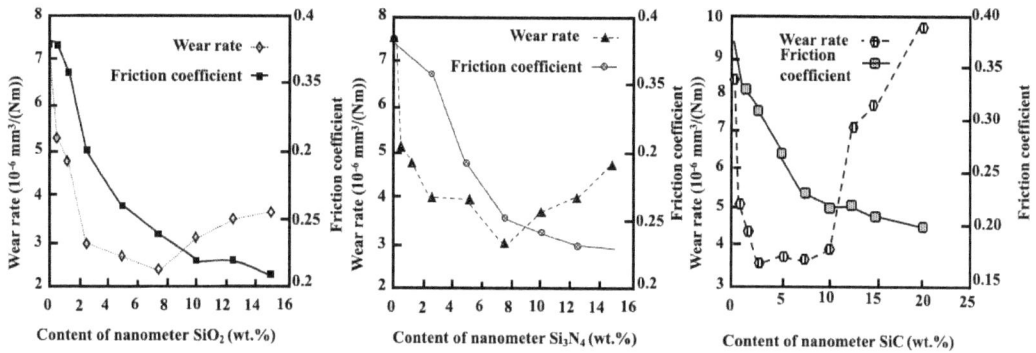

Figure 6.3 Effect of nanometer (a) SiO_2 (b) Si_3N_4 and (c) SiC content on the coefficient friction and wear rate of PEEK nanocomposite (load: 196 N; sliding velocity: 0.445 m/s). (Reprinted with permission from: Shi 2015. The recent progress of tribological biomaterials. *Biosurface and Biotribology* 1,2(2015): 81–97).

3.2.3 Ultra-High-Molecular-Weight-Polyethylene (UHMWPE)-based Biomaterials

The UHMWPE-based nanocomposite is the most extensively used material for bio-tribo applications. High biocompatibility and wear resistance are the core properties responsible for its widespread acceptance. For most bio-implants, the wear debris produced during interaction can cause aseptic loosening, leading to failure and repair requirements. Due to the widespread use of UHMWPE, a significant number of investigations on increasing the wear resistance without compromising other desirable properties have been reported. Methods such as ion irradiation, ion implant, filler addition, surface modifications are mainly used for improving their tribo-properties.

3.3 Applications of Metallic, Ceramic and Polymer Nanocomposites in Artificial Joints

The desirable properties of bone substitutes include biocompatibility, non-toxicity, mouldability, durability, availability and affordability. Metal-based implants are not popular due to varied success rates in the past. Metal-on-metal contact joints are sensitive to lubrication and cause higher wear under adverse lubrication. Ceramic-on-ceramic combinations, on the other hand, showed the least wear among all the materials combinations. Due to this reason, zirconia toughened alumina ceramics are used for adverse edge contact conditions; where high wear can cause loosening of the prosthetic components (Halma et al. 2014).

Polymeric nanocomposites are extensively used for artificial joints and bone replacements. The UHMWPE and PEEK are the most preferred polymers in prosthetics. It is possible to manufacture polymer nanocomposites with properties close to that of natural bones with high dimensional accuracy. UHMWPE is used to fabricate hip, knee and shoulder prostheses, whereas PEEK finds its application in bone replacement prosthetics. However, these polymers are bioinert and do not interact with bone material. To counteract this problem, Hydroxyapatite (HA), a component in natural bone, is added to polymeric biomaterials to reduce the healing time after surgery. In these prosthetics, the polymer nanocomposites provide the required mechanical properties and HA helps in better integration with the bio-tissues.

The formation of wear debris during sliding interaction has a detrimental effect on the success of prosthetic surgery. It can lead to the inclusion of wear particles into the bone and eventually loosening the implanted joints. Figure 6.4 shows the effect of wear debris formation on the failure of an artificial joint made using a metal and polymer (UHMWPE).

Surgical techniques and patient-specific variation of anatomy and loading/motion largely determine the success of an implant surgery. Due to this reason, it is important to integrate tribological studies with biological and clinical studies when designing implants. For instance, the most important parameter in the hip joint is the femoral head radius. A larger femoral head radius helps in terms of biomechanics, but it increases the sliding distance leading to more wear. Also, conformity between the two articulating surfaces in the hip joint mainly depends on the radial clearance between the femoral head and the acetabular cup. In the case of knee implants, this conformity depends on the radii of the femoral and tibial bearing surfaces (Wang et al. 2001). It is challenging to address these factors in tribological studies alone. The artificial joints used for hip and knee implants are shown in Fig. 6.5.

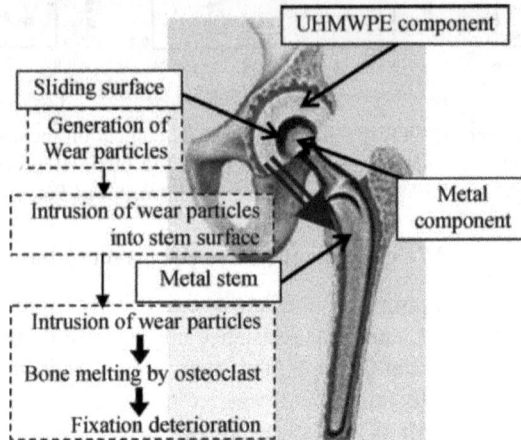

Figure 6.4 Effect of wear debris on artificial joint. (Reprinted with permission from: Shen et al. 2018. Tribological performance of bio implants: A comprehensive review. *Nanotechnology and Precision Engineering* 1,2(2018): 107–122).

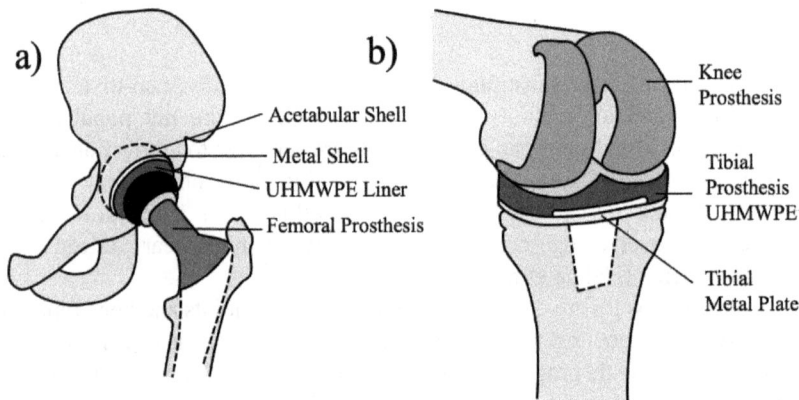

Figure 6.5 Artificial Hip (a) and knee joint (b) using UHMWPE. (Reprinted with permission from: Cobelli et al. 2011. Mediators of the inflammatory response to joint replacement devices. *Nature Reviews Rheumatology* 7,10(2011): 600–608).

3.4 Applications of PEEK-based Nanocomposites in Medicine

PEEK becomes popular due to its elastic and mechanical properties similar to human bone (Scholes and Unsworth 2009; Najeeb et al. 2016; Panayotov et al. 2016). PEEK with dispersed Hydroxyapatite (HA) assists in a speedy recovery. It is also easy to manufacture and implement during surgery with high dimensional accuracy and flexibility. PEEK is one of the most sort out bio implant materials due to its high success rate in operations. It is used in orthopaedic and spinal implants, also in various surgical and dental equipment (Rahmitasari et al. 2017, Ferguson et al. 2006).

Due to its high vacuum and pressure holding capability, PEEK tubing is used in numerous medical devices. It can replace titanium and ceramics in orthopedic, maxilla-facial, cranial and spine surgeries (Gallagher et al. 2018). Tables 6.1 and 6.2 show the desirable properties of PEEK implants and their application areas in the medical field respectively.

Table 6.1 Properties of PEEK implants; (*Source:* Data from: Haleem and Javaid 2019. Polyether ether ketone (PEEK) and its 3D printed implants applications in medical field: An overview. Clinical *Epidemiology and Global Health* 7,4 (2019): 571–577)

Excellent biocompatibility	• PEEK material is biocompatible • It can interact with the human body and increases the success rate of surgeries • PEEK-based nanocomposite implants can be designed to perform required functions concerning medical therapy that helps in accelerated recovery
Less weight	• Implants manufactured by PEEK-based nanocomposites have less weight as compared with other traditional materials • These implants provide a natural look and feel with excellent safety characteristics and enhance comfort
Good mechanical properties	• PEEK implants have good mechanical strength to withstand the loadings on the whole body like bones • Its functionality is unaffected by variation of body temperature • It has high rigidity, modulus, stability and chemical resistance
High Tensile Strength	• These implants have high tensile strength under the action of the load • Its tensile strength is between 90–100 MPa
Low moisture absorption	• This material has a low moisture absorption • It needs less drying time and helps in easy handling during surgery
High flexibility	• PEEK material is capable of bending easily without breaking
Suitable for high vacuum and pressure applications	• Used in medical applications for the manufacturing of pressure or vacuum devices • It increases the performance of devices that have highly resistant to most chemicals
Stable high temperature	• PEEK is stable high temperature hence beneficial for medical and engineering applications • It melts at 343 °C, a high melting temperature, as compared with most other thermoplastics • It has good toughness, low toxicity and excellent abrasion resistance

Table 6.2 Application areas of PEEK in the medical field; (*Source:* Data from: Haleem and Javaid 2019. Polyether ether ketone (PEEK) and its 3D printed implants applications in medical field: An overview. Clinical *Epidemiology and Global Health* 7, 4 (2019): 571–577)

Orthopaedic surgery	• Due to high strength, PEEK is used for various orthopaedic applications like hip replacements, hip resurfacing and construction of the femoral component • Artificial implant in the bone to assist with the load of the body
Medical tubing	• It is a high purity, organic polymer with good resistance • It is also have required flexural modulus and tensile strength • PEEK tubing is used medical applications where high rigidity is required
Spinal implants	• For spinal implants, PEEK offers many advantages as compared to metal. • Include modulus close to bone which is the ultimate solution for the surgeon • Applications of this material are expanded significantly for the manufacturing of intervertebral fusion cage
Spinal fusion	• Applications of this material are expanded significantly for the manufacturing of intervertebral fusion cage
Spinal arthroplasty devices	• PEEK materials are used innovatively for spinal arthroplasty devices • These are extremely hard and capability to withstand high temperatures

Bone screws and pins	• For the manufacturing of screws and pins, PEEK has excellent capability in required strength used to hold the bone in place. • This increases the recovery chance of patient in less time
The smooth motion of the spine	• PEEK helps in the smooth motion of spine which is crucial in the success of spine surgery
Minimally invasive fusion surgery	• Used for innovation in spine surgery with minimally invasive fusion • It also reduces nerve root retraction and tissue dissection
Used for Stabilization devices	• PEEK materials are used for manufacturing unique instrumentation for medical applications • Used for soft stabilisation and flexible stabilization for complex treatments
Face reconstruction	• Applicable for the reconstruction of the face of the individual patient with high strength, durability and stiffness • These implants are easily workable, non-porous for facial reconstruction
Heart valve and stents	• Used in the manufacture of the heart valve, stents with high durability • Efficient for the cardio patient during surgery and performed high comfort as compared to other materials
Dental implants	• In dentistry, PEEK also shows an excellent contribution to the manufacturing of missing teeth to enhance the comfort of the patient. • Used for construction of partial dentures, crowns and bridges due to its lightweight and strength

Using 3D printing technology, it is possible to manufacture polymer nanocomposite with high dimensional accuracy and flexibility. 3D printed PEEK is used for facial reconstruction of an individual patient with the required strength, durability and stiffness. In cardiac surgery, PEEK finds application in manufacturing heart valve prostheses, leaflet heart valves and stents. Another application in cardiology includes the manufacture of the rotor of a micro axial pump. In dentistry, PEEK implants are used for tooth replacement. PEEK has a wide variety of applications in bone tissue engineering, post teeth bleaching, spinal implants, joint replacement and restoration of periodontal defects.

3.5   Applications of PMMA Nanocomposites in Dentistry

PMMA is used in dental applications as orthodontic retainers, dentures, relining and for the fabrication of artificial teeth. It has low density, aesthetics, ease of manipulation and tailorable physical and mechanical properties. Figure 6.6 shows the desirable properties of PMMA in denture applications. However, pure PMMA possesses low impact and flexural strength and is also susceptible to fracture due to water absorption. To counteract this problem, reinforcements with natural fibers, nanoparticles and nanotubes are mostly employed. For fiber reinforced PMMA, it was determined that fiber diameter, length, orientation and interfacial strength greatly determine its mechanical properties. For instance, adding glass fibers shorter than a critical length of 0.5–1.6 mm negatively affects their mechanical properties.

The addition of metal and ceramic nanoparticles also showed enhancement in properties. Metallic nanocomposites enhanced the thermal conductivity of the denture base; helping in better judgment of food temperature. CNT-based nanocomposites are also extensively studied because of their improved electrical and mechanical properties. In most research on PMMA/CNT nanocomposites, it was observed that an optimum of CNT nanoparticles concentration has to be determined for better and improved strength and tribo-properties. CNT addition reduced polymerization shrinkage, but with an increase in concentration beyond optimum, the properties deteriorated due to insufficient dispersion and amalgamation.

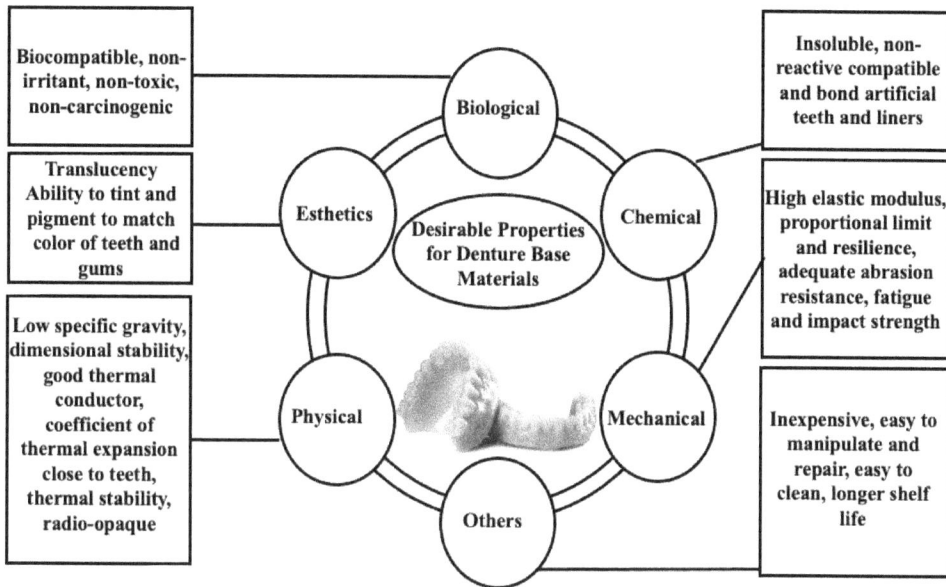

Figure 6.6 Ideal properties of PMMA for denture base applications. (Reprinted with permission from: Zafar 2020. Prosthodontic applications of polymethyl methacrylate (PMMA): an update. *Polymers* 12,10 (2020): 2299(1–35)).

4 APPLICATIONS OF NANOCOMPOSITES AND NANOSTRUCTURED MATERIALS IN SPACE TRIBOLOGY

Aeronautical and astronomical applications are subject to extreme working conditions where temperature, pressure and atmospheric influences change drastically in a short time. For example, aircraft bearing surfaces are subject to contact stresses in the order of 10^7–10^{10} Pa with sliding velocities vary in the range of 0 to 20 meters per second. These loads are typically due to high vibrations and can create significant wear and fatigue of the bearing surfaces. They are also prone to environmental cycling, with humidity, temperature and pressure varying between flights and landings. The dust experienced by aircrafts also adds to the problems of wear in bearing surfaces. In spacecraft bearings, they may be exposed to atomic oxygen, which creates electrostatically attracting dust causing massive abrasive wear. The bearings for these applications are expected to provide a life of 10^9–10^{10} cycles. This is challenging with most industrial materials.

These mechanical, environmental and endurance needs challenge scientists to develop novel methods to cope with the requirements. Widely available liquid lubricants are inadequate for space applications because of their limited environmental stability and challenges imposed by low pressure or vacuum in the design of such systems. Materials with a layered structure (2D), soft metals and metal oxide-based nanocomposite coatings are used to reduce wear and friction in these applications. Transition metal dichalcogenides-based solid lubricants such as MoS_2, WS_2, $NbSe_2$ are the most extensively used tribo-coatings (Manu et al. 2021). However, the wear resistance of these pure solid lubricants is low and inadequate to handle extreme working conditions. To counteract this problem, self-lubricating composites, where the solid lubricants are mixed into a supporting matrix, with nanoparticle dispersion that provides improved wear and strength properties are developed. In most space applications, MoS_2 is the preferred and used solid lubricant. With the advancements in vacuum deposition techniques, it is now possible to manufacture nanocomposites with greater control over chemistry, structure, morphology and thickness. This helps in creating tribo-coatings with more reliability; over a broader working range.

For composite materials, the size of the reinforcement particles greatly influences properties. With an increase in solid lubricant particle size (micro-sized); the friction between the sliding surfaces decreases, but it also reduces the strength. The reduction of strength is due to the weakening of matrix-filler interfacial strength. Nanosized particles possess a large surface area at the fillers-matrix interface, resulting in enhanced bonding (Wetzel et al. 2003; Dasari et al. 2009). They also increase the nucleation capability and rigidity by restricted mobility of the matrix at the vicinity of the fillers. For polymer nanocomposites, material removal also decreases. This is due to the similar size of the nanofiller and surrounding polymer chains.

4.1   Applications of Polymer Nanocomposites in Aerospace

Polymeric materials such as PTFE, PEEK and PMMA are known for their low friction coefficient, however, they are subject to high wear. The challenge towards the use of polymer nanocomposite tribo-coatings is reducing the wear. The type of composite filler material influences the lubricating and wear properties. For instance, graphite, PTFE, molybdenum disulfide (MoS_2) decreases friction at the expense of films strength. Glass and carbon fibers fillers increase the strength but also increases the abrasiveness.

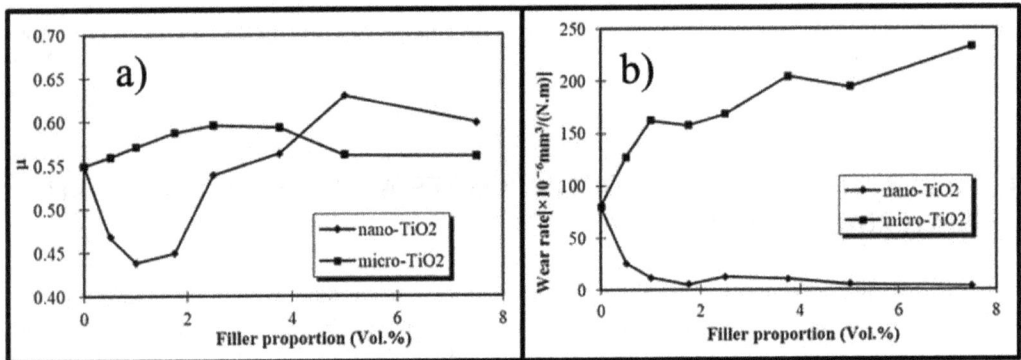

Figure 6.7   Variation in properties of nano- and micro-TiO_2-reinforced PPESK with different filler volume proportions: (a) coefficient of friction and (b) wear rate. (Reprinted with permission from Rathod et al. 2017. Polymer and ceramic nanocomposites for aerospace applications. *Applied Nanoscience* 7,8(2017): 519–548).

The size and concentration of fillers in the matrix also determines the tribo-performance. A filler's optimum concentration for a specific application depends on the process parameters and working conditions (Shao et al. 2004). Figure 6.7 shows the effect of the size and volume concentration of TiO_2 particles in the PPESK copolymer matrix. Beyond a filler concentration of 2 vol.% the friction coefficient drastically increased for nano-TiO_2, but the friction coefficient of micro-TiO_2 remained almost constant (~0.55). On the other hand, wear rate significantly decreased with nano-TiO_2 filler concentration, but for micro-TiO_2 a gradual increase in wear was observed with an increase in filler concentration. A filler concentration of ~2 vol.% nano-TiO_2 will provide the best tribo-performance for this film under these loads and working conditions. It is worth noting that nanofillers also influence the transfer films, by increasing their stability and reducing wear (Kong and Ashby 1992). The decrease in wear rate in nano-dispersed PPESK/TiO_2 composite films is due to the formation of stable tribo-films.

4.1.1   Application of Polymer Nanocomposites in Aerospace Bearings

The polymer nanocomposite coatings are extensively used in spacecraft and aircraft bearings. The properties of some nanocomposites used in these applications are shown in Table 6.3.

These nanocomposites are lightweight, possess low friction and wear rates. Polyimide/CNT nanocomposite shows the lowest wear rate. CNT fillers also impart thermal conductivity to the films and remove the heat built up during operation. Si_3N_4 addition to thermoplastic or thermosetting matrix provides hydrodynamic lubrication (Shi et al. 2003, Wang et al. 1996). This is due to tribo-chemical wear leading to the formation of SiO_2 in the tribo-films and thus protecting the sliding surfaces.

Table 6.3 Tribological properties of some polymer nanocomposites; (*Source:* Data from Rathod et al. 2017, "Polymer and ceramic nanocomposites for aerospace applications," *Applied Nanosci* 7,8 (2017): 519–548)

Material systems	Wear rate (10^{-6} mm^3/km)	Toughness (kJ/m^2)	Hardness (MPa)
Epoxy	38	9.3	19
Epoxy/4 vol.% TiO_2	15	46	—
Epoxy/0.9 vol.% nano-Si_3N_4	17	13.2	20
Epoxy/2 vol.% Al_2O_3	3.9	55	—
Epoxy/3 vol.% SiO_2	22	—	208
Polyimide/10 wt.% CNT	0.1	—	380
Polyimide/10 wt.% graphite	0.7	—	300
PEEK/7.5 wt.% Si_3N_4	1.3	—	133
PEEK/7.5 wt.% SiO_2	1.4	—	—
PEEK/2.5 wt.% SiC	3.4	—	—
PEEK/10 wt.% ZrO_2	3.9	—	—

4.1.2 Application of Epoxy-based Nanocomposites in Aircraft Body

Epoxy nanocomposites are becoming more popular than conventional fiber-reinforced composites. Their excellent properties include ease of fabrication, low filler requirement and low cost of manufacture (Prusty et al. 2017). Fillers such as silica, aluminum oxide, titanium oxide, nano clay, carbon nanotubes and graphene have been used to enhance various properties; with carbon nanotubes (CNTs) and Graphene (Ga) mixed with epoxy nanocomposites being popular. Ga has a large surface area that helps in increasing the interfacial adhesion. On the other hand, CNTs possess a high aspect ratio that both helps in better interfacial strength between filler and matrix. Ga/CNTs/epoxy hybrid nanocomposite is the go-to material for the manufacture of control surfaces of sub-sonic fixed-wing aircraft (King et al. 2013).

4.2 Ceramic Nanocomposites in Aerospace Tribology

Ceramics possess high chemical and temperature stability. Hence, they are widely used as a matrix material in tribo-applications in space programs. Solid lubricants and nano-fillers are usually dispersed in them to improve the lubricating and strength properties of these composites. Diamond-Like-Carbon (DLC) fillers are highly considered in space exploration applications. However, under high load or speed, graphitization of DLC takes over time, thus depleting its performance. Hydrogenated DLC is known to increase the service life, hydrogen desorption again leads to deteriorated performance. Dichalcogenide and carbides incorporated DLC (carbide/DLC/dichalcogenide composite) coating is known to improve its service life. Nano-crystalline TiC, WC, WS_2 and laser processed MoS_2 reservoirs in amorphous DLC coating demonstrated an order of magnitude improvement in toughness. Single-phase carbides with DLC addition showed a low friction coefficient in cycling from dry to humid environments, with a long life in both terrestrial and space environments.

Advanced vacuum deposition techniques are widely used in nanocomposite manufacture. It allows the creation of hard and lubricious phases in nanocomposite coatings with precise control of morphology, structure, chemistry. Vacuum deposition processes are widely employed for producing MoS_2, WS_2, $(NbSe_2)$-based nanocomposites. MoS_2 or WS_2 embedded in carbide/DLC/WC dichalcogenide composite lasts more than a million cycles in a space environment with humid/dry cycling. Ceramic-based adaptive PbO/MoS_2, ZnO/MoS_2 and ZnO/WS_2 nanocomposites are used in applications involving large fluctuation in temperature.

4.2.1 Applications of Ceramic Nanocomposites for Propulsion and Exhaust Washed Structures

For propulsion and exhaust-washed structures, titanium alloys or nickel-based superalloys were predominantly used in the past. Recently, the prospects of ceramics and their composites have gained significant attention among engine manufacturers. Their high-temperature stability, less cooling air requirements and weight reduction capabilities have resulted in this interest. Among various ceramics available for these applications' silicon carbide and alumina-fibre reinforced aluminosilicate matrix composite have found commercial applications. Alumina-fiber reinforced aluminosilicate recently replaced traditional materials for center body and exhaust nozzle in commercial aircraft. Figure 6.8 shows an exhaust ground test demonstrator of a 1.60-m-diameter nozzle and 1.14 m diameter by 2.34 m conical center body made of metal oxide-based ceramic matrix composites.

Figure 6.8 Oxide ceramic matrix composite exhaust ground test demonstrator consisting of a 1.60-m-diameter nozzle (reveal showing the underneath side of the titanium-alloy faring) and 1.14 m diameter by 2.34 m conical center body with titanium end cap inspection portal. (Reprinted with permission from: Steyer 2013. Shaping the future of ceramics for aerospace applications. *Int. J. App. Cer. Tech.* 10,3(2013): 389–394).

4.2.2 Applications of Ceramic Nanocomposites for Drag Reduction

Hypersonic flights have a sharp, wedge-shaped profile for reducing air friction. These vehicle's noses and leading edges are subjected to high shear and temperatures. For these super and hypersonic applications, materials with superior thermal and shear properties are required. Ultra-High-Temperature Ceramics (UHTCs) are a promising class of materials for these vehicles (Opeka et al. 2004; Fahrenholtz et al. 2007). The refractory nature of carbides, borides and nitrides makes them attractive candidates for the high heat flux, heat load and shear resistance. Particulates, whiskers, chopped fibers and continuous fibers reinforcements are used to improve the mechanical properties of these composites for high-temperature drag reduction applications.

4.2.3 Ceramic Nanocomposites Coatings for High Contact Load and Velocity Applications

The hardness of the surface is greatly influential in extreme temperature and contact pressure tribo-applications. Ductile materials working under these conditions are coated with nanocomposites of high-yield strength and fracture toughness. Usually, these nanocomposite coatings consist of crystalline nanoparticles dispersed in an amorphous matrix. When crystalline nanoparticles are dispersed in an amorphous matrix, the volume of matrix-particle grain boundaries increases. This limits the size of the initial crack and terminates crack propagation. These nanocomposites also possess good ductility, credited to an increase in grain boundary diffusion and grain boundary sliding. Equiaxial grain shapes, high angle grain boundaries, low surface energy and the presence of an amorphous boundary phase facilitate grain boundary sliding and hence ductility. The addition of solid lubricant into the matrix also promotes the self-lubricating property. Examples of such coatings used in high contact load or velocity applications include nanocrystalline carbides dispersed in an amorphous DLC matrix such as TiC/DLC and WC/DLC. Yttria-Stabilized Zirconia (YSZ) nanocrystals grains in a mixed YSZ–Au amorphous matrix. The increase in ductility, toughness and crack termination due to an increase in interfacial area is the underlying mechanism responsible for its superior performance.

4.3 Ceramic Adaptive Solid Lubricants

The drawback of solid lubricants is their sensitivity to the working conditions; most solid lubricants can perform efficiently on a relatively single environmental condition. Adaptive lubrication is used for tribo applications that work on a broad range of operating conditions. These nanocomposites provide lubrication by changing surface chemistry and microstructures with operating conditions (Muratore and Voevodin 2009). Adaptive lubricants are multicomponent lubricant composites, where different lubricants provide lubrication corresponding to different anticipated environmental conditions. They also possess high hardness and toughness, imparted mostly by nanocrystallites, amorphous-oxides and carbides (Sliney 1979). An insight into dry/humid adaptive and broad temperature adaptive solid lubricating nanocomposites are given next.

4.3.1 Dry/Humid Adaptive Solid Lubricating Nanocomposites

Aerospace components are subject to humid environments before launch. Hence for these applications, the solid lubricants need to have both dry and humid condition functionality. TMDs function better in vacuum environments, on the other hand, DLC, exhibits low friction in humid environments. $WC/DLC/WS_2$ nanocomposite films produced by magnetron-assisted pulsed laser deposition were one of the first solid lubricants developed for dry/humid compatibility (Voevodin et al. 1999). A graphite-like transfer film provided lubrication in humid conditions and also sealed the WS_2 dispersed particles from oxidation. As the condition changes to vacuum or dry air conditions, graphite particles wear off exposing the WS_2; which crystallizes and reorient to provide lubrication (Voevodin et al. 2000). WC nanoparticles provide the required hardness and toughness to the film demanded by the application need. These films provided an extremely low COF of 0.007 in dry nitrogen with a film hardness of 7–8 GPa. The mechanism of adaptive lubrication for $WC/DLC/WS_2$ nanocomposite films is shown in Fig. 6.9.

Other nanocomposites developed for adaptive dry/humid applications include WS_2/amorphous carbon (Cao et al. 2018), C–Ti/MoS$_2$ (Li et al. 2020), Mo–Se–C (Vuchkov et al. 2020) and Al_2O_3/DLC/Au/MoS$_2$. In the Al_2O_3/DLC/Au/MoS$_2$ system, COFs of 0.02–0.03 in dry nitrogen and 0.1–0.15 in humid air under 100,000 cycles varying environments were observed (Baker et al. 2006). A MoS_2–Sb_2O_3–C coating produced by the plasma electrolytic oxidation process showed a film hardness of 11–12 GPa with COFs of 0.10–0.15 in humid air and 0.06–0.09 in dry nitrogen (Liu et al. 2018).

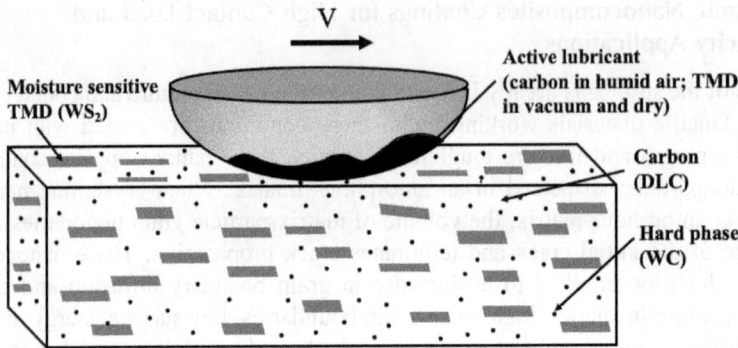

Figure 6.9 The mechanism of reversible dry/humid adaptive lubricating materials. (Reprinted with permission from: Gong et al. 2020. Intelligent lubricating materials: A review. *Composites Part B: Eng.* (2020): 1078450 (1–18)).

4.3.2 Broad Temperature Adaptive Solid Lubricating Nanocomposites

Temperature adaptive solid lubricants are used in high-temperature engines where the temperature varies from room to ~900°C (Sliney 1979). Soft metals, TMD's and metal oxides and fluorides are known to perform effectively in high-temperature applications (Torres et al. 2018). Creating adaptive composites that provide lubrication both at room temperature and high temperature is crucial for these applications. Examples of broad temperature adaptive solid lubricants nanocomposites include MoVN-Ag and steel/ZnO/MoS$_2$.

The MoVN–Ag adaptive lubricants are used in a temperature range of 25 to 700°C. An increase in silver decreases friction but also deteriorates the film's mechanical properties. The MoVN film with 45.6 at % Ag showed low COFs of about 0.19 and 0.28 at 500°C and 700°C respectively (Wang et al. 2020). For M50 steel/ZnO/MoS$_2$ composites good tribo-properties were observed in the 25 to 800°C range. In an experiment at 800°C (sliding speed = 0.2 m/s; load = 12 N) the COF of 0.17 and a wear rate of 2.88 × 10^{-5} mm^3/Nm were obtained for this nanocomposite (Essa et al. 2017). MoS$_2$ provides low-temperature lubrication. At temperatures above 400°C, ZnO and the formation of the ZnMoO$_4$ complex improves the tribological properties. A 10 wt.%, TiO$_2$ and 5 wt.% graphene dispersed in steel matrix gave good antifriction/wear properties from 25 to 450°C (Ali and Xianjun 2019).

Most broad temperature adaptive lubricants undergo irreversible reactions with temperature increase. Soft metals diffuse out of the coating after heating, whereas, medium-temperature lubricants undergo oxidation at high temperatures. Multi-layer diffusion barrier deposition has been shown to improve reusability in some temperature adaptive lubricants (Muratore and Voevodin 2009). They were effective in maintaining lubrication over multiple high/low-temperature cycles. For instance, a TiN barrier layer with a random array of pinholes was able to limit the diffusion of noble metals in TiN/YSZ-Ag-Mo coatings. (Muratore and Voevodin 2007). Annealed chromium oxide coating with self-assembled mesh-like heave structures helped in making reusable nanocomposite coating in a temperature range from 25 to 1000°C (He et al. 2017).

4.3.3 'Chameleon' Adaptive Solid Lubricating Nanocomposites

Chameleon coatings are reversible self-adjusting coatings where the tribo-properties are dependent on the applied load and the operating environment. They usually consist of crystalline nanoparticles and solid lubricant particles dispersed in the amorphous matrix for improved tribo and strength properties. Chameleon coatings are designed to provide lubrication over a wider range of working conditions by changing the mechanism of transfer film generated based on the operating conditions.

Figure 6.10 presents a schematic of a nanocomposite design for the fabrication of YSZ/Au/DLC/MoS$_2$ and WC/DLC/WS$_2$ chameleon coating on an amorphous matrix. Here crystalline nanoparticles (YSZ or WC) provide the required mechanical and strength properties. Au, MoS_2, and DLC provide chemical and structural adjustments for stable transfer films based on working conditions. Under dry air or nitrogen, MoS_2 or WS_2 transfer films provides lubrication, whereas DLC forms graphite-like carbon transfer films in humid air. Soft metals or low melting ceramics provide lubrication around the temperature range of 500–600°C.

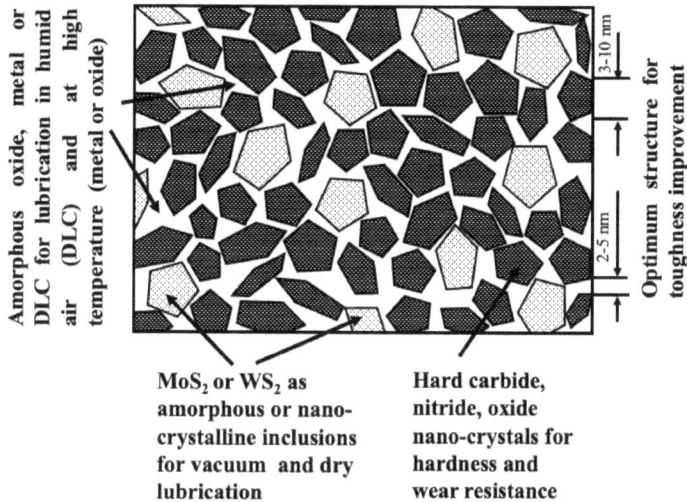

Figure 6.10   Schematic of a conceptual design of the $YSZ/Au/MoS_2/DLC$ tribological coating with chameleon-like surface adaptive behavior. (Reprinted with permission from: Voevodin and Zabinski 2005. Nanocomposite and nanostructured tribological materials for space applications. *Composites Sci. and Tech.* 65,5(2005): 741–748).

5   TRIBOLOGICAL APPLICATIONS OF NANOCOMPOSITES IN RADIATIVE ENVIRONMENTS

Nuclear energy is one of the most studied fields due to its importance in defence programs. Tribo materials in these applications are subject to both high temperature, vacuum pressure and radiations. Hence it is important to understand the effect of these parameters on the tribo-properties of materials used in these applications. Most liquid lubricants decompose around 350°C. They also possess high vapor pressure, restricting their use in high temperature and vacuum conditions.

In these conditions, nanocomposite coatings with solid lubricants, polymers or soft metals are widely used. Two-dimensional (2D) layered materials such as graphite, Hexagonal Boron Nitride (HBN) and transition metal dichalcogenides are used in radiative environments. Graphite is used in a vacuum (less than 10^{-6} torr) and up to temperatures of 550°C; MoS_2 provides lubrication up to 400 to 500°C (Huai et al. 2020).

Polymers such as polyimide, PE, nylon, PTFE; soft metals, metal salts and metal oxides-based nanocomposites have also been used in these applications. Carbon fiber reinforced polymer nanocomposites possess good thermal stability, wear resistance, effectiveness in a vacuum and nuclear environment. Soft metals and alkali, alkaline earth metal fluorides and lubricious metal oxides-based nanocomposites are used in high-temperature rolling contact applications. For instance, CaF_2, LiF, and BaF_2-based solid lubricants are used in a temperature range of 400 to 800°C, LaF_3 and CeF_3 are effective from 540 to 1000°C (Giltrow 1973).

Oxides such as PbO, MoO_3, K_2MoO_4, Co_2O_3 and $NiMoO_4$ also find application in the radiative environment (Sliney 1982). A combination of two oxides with nanoparticle inclusion helps in increasing the working range. For example, combining PbO and SiO_2 prevents PbO from oxidation and their composite is used for lubrication in a temperature range of 350 to 500 °C in air. Nanocrystalline diamond films are used for the lubrication of CRDM systems in nuclear reactors. The presence of diamond nano crystallites provides lubrication during dry sliding, whereas sp^3 C-rich grain boundaries help in irradiation tolerance (Podgursky et al. 2019; Peng et al. 2021).

6 APPLICATIONS OF NANOCOMPOSITES IN INDUSTRY

Nanocomposite coatings are widely used in industrial applications to reduce friction and wear of moving parts. The piston ring of the internal combustion engine has the most energy loss in an automobile. nc-TiC/a-C(Al) film on piston rings is widely used to improve the overall efficiency. CrAlN nanocomposite films have been shown to have high adhesion to stainless steel and are employed as tribo-coating in gears, turbine blades, cutting tools. Figure 6.11 shows different industrial components with nanocomposite coatings.

Figure 6.11 Images of several components coated with nanocomposite film.

The nc-(Al$_{1-x}$Ti$_x$) N/a-Si$_3$N$_4$ nanocomposite on metal cutting tools showed two to four times better performance compared with the conventional state-of-the-art (Al$_{1-x}$Ti$_x$)N. MoS$_2$/titanium composite (Jilek et al. 2004). It also showed an increase in tool life, reduction in milling force with an improvement of surface finish. Ceramic solid lubricants are widely used in engineering applications such as cylinder liner, piston ring, car engine cam and tappet (Nakamura and Hirayama 1989; Grzesik and Małecka 2011). Ceramics such as Al$_2$O$_3$ ceramics (Xu-Guo et al. 2017), ZrO$_2$ ceramics (Kong et al. 2012), SiC ceramics (Sharma et al. 2015; Zhang et al. 2020), Si$_3$N$_4$ ceramics have been extensively studied for their tribological properties. B$_4$C ceramics are used as wear-resistant coating mechanical seals, bearings, cutting tools, wheel dressing tools and blast nozzles (Maros et al. 2016). Compared with hard metals, B$_4$C ceramics are also used as abrasive water-jet nozzle materials when using hard abrasives such as Al$_2$O$_3$ (Schwetz et al. 1995).

Polymers are widely used as a coating on bearings and brakes to reduce friction. Pure polymers are prone to surface adhesion, shear and deformation; which affects their tribological performance (Myshkin et al. 2005). The addition of fillers has shown an improvement in the tribo-performance of these polymers (Myshkin et al. 2005; Pesetskii et al. 2008). Polymer nanocomposites with a thermoplastic matrix possess self-lubricating properties. For instance, PTFE impregnated with bronze nanocomposites coatings is widely used for dry film lubrication of bearings. Polyamides-based nanocomposites coatings are used for the lubrication of aluminum-based gears. Thermosetting polymers such as polyimides are used for high-temperature applications; mostly in the temperature range of 220–260°C. Polyformaldehyde, polycarbonate and polyacrylates are used in lubricating gears, bushings and sliding bearings. Epoxy and phenol resins with solid lubricant fillers can protect machine guide linings. Composites with thermosetting polymer matrices are used in the manufacture of brakes, clutches and other frictional units. Rubbers and polyurethanes are used to create abrasion-resistant linings on metal components. Polymer nanofilms, monolayers and grafted structures find applications in MEMS and NEMS systems.

SUMMARY

Nanocomposite coatings are extensively used in various applications to reduce wear and tear. They find applications in industries, medical and extreme conditions. Depending on the matrix material, nanocomposites can be divided into metal, ceramic or polymer-based. In industries, they are used as coatings on pistons, fasteners gear among others to improve the overall efficiency. They are also used as a superlubric coating on bio implants. The biocompatibility and chemical inertness make it a preferred material for various medical applications. Another major application of nanocomposite coating is in aeronautical and astronomical fields, where the components are subjected to radiations, high variations in temperatures and pressure. The high inertness to chemical reactants of ceramics and polymers are exploited in marine and chemical industries. On the other hand, metal matrix composites are extensively used in automotive industries to manufacture tribo-components that help in reducing weight.

References

Ali, M.M., S.G.S. Raman, S.D. Pathak and R. Gnanamoorthy. 2010. Influence of plasma nitriding on fretting wear behaviour of Ti–6Al–4V. Tribology International 43: 152–160.

Ali, M.K.A. and H. Xianjun, H. 2019. M50 matrix sintered with nanoscale solid lubricants shows enhanced self-lubricating properties under dry sliding at different temperatures. Tribology Letters 67: 1–16.

Baker, C.C., J.J. Hu and A.A. Voevodin. 2006. Preparation of Al_2O_3/DLC/Au/MoS_2 chameleon coatings for space and ambient environments. Surface and Coatings Technology 201: 4224–4229.

Cao, H., F. Wen, J.T.M. De Hosson and Y.T. Pei. 2018. Instant WS2 platelets reorientation of self-adaptive WS2/aC tribocoating. Materials Letters 229: 64–67.

Cobelli, N., B. Scharf, G.M. Crisi, J. Hardin and L. Santambrogio. 2011. Mediators of the inflammatory response to joint replacement devices. Nature Reviews Rheumatology. 7: 600–608.

Dante, R.C. and C.K. Kajdas. 2012. A review and a fundamental theory of silicon nitride tribochemistry. Wear 288: 27–38.

Dasari, A., Z.Z. Yu and Y.W. Mai. 2009. Fundamental aspects and recent progress on wear/scratch damage in polymer nanocomposites. Materials Science and Engineering: R: Reports 63: 31–80.

Essa, F.A., Q. Zhang, X. Huang, M.K.A. Ali, A. Elagouz and M.A. Abdelkareem. 2017. Effects of ZnO and MoS_2 solid lubricants on mechanical and tribological properties of M50-steel-based composites at high temperatures: Experimental and simulation study. Tribology Letters 65: 1–29.

Fahrenholtz, W.G., G.E. Hilmas, I.G. Talmy and J.A. Zaykoski. 2007. Refractory diborides of zirconium and hafnium. Journal of the American Ceramic Society 90: 1347–1364.

Ferguson, S.J., J.M. Visser and A. Polikeit. 2006. The long-term mechanical integrity of non-reinforced PEEK-OPTIMA polymer for demanding spinal applications: experimental and finite-element analysis. European Spine Journal 15: 149–156.

Freed, L.E., G. Vunjak-Novakovic, R.J. Biron, D.B. Eagles, D.C. Lesnoy and R. Langer. 1994. Biodegradable polymer scaffolds for tissue engineering. Biotechnology 12: 689–693.

Fu, Y., N.L. Loh, A.W. Batchelor, D. Liu, X. Zhu, J. He, et al. 1998. Improvement in fretting wear and fatigue resistance of Ti–6Al–4V by application of several surface treatments and coatings. Surface and Coatings Technology 106: 193–197.

Gallagher, E.A., S. Lamorinière and P. McGarry. 2018. Multi-axial damage and failure of medical grade carbon fibre reinforced PEEK laminates: Experimental testing and computational modelling. Journal of the Mechanical Behavior of Biomedical Materials 82: 154–167.

Giltrow, J.P. 1973. Friction and wear of some polyimides. Tribology 6: 253–257.

Golden, P.J., A. Hutson, V. Sundaram and J.H. Arps. 2007. Effect of surface treatments on fretting fatigue of Ti–6Al–4V. International Journal of Fatigue 29: 1302–1310.

Gong, H., C. Yu, L. Zhang, G. Xie, D. Guo and J. Luo. 2020. Intelligent lubricating materials: A review. Composites Part B: Engineering 108450: 1–18.

Gonzalez, P., J. Serra, S. Liste, S. Chiussi, B. Leon and F.M. Varela-Feria. 2003. New biomorphic SiC ceramics coated with bioactive glass for biomedical applications. Biomaterials 24: 4827–4832.

Grzesik, W. and J. Małecka. 2011. Documentation of tool wear progress in the machining of nodular ductile iron with silicon nitride-based ceramic tools. CIRP Annals 60: 121–124.

Haleem, A. and M. Javaid. 2019. Polyether ether ketone (PEEK) and its 3D printed implants applications in medical field: An overview. Clinical Epidemiology and Global Health, 7: 571–577.

Halma, J.J., J. Senaris, D. Delfosse, R. Lerf, T. Oberbach and A. de Gast. 2014. Edge loading does not increase wear rates of ceramic-on-ceramic and metal-on-polyethylene articulations. Journal of Biomedical Materials Research Part B: Applied Biomaterials 102: 1627–1638.

He, N., H. Li, L. Ji, X. Liu, H. Zhou and J. Chen. 2017. Reusable chromium oxide coating with lubricating behavior from 25 to 1000°C due to a self-assembled mesh-like surface structure. Surface and Coatings Technology 321: 300–308.

Hu, T., L. Hu and Q. Ding. 2012. Effective solution for the tribological problems of Ti-6Al-4V: Combination of laser surface texturing and solid lubricant film. Surface and Coatings Technology 206: 5060–5066.

Huai, W., C. Zhang and S. Wen. 2020. Graphite-based solid lubricant for high-temperature lubrication. Friction 9, 1660–1672.

Jilek, M., T. Cselle, P. Holubar, M. Morstein, M.G.J. Veprek-Heijman and S. Veprek. 2004. Development of novel coating technology by vacuum arc with rotating cathodes for industrial production of nc-$(Al_{1-x}Ti_x)$ N/a-Si_3N_4 superhard nanocomposite coatings for dry, hard machining. Plasma Chemistry and Plasma Processing 24: 493–510.

King, J.A., D.R. Klimek, I. Miskioglu and G.M. Odegard. 2013. Mechanical properties of graphene nanoplatelet/epoxy composites. Journal of Applied Polymer Science 128: 4217–4223.

Kong, H. and M.F. Ashby. 1992. Wear mechanisms in brittle solids. Acta Metallurgica et Materialia 40: 2907–2920.

Kong, L., Q. Bi, S. Zhu, J. Yang and W. Liu. 2012. Tribological properties of ZrO_2 (Y_2O_3)–Mo–BaF_2/CaF_2 composites at high temperatures. Tribology International 45: 43–49.

Li, J., X. Zeng, T. Ren and E. Van der Heide. 2014. The preparation of graphene oxide and its derivatives and their application in bio-tribological systems. Lubricants 2: 137–161.

Li, L., Z. Lu, J. Pu, H. Wang, Q. Li, L. Wang, et al. 2020. The superlattice structure and self-adaptive performance of C–Ti/MoS_2 composite coatings. Ceramics International 46: 5733–5744.

Liu, Y.F., T. Liskiewicz, A. Yerokhin, A. Korenyi-Both, J. Zabinski and A.A. Voevodin. 2018. Fretting wear behavior of duplex PEO/chameleon coating on Al alloy. Surface and Coatings Technology 352: 238–246.

Manu, B.R. and A.H. Jayatissa. 2020. Effect of humidity on friction of molybdenum disulfide films produced by thermal evaporation on titanium substrates. Chemical Science International Journal 29: 25–33.

Manu, B.R., A. Gupta and A.H. Jayatissa. 2021. Tribological properties of 2D materials and composites—A review of recent advances. Materials 14: 1630.

Maros, M., A.K. Németh, Z. Károly, E. Bódis, Z. Maros, O. Tapasztó and K. Balázsi. 2016. Tribological characterisation of silicon nitride/multilayer graphene nanocomposites produced by HIP and SPS technology. Tribology International 93: 269–281.

Moseke, C. and U. Gbureck. 2010. Tetracalcium phosphate: Synthesis, properties and biomedical applications. Acta Biomaterialia 6: 3815–3823.

Mota, R.C.D.A.G., E.O. da Silva and L.R. de Menezes. 2018. Polymer nanocomposites used as scaffolds for bone tissue regeneration. Materials Sciences and Applications 9: 679–697.

Muratore, C., J.J. Hu and A.A. Voevodin. 2007. Adaptive nanocomposite coatings with a titanium nitride diffusion barrier mask for high temperature tribological applications. Thin Solid Films 515: 3638–3643.

Muratore, C. and A.A. Voevodin. 2009. Chameleon coatings: Adaptive surfaces to reduce friction and wear in extreme environments. Annual Review of Materials Research 39: 297–324.

Muratore, C., J.J. Hu and A.A. Voevodin. 2009. Tribological coatings for lubrication over multiple thermal cycles. Surface and Coatings Technology 203: 957–962.

Myshkin, N.K., M.I. Petrokovets and A.V. Kovalev. 2005. Tribology of polymers: Adhesion, friction, wear, and mass-transfer. Tribology International 38: 910–921.

Najeeb, S., M.S. Zafar, Z. Khurshid and F. Siddiqui. 2016. Applications of polyetheretherketone (PEEK) in oral implantology and prosthodontics. Journal of Prosthodontic Research 60: 12–19.

Nakamura, Y. and S. Hirayama. 1989. Wear tests of grey cast iron against ceramics. Wear 132: 337–345.

Nath, S., N. Sinha and B. Basu. 2008. Microstructure, mechanical and tribological properties of microwave sintered calcia-doped zirconia for biomedical applications. Ceramics International 34.1509–1520.

O'brien, F.J. 2011. Biomaterials & scaffolds for tissue engineering. Materials Today 14: 88–95.

Opeka, M.M., I.G. Talmy and J.A. Zaykoski. 2004. Oxidation-based materials selection for 2000 C+ hypersonic aerosurfaces: Theoretical considerations and historical experience. Journal of Materials Science 39: 5887–5904.

Panayotov, I.V., V. Orti, F. Cuisinier and J. Yachouh. 2016. Polyetheretherketone (PEEK) for medical applications. Journal of Materials Science: Materials in Medicine 27: 1–11.

Peng, J., C. Xiong, J. Liao, J. Liao, L. Yuan and L. Li. 2021. Study on the effect of Ar-containing work gas on the microstructure and tribological behaviour of nanocrystalline diamond coatings. Tribology International 153: 106667.

Pesetskii, S.S., S.P. Bogdanovich and N.K Myshkin. 2008. Tribological behaviour of polymer nanocomposites produced by dispersion of nanofillers in molten thermoplastics. Tribology and Interface Engineering Series 55: 82-107.

Podgursky, V., A. Bogatov, M. Yashin, M. Viljus, A.P. Bolshakov and V. Ralchenko. 2019. A comparative study of the growth dynamics and tribological properties of nanocrystalline diamond films deposited on the (110) single crystal diamond and Si (100) substrates. Diamond and Related Materials 92: 159–167.

Prusty, R.K., D.K. Rathore, S. Sahoo, V. Parida and B.C. Ray. 2017. Mechanical behaviour of graphene oxide embedded epoxy nanocomposite at sub-and above-zero temperature environments. Composites Communications 3: 47–50.

Rahmitasari, F., Y. Ishida, K. Kurahashi, T. Matsuda, M. Watanabe and T. Ichikawa. 2017. PEEK with reinforced materials and modifications for dental implant applications. Dentistry Journal 5: 35.

Ramli, M.S., M.S. Wahab, M. Ahmad and A.S. Bala. 2016. FDM preparation of bio-compatible UHMWPE polymer for artificial implant. ARPN Journal of Engineering and Applied Sciences 11: 5473–5480.

Rathod, V.T., J.S. Kumar and A. Jain. 2017. Polymer and ceramic nanocomposites for aerospace applications. Applied Nanoscience 7: 519–548.

Salgueiredo, E., M. Vila, M.A. Silva, M.A. Lopes, J.D. Santos, M.H. Fernandes, et. al. 2008. Biocompatibility evaluation of DLC-coated Si_3N_4 substrates for biomedical applications. Diamond and Related Materials 17: 878–881.

Scholes, S.C. and A. Unsworth. 2009. Wear studies on the likely performance of CFR-PEEK/CoCrMo for use as artificial joint bearing materials. Journal of Materials Science: Materials in Medicine 20(1): 163.

Schwetz, K.A., L.S. Sigl, J. Greim and H. Knoch. 1995. Wear of boron carbide ceramics by abrasive waterjets. Wear 181–183: 148–155.

Shao, X., W. Liu and Q. Xue. 2004. The tribological behavior of micrometer and nanometer TiO_2 particle-filled poly (phthalazine ether sulfone ketone) composites. Journal of Applied Polymer Science 92: 906–914.

Sharma, S.K., B.V.M. Kumar and Y.W. Kim. 2015. Effect of WC addition on sliding wear behavior of SiC ceramics. Ceramics International 41: 3427–3437.

Shi, G., M.Q. Zhang, M.Z. Rong, B. Wetzel and K. Friedrich. 2003. Friction and wear of low nanometer Si_3N_4 filled epoxy composites. Wear 254: 784–796.

Shi, L., Z.G. Guo and W.M. Liu. 2015. The recent progress of tribological biomaterials. Biosurface and Biotribology 1: 81–97.

Shen, G., F. Fang and C. Kang. 2018. Tribological performance of bio implants: A comprehensive review. Nanotechnology and Precision Engineering 1: 107–122.

Sliney, H.E. 1979. Wide temperature spectrum self-lubricating coatings prepared by plasma spraying. Thin Solid Films 64: 211–217.

Sliney, H.E. 1982. Solid lubricant materials for high temperatures—A review. Tribology International 15: 303–315

Soppimath, K.S., T.M. Aminabhavi, A.R. Kulkarni and W.E. Rudzinski. 2001. Biodegradable polymeric nanoparticles as drug delivery devices. Journal of Controlled Release 70: 1–20.

Steyer, T.E. 2013. Shaping the future of ceramics for aerospace applications. International Journal of Applied Ceramic Technology 10: 389–394.

Torres, H., M. Rodríguez Ripoll and B. Prakash. 2018. Tribological behaviour of self-lubricating materials at high temperatures. International Materials Reviews 63: 309–340.

Varela-Feria, F.M., J. Martínez-Fernández, A.R. de Arellano-Lopez and M. Singh. 2002. Low density biomorphic silicon carbide: microstructure and mechanical properties. Journal of the European Ceramic Society 22: 2719–2725.

Vaz, F., L. Rebouta, P. Goudeau, J. Pacaud, H. Garem, J.P. Riviere, et. al. 2000. Characterisation of Ti1−xSi$_x$N$_y$ nanocomposite films. Surface and Coatings Technology 133: 307–313.

Voevodin, A.A., J.P. O'neill and J.S. Zabinski. 1999. WC/DLC/WS2 nanocomposite coatings for aerospace tribology. Tribology Letters 6: 75–78.

Voevodin, A.A. and J.S. Zabinski. 2000. Supertough wear-resistant coatings with 'chameleon' surface adaptation. Thin Solid Films 370: 223–231.

Voevodin, A.A. and J.S. Zabinski. 2005. Nanocomposite and nanostructured tribological materials for space applications. Composites Science and Technology 65: 741–748.

Vuchkov, T., T.B. Yaqub, M. Evaristo and A. Cavaleiro. 2020. Synthesis, microstructural and mechanical properties of self-lubricating Mo-Se-C coatings deposited by closed-field unbalanced magnetron sputtering. Surface and Coatings Technology 394: 125889.

Wang, Q., J. Xu, W. Shen and W. Liu. 1996. An investigation of the friction and wear properties of nanometer Si$_3$N$_4$ filled PEEK. Wear 196: 82–86.

Wang, A., A. Essner and R. Klein, R. 2001. Effect of contact stress on friction and wear of ultra-high molecular weight polyethylene in total hip replacement. Proceedings of the Institution of Mechanical Engineers, Part H: Journal of Engineering in Medicine 215: 133–139.

Wang, W., F. Salvatore and J. Rech. 2020. Characteristic assessment and analysis of residual stresses generated by dry belt finishing on hard turned AISI52100. Journal of Manufacturing Processes 59: 11–18.

Werner, Z., M. Barlak, M. Grądzka-Dahlke, R. Diduszko, W. Szymczyk, K. Borkowska, et al. 2007. The effect of ion implantation on the wear of Co–Cr–Mo alloy. Vacuum 81: 1191–1194.

Wetzel, B., F. Haupert and M.Q. Zhang. 2003. Epoxy nanocomposites with high mechanical and tribological performance. Composites Science and Technology 63: 2055–2067.

Xu-Guo, H., C. Zi-Shang, L. Xiao-Ping and B. Sheng. 2017. Effect of sintering additives size on the microstructure and wear properties of Al$_2$O$_3$ ceramics. Ferroelectrics 521: 101–107.

Yilbas, B.S. and M.S.J. Hashmi. 2000. Laser treatment of Ti–6Al–4V alloy prior to plasma nitriding. Journal of Materials Processing Technology 103: 304–309.

Zafar, M.S. 2020. Prosthodontic applications of polymethyl methacrylate (PMMA): An update. Polymers 12: 2299.

Zhang, W., S. Yamashita and H. Kita. 2020. Progress in tribological research of SiC ceramics in unlubricated sliding–A review. Materials & Design 190: 108528.

Chapter 7

Fiber Reinforced Polymer and Recycling using Biomaterials

Keiichiro Sano

Department of Symbiotic Design, College of Inter-human Symbiotic Studies,
Kanto Gakuin University, 1-50-1, Mutsuura-higashi, Kanazawa, Yokohama 236-8503, Japan

E-Mail: keisano@kanto-gakuin.ac.jp

1   INTRODUCTION

In this paper, the technological development and practical application of carbon fiber reinforced plastics, natural fiber-reinforced plastics and GFRP recycling in Japan are described. First, the Carbon Fiber Reinforced Polymer (CFRP) parts having high-strength and lightweight using carbon fiber and epoxy resin were developed and its practical use expanded. The CFRP has been used in airplanes and automobiles improving safety and fuel efficiency by reducing weight. Second, to reduce the consumption of petroleum resources, Natural Fiber Reinforced Polymer (NFRP) using natural fiber of reinforcement and Polylactic Acid (PLA) of base polymer made from biomass was developed for German and Japanese car parts. The authors explain in detail the feedstock recycling of GFRP using vegetable oil as the solvent for polymer dissolution, which was developed in collaboration with Japanese industries. Thermal recovery such as recycling, and landfill disposal are the mainstream for disposal of GFRP waste in Japan. In the future, the material selection of FRP composite and the recycling methods will contribute to the maintenance of environmental sustainability.

2   FIBERS AND RECYCLING OF FIBER REINFORCED POLYMER (FRP)

Harvesting and conversion of fibers originating from plants or animals into clothing for humans are one of the oldest crafts of humankind. During development, especially during the industrialization of the 19th and 20th centuries, not only natural fibers were used but natural raw materials were chemically converted, e.g., cellulose, and later also by using polymer composites. Parallel to the manufacturing processes new application areas developed—other than in clothing textiles—

especially the area of home and household, automotive, and currently the area of technical textiles (Reate Luzkendorf 2009).

In Germany, flax has been cultivated in large quantities for a long time, and the oil extracted from seeds has been used for food and industrial raw materials. Natural fibers made by processing the surplus flax stalks have been used for daily necessities, furniture and industrial products. After World War II, in East Germany, natural fibers of flax and cotton were used as the inner material of the seat in the Trabant of domestic cars, and NFRP of flax fiber and phenol resin was used as the outer panel of the car body.

Even after the unification of East and West Germany, the ancient industrial technology of flax fiber continues to be utilized in the parts of the latest German cars to protect the global environment.

In Japan, there has long been an industry in which cultivated hemp is processed and used for clothing. However, after World War II, Japan discontinued its natural fiber industry and switched to a chemical fiber industry using petroleum. On the other hand, in Japan, the industry of processing used cotton clothing and recycling it into industrial felt continues even after the war. This is a rare case. Felt made from used natural fibers continues to be used in the latest Japanese car parts.

The types of industrial fibers are roughly classified into natural fibers and man-made fibers. From ancient times, fiber materials have been further selected from both fibers in consideration of product performance and cost and used in products. From the end of the 20th century to the present, the types of fibers and base materials have begun to be selected in consideration of product performance and the environment. The types of industrial polymer are classified into petroleum-based resin and biomass base resin. In the distant future, FRP using both fiber and polymer materials made from biomass must be the most environmentally friendly composite.

2.1 Classification and Application of Fibers

Textile materials today are subject to different classifications criteria. They are classified according to the kind of textile fibers, the spinning process, the fiber geometry, the construction of a textile surface structure or also of their application areas. Figure 7.1 shows as an example the classification of fiber types. The world production of all textile fibers was 70.6 million tons in 2007. Of that 63% were chemical fibers, 35% cotton and 2% wool. The USA produced 8% of the chemical fibers, West Europe 7%, Japan 3% and 82% were produced in other parts of the world, especially in China, India, and South America (Reate Luzkendorf 2009).

In recent years, research and development of high-strength nanocomposites prepared cellulose of wood and crop have been active.

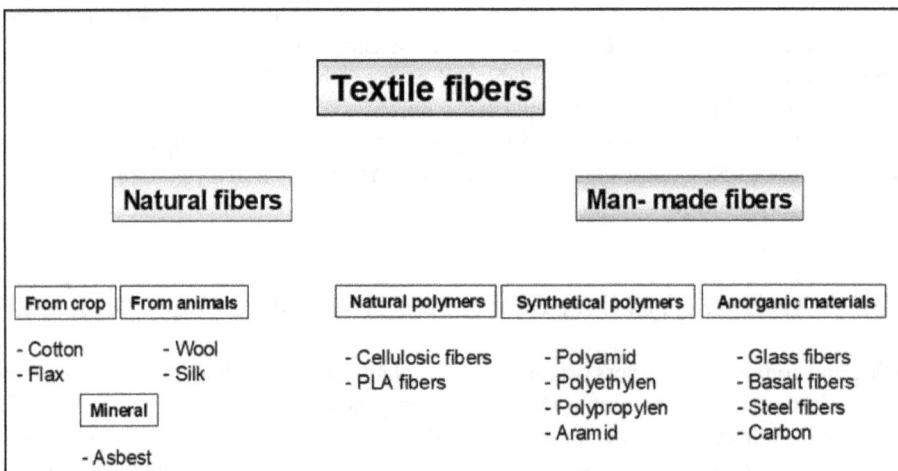

Figure 7.1   Classification of fiber types (Renate Luzkendorf 2009).

2.2   Materials and Recycling Methods for Fiber Reinforced Polymer

Figure 7.2 shows the theory of FRP material design. Conventionally, when manufacturing FRP products, the types of fibers and resins are selected in consideration of cost and performance such as strength, lightness and durability. In the future, it will be important to select fibers and resins in consideration of the environmental impact and recycling methods.

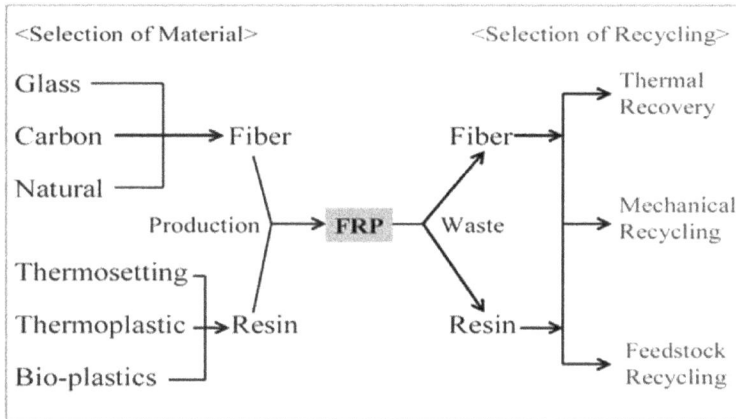

Figure 7.2   Fundamental theory to design Fiber Reinforced Polymer (FRP) for environment.

In summary, it is clear that textile fiber material and FRP over the centuries and especially in the age of industrialization has gained entrance into a multitude of areas of daily life. There is a steady increase in the number of fibers produced worldwide as well as fiber types and their appearance. The development of appropriate strategies for recycling and reuse is now all the more urgent.

3   DEVELOPMENT AND RECYCLING OF FIBER REINFORCED POLYMER (FRP)

Many developments and practical uses of new FRP products are proceeding from the viewpoint of lightweight, toughness and durability in Japan. Simultaneously, the research and the realization of FRP recycling are important for the environment. However, these recycling efforts are not successful enough in all countries, including Japan. In this research, the trend of development and the practical use for new FRP in Japan were analyzed and the situation of FRP recycling was studied. Finally, the fundamental theory for the design and development of new FRP composite and FRP recycling has been proposed. In all countries, FRP products composed of UP thermosetting resin and glass fiber are used widely as the most general materials from before. The reason to use GFRP is its lightweight, high strength and durability, cheapness.

Recently, the development and practical use of the CFRP which used carbon fiber and epoxy resin were examined in depth in Japan. In the U.S.A., earlier the CFRP was used for the large size parts of airplanes. In Japan, the application of CFRP for the automobile industry is developed actively by the influence of the U.S.A. The reason to use CFRP for cars is more lightweight and a thinner part to reach a high performance of fuel mileage and power.

The NFRP composed from natural fiber and thermoplastic had developed and used for a few car parts in Japan. In Germany, the development and use of NFRP are more active than in Japan. The cultivation and the industry of flax and hemp fiber are prospering in Germany. In NFRP, the use of plastic made from fossil oil can be decreased by adding natural fiber. The NFRP using bio-plastics have developed and used for a few car parts and electric appliances in Japan. In NFRP using bio-plastics, the use of plastic made from fossil oil can be decreased by adding bio-plastics. The last purpose is that consumption of plastic from fossil oil becomes zero.

The world is in an economic depression now, but the amount of FRP production will increase in the world by the normal economic growth of China and Indian. Recently in Japan, the total volume of FRP production was about 360 thousand tons for one year. And following it, the quantity of the FRP waste from a building or a transportation machine is about 400 thousand tons every year. It is predicted that the production of FRP will? slightly decreases by the recent global recession, but the waste from the past does not decrease for a long time.

From the old days, a lot of pleasure FRP boats from UP and glass were dumped illegally in water and bank of seas and lakes in Japan. The FRP waste of bathtubs and mannequins is disposed into landfills. Therefore, the Japanese government is now promoting FRP recycling. Furthermore, some companies are developing FRP recycling. However, all the FRP recycling does not spread in Japan from the cost and energy, infrastructure. The FRP recycling does not spread or succeed in the world either.

In this study, the important points for the development and recycling of FRP were investigated from the situation of FRP development and FRP recycling.

3.1   Situation of Development for Fiber Reinforced Polymer (FRP)

3.1.1   Carbon Fiber Reinforced Polymer (CFRP)

Toyota motor corp. exhibited two near-future cars of 1/X and LF-A using CFRP in the Tokyo motor show 2007. At first, the 1/X was light-weight to 420 kg from 1200 kg of a general model by using a lot of CFRP body parts. Figure 7.3 shows the appearance of 1/X.

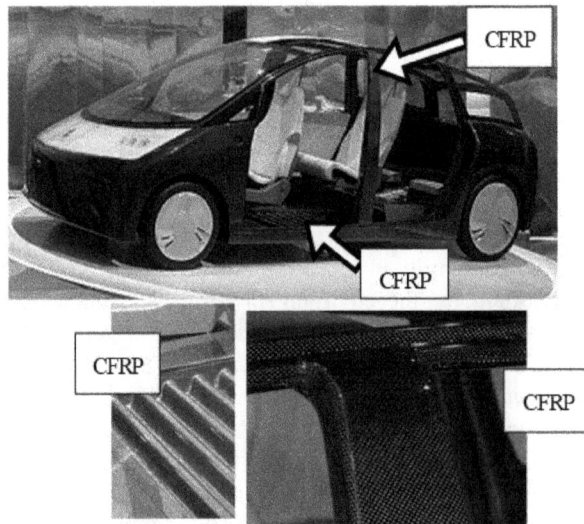

Figure 7.3   Toyota's 1/X of near future car using Carbon Fiber Reinforced Polymer.

The body of the light weight is very tough because it has the CFRP floor of rib shape. This car has an engine of a hybrid system. Hence, the fuel mileage of this car is very good (Tokyo motor show 2007). Next, the LF-A model has a lightweight body of CFRP and carries a V10 engine (Green car congress 2007). Therefore, this car is a very high-performance car like a Porsche. The CFRP can be used already for various products. On the other hand, cost reduction by productivity improvement and the development of recycling methods are required.

3.1.2 Natural Fiber Reinforced Polymer (NFRP)

All automotive companies in Germany use natural fiber composites for interior parts. One of the pioneers of this development was Mercedes Benz. In the example of the Mercedes-Benz S-Class, the use of natural materials can be demonstrated. Today the Mercedes-Benz S-Class contains 27 parts with a total weight of about 43 kg. In the earlier of this model, the amount of natural fiber composites was about 25 kg. The parts inside the car consist of different materials. In addition, a big NFRP part was used in Smart four two coupe of German car produced from 2007. Figutre 7.4. shows the instrumental panel of NFRP using flax fiber and polypropylene (PP). The use of NFRP in German cars is spreading (Reussmann and Luetzkendorf 2008).

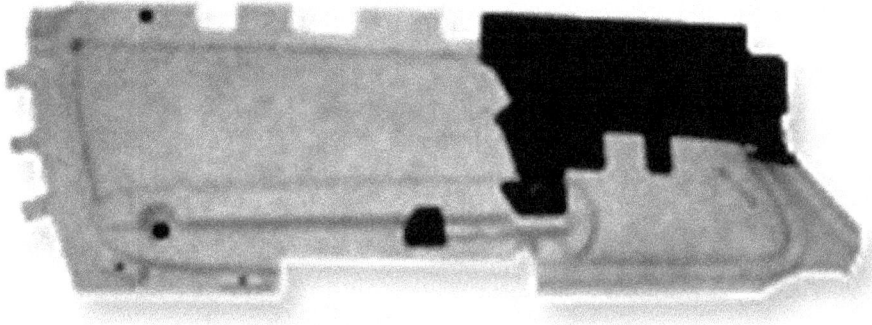

Figure 7.4 Instrumental panel for German car using Natural Fiber Reinforced Polymer (NFRP) (Faurecia S.A.).

In Toyota Celsior of a luxury sedan which begun production in 2000, the composite material of kenaf fiber and PP is used for the inner board of the door panel. Figure 7.5 shows the door inner using NFRP. Toyota had developed the roof lining which molded the mixture mat of carbon fiber and sisal fiber from Brazil by using a petrochemical binder in 2003. The roof lining was used for the Crown of a luxury sedan in Japan (Takeru 2014).

Figure 7.5 Inner side of door using Natural Fiber Reinforced Polymer (NFRP) of door of Toyota Lexus (Takeru, N. 2014).

In Germany, development and practical use for NFRP are more active than in Japan as was written above. The reason is a difference of country policy in the government and companies users. The recycling method of the NFRP will be demanded in the future. In this study, it had been found that the development and use of NFRP in Japan are difficult as compared with Germany because all the cultivation and industry of natural fiber discontinued their business after World War II. Furthermore, the recognition of the use of natural fiber is necessary as an environmental policy. In addition, there is a problem that the Japanese are sensitive to the original odor emitted from natural fiber. The problem of odor from NFRP should be improved in the future (Keiichiro Sano et al. 2010).

3.1.3   Natural Fiber Reinforced Polymer (NFRP) using Polylactic Acid (PLA)

In Raum of Toyota, the composite of kenaf fiber and PLA is already used for a cover board of the spare tire. However, the use of the NFRP using bio-plastic does not spread because of the bad smell from natural fiber and bio-plastics that are emitted. In Toyota's 1/X, all-natural material which molded PLA involving kenaf fiber and lamie fiber is used for roof lining (Green car congress 2007). Figure 7.6 shows the rooflining of NFRP using PLA. In this composite, more improvement of performance durability will be necessary for practical use in the future. The recycling method of the NFRP will also be demanded in the future. Furthermore, the improvement of the bad smell from natural fiber and bio-plastics in the composite will be necessary.

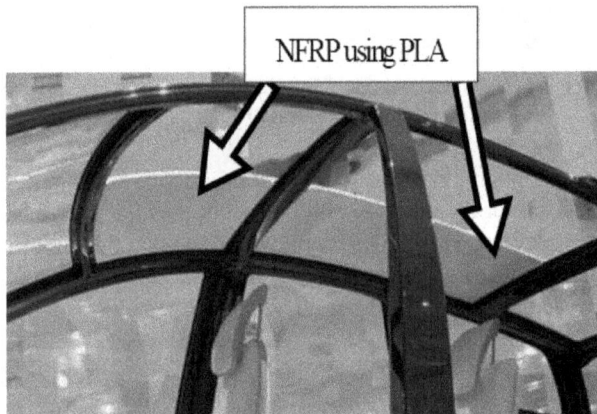

Figure 7.6   Roof lining of 1/X applied Natural Fiber Reinforced Polymer (NFRP) using Polylactic Acid (PLA) resin.

As another example, NEC and UNITIKA had developed the mobile telephone using improved PLA involving kenaf fiber, and sold it experimentally. Now, NEC and UNITIKA are improving moldability by using original additives (Topics of NEC).

3.2   Recycling of Glass Fiber Reinforced Polymer (GFRP) in Japan

3.2.1   Thermal Recovery in Cement Kiln

The Ministry of Land, Infrastructure, Transport and Tourism (MLIT) is acting to promote the appropriate treatment of end-of-life vehicles using GFRP (Glass Fibre Reinforced Polymer) and thus prevent illegal dumping with the implementation of the vehicle recycling system and the enforcement of the Road Transport Vehicle Law. Figure 7.7 shows the boat waste which was dumped on the shore. The ministry is researching the development of a recycling system for the FRP pleasure boats. Each year, the ministry plans to turn 10 thousand FRP boats into the cement kiln. Figure 7.8 shows the FRP boat recycling process.

Figure 7.7 Waste of Glass Fiber Reinforced Polymer (GFRP) boat on the shore.

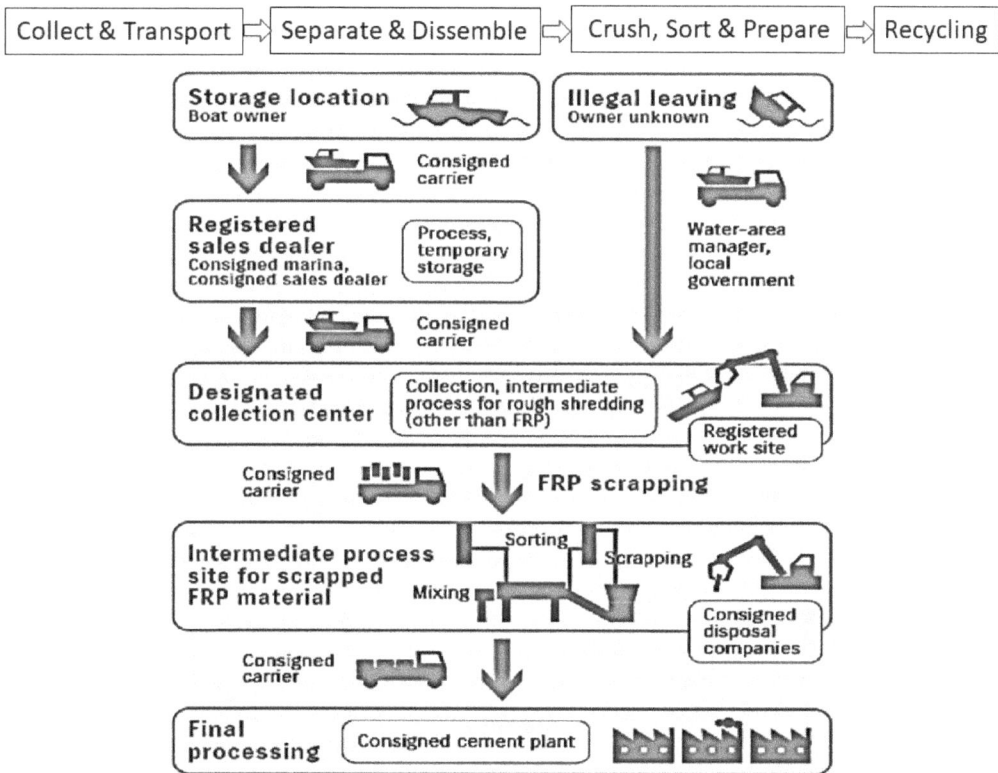

Figure 7.8 Recycling process of Glass Fiber Reinforced Polymer (GFRP) boat planned by the Japanese ministry.

Under the program, boat owners will be asked to bring the boat waste to a nearby shipyard. The FRP will be separated from other valuable materials. Then, the FRP goes through intermediate treatment at a waste-processing center. The FRP is crushed finely, mixed with waste oil, sent to the cement plant, incinerated in a kiln to obtain thermal energy and cement material.

In 2005, the recycling system operation was limited to two areas of Western Seto inland sea and the Northern Kyushu areas. The plan is continuing to expand areas covered nationwide. However, it does not advance as scheduled because there is the problem of infrastructure for transportation and the costs.

3.2.2 Mechanical Recycling of Glass Fiber

Hitachi Chemical had developed the recycling method which decomposed GFRP under the atmosphere by using benzyl alcohol solvent adding tripotassium phosphate. In this method, the long fibers for mechanical recycling can be obtained from decomposed waste. The reusable GFRP using the recovered fiber has a strength of about 70% as compared with the new original GFRP. The feedstock recycling for decomposed plastic is now expected .

3.2.3 Feedstock Recycling of GFRP Resin using Sub-critical Water

Panasonic Electric Works Ltd. had succeeded in the technology development of feedstock recycling based on sub-critical water for hydrolysis and regenerating FRP resin. Figure 7.9 shows the recycling process of GFRP by Panasonic. Sub-critical water can be made by high pressure and high-temperature conditions. Styrene-fumaric-acid copolymer, a functional polymer recovered from this process using this method, holds promise as an added polymer material in a wide spectrum of applications. The value of this functional polymer has been evaluated to be 5–10 times higher price than the market styrene and organic acid which are original components of the FRP resin. Furthermore, the new material exceeded the value of the original raw materials. Panasonic is conducting this research for the commercialization of this added functional polymer. The bench-scale process with 40 kg capacity can successfully perform sub-critical water hydrolysis and inorganic material separation. However, this method has a problem of high cost by high-pressure processing.

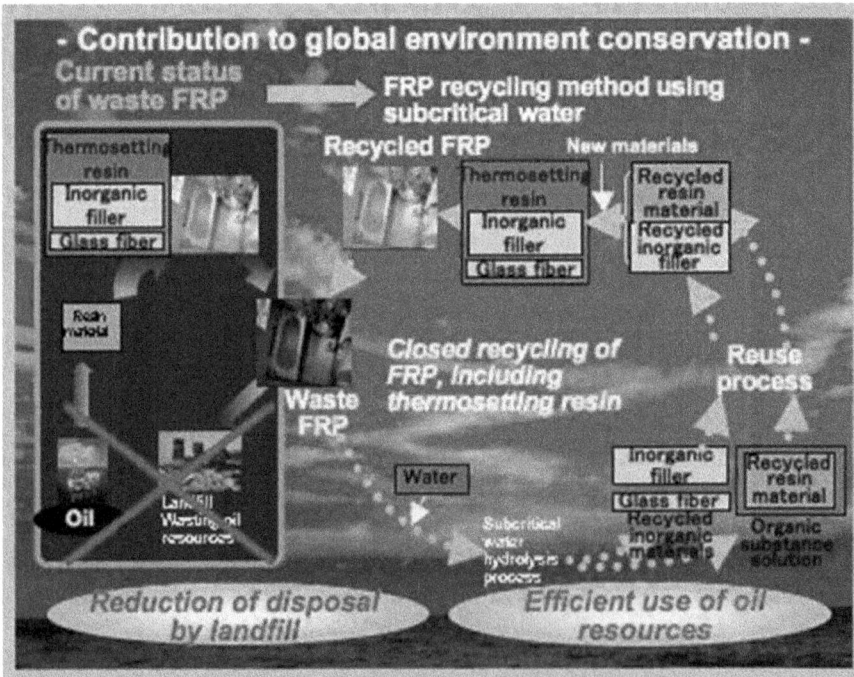

Figure 7.9   Recycling process of Glass Fiber Reinforced Polymer (GFRP) by Panasonic Electric Works (Panasonic Electric Works 2008).

GFRP has been widely used since ancient times. The use of GFRP is increasing from the viewpoint of product weight reduction. In addition, the use of NFRP and PLA will gradually increase from the viewpoint of environmental protection. Furthermore, various recycling methods for GFRP and CFRP are required, and it is necessary to develop their technologies.

4 DEVELOPMENT OF GLASS FIBER REINFORCED POLYMER (GFRP) RECYCLING USING VEGETABLE OIL

It is difficult to recycle GFRP waste because the UP resin of thermosetting resin and inorganic fibers of filler is used. In this study, a new thermal degradation method of FRP waste in vegetable oil was examined for recycling. In this method, the UP degradation and the separation of glass fiber from degraded UP were achieved very simply. Further, it has been found that ortho-phthalic acid of UP raw material was precipitated in degradation oil when FRP degraded under sealed pressurization, on the other hand, the needle-shaped crystal which mixed the ortho-phthalic acid and the phthalic anhydride was generated when the vaporized gases were cooled at the FRP degradation reaction under atmospheric pressure. These decomposition materials extracted from the UP degradation reaction have a possibility of being chemically recyclable. Some fundamental data of FRP degradation and the new recycling process were investigated.

The FRP is used in various structural materials including bathtubs, pleasure-boat, trains and airplanes due to its lightweight, fastness and corrosion resistance. However, an epoch-making technology of FRP recycling has not been invented in the world, as many of the FRP comprises thermosetting resin such as UP or epoxy resin and contain glass fiber or carbon fiber. Most FRP wastes are treated by means of incineration or landfill processes which inflict damage on the environment. At the present time, some studies have been reported for FRP recycling such as liquefaction by high-temperature degradation under high pressure, energy recovery by incineration of the waste, and reusing of the residual substance as cement raw material. In addition, material recycling such as reusing technology of the crushed FRP as plastic fillers has been developed. However, an innovative recycling method with the effective matching of the cost and the environmental preservation has not been developed. Therefore, FRP recycling has not been spread yet in the world. Especially, the study of chemical recycling such as extraction of chemical materials from the thermosetting resin is just going to be late. Several years earlier, we reported the thermal degradation method of thermosetting resin in vegetable oil under the atmospheric condition to liquefy the resin. In this method, various plastics can be degraded in a short time and the consumption energy is less as compared with other high-temperature method. As a good feature, the various vegetable oil manufactured from the crops grown in the fields can be used as a solvent of plastic waste, moreover, the used oil for cooking can also be used as a solvent of plastics. Hence, it can be called the environment-friendly degradation method of plastic waste. Moreover, since vegetable oil has the highest boiling temperature and ignition temperature compared with other industrial solvents, the heating reaction (300–350°C) of plastics can be safely carried out under the atmosphere. The main purposes of this research are a development of an easy degradation method of the FRP waste, and an easy recycling method of chemical materials extracted from the degradation substance. In this study, we experimented that the FRP degradation in heated vegetable oil, the removal of glass fiber filler from the degraded FRP waste by a centrifuge, and the extraction of phthalic materials from the FRP degradation. Finally, the new FRP recycling process was proposed.

4.1 Experimental Methods

4.1.1 Specimen

Figure 7.10 shows the chemical structure of typical UP. Figure 7.11 shows the structure of UP cross-linked by styrene. Table 7.1 shows the composition of raw materials in cross-linked UP. The FRP resin was prepared by adding methyl ethyl ketone peroxide as a cross-linking initiator to UP resin-based ortho-phthalic acid mixed with styrene (Japan Composite Co., Ltd.). The UP solution was introduced into molds containing glass fiber chopped-stacked mat (fiber diameter of 11 μm, fiber length of 50 mm, mat weight of 450 g/m²), then it was cross-linked under atmosphere to manufacture a flat FRP substrate (density of 1.2 g/cm³, glass fiber content of 30 wt.%). The FRP substrate was crushed. Further, the crushed grain was prepared to about 5 mm by using a grinder and the specimen was obtained.

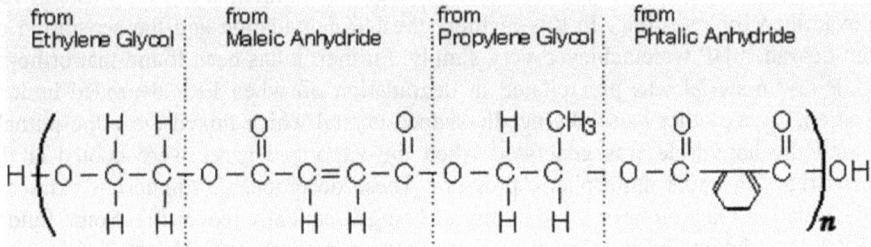

Figure 7.10   Chemical structure of Unsaturated Polyester (UP).

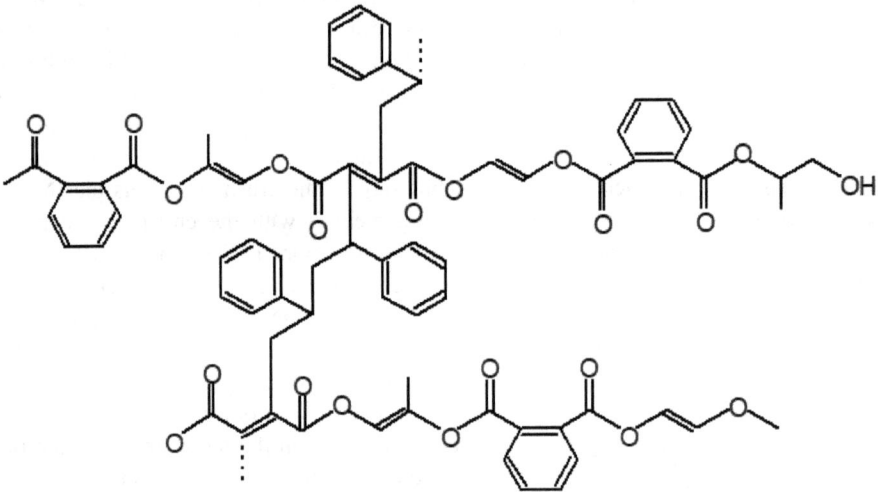

Figure 7.11   Unsaturated Polyester (UP) cross-linked by styrene.

Table 7.1   Raw Materials of Cross-linked UP
(Keiichiro Sano and Masaaki Takayanagi 2003).

Phthalic Anhydride	25.8 (wt.%)
Maleic Anhydride	11.4
Propylene Glycol	14.3
Ethylene Glycol	17.6
Styrene	45.0
Other	2.3

Purified vegetable oil of soybean (Keiichiro Sano and Masaaki Takayanagi 2003) and rapeseed (Keiichiro Sano 2005) made in Nisshin OilliO Group, Ltd. was used as a solvent to dissolve the UP material. Figure 7.12 shows the triglyceride structure of vegetable oil. Table 7.2 shows the weight ratio of fatty acid-containing in the rape oil.

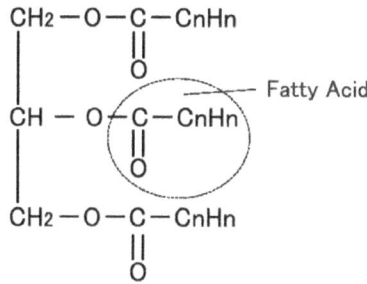

Figure 7.12   Triglyceride.

Table 7.2   Composition of fatty acid in rape oil* (Keiichiro Sano 2005)

Kind of Acid	(C : =)**	Radio (wt.%)
Oleic Acid	(18 : 1)	62.7
Linoleic Acid	(18 : 2)	19.7
Lionolenic Acid	(18 : 3)	8.8
Palmitin Acid	(16 : 0)	4.1
Stearic Acid	(18 : 0)	1.9
Other		3.0

*Iodine Number = 100 – 120; Double Bonds Number = 3.8/molecule
**Number of Carbon and double bonds.

4.1.2   Thermal Degradation of Glass Fiber Reinforced Polymer (GFRP)

Figure 7.13 shows a schematic diagram of the autoclave unit with a heater and mixer for GFRP. This unit was used for the degradation of cross-linked UP in FRP material. First, a suitable amount of FRP and rape oil (FRP: oil = 25 g: 75 g) were added to a stainless chamber (diameter of 4 cm, height of 15 cm). By using the autoclave equipped with a temperature controller, the rape oil was heated from 310 °C to 340 °C to measure the UP dissolution rate under pressurization (3–4 MPa) introduced nitrogen gas and under atmospheric condition. The UP dissolution rate was calculated from the elapsed time when the solid UP dissolved completely (Keiichiro Sano and Masaaki Takayanagi 2003).

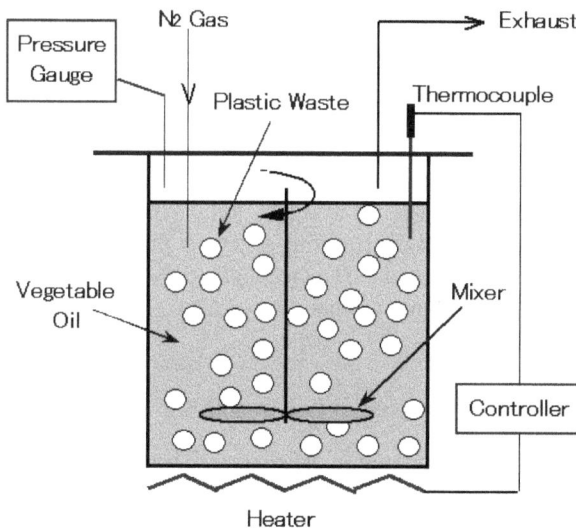

Figure 7.13   Schematic diagram of autoclave for Glass Fiber Reinforced Polymer (GFRP) degradation.

Figure 7.14 shows a schematic diagram of the cooling deposition unit for the vaporized gases from degraded polymer involving nitrogen carrier gas from the autoclave unit of Fig. 7.13 under atmospheric pressure. The vaporized gases that passed the pyrex flask (25 ml) soaked in the ice water bath were cooled, and the needle-shaped crystal of sublimation nature is reduced in the pyrex flask. In addition, the glass fiber was fully removed from dissolved FRP by the centrifuge (TANABE WILLTEC Inc.) using a filter (50 mesh/inch made from stainless steel) (Keiichiro Sano 2005).

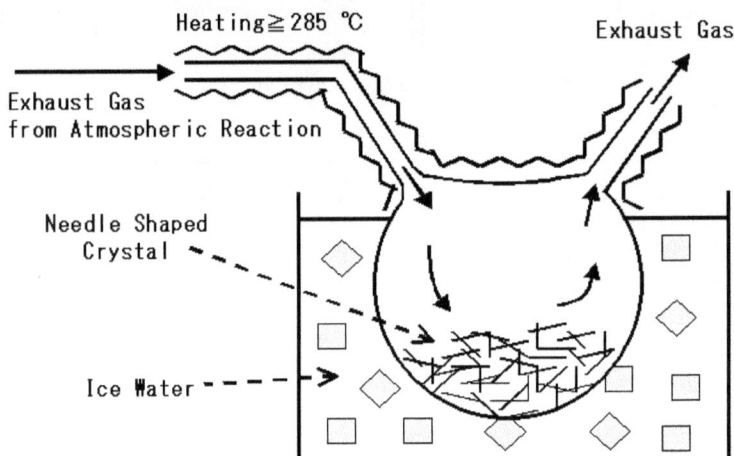

Figure 7.14 Cooling deposition unit for vapor from degradated polymer.

4.1.3 Chemical Analysis

The molecular weight distribution of dissolved UP was measured by gel permeation chromatography (GPC, Tosoh Corp.). The GPC machine was operated with the column and detector temperature at 50°C and at a flow rate of 0.8 mL/min. by using tetrahydrofuran as an elution solution. Further, the calibration curve was obtained from standard polystyrene having a molecular weight of 266 to 190×10^3.

The structures of a solution, precipitation, and needle-shaped crystal from the degraded UP were analyzed by the Fourier transform infrared spectrometry (FTIR, Spectrum One, Perkin Elmer Instruments). In this FTIR measurement, the Attenuated Total Reflection (ATR) method was used. In addition, the precipitation and the needle-shaped crystal were identified by the liquid chromatography (LC-10A, Shimadzu Corp.). The operating conditions of this chromatography are shown below. Elution liquid, pure water of 67 vol.% and methyl alcohol of 30 vol.%, acetic acid of 3 vol.%; Column of Waters Corp., ODS μ-bond sphere 5 μm C_{18}-100 Å, column temp. of 50°C; Flow rate of 0.5 mL/min; Detector of Tosoh Corp., UV-8000, detection of 234 nm.

The resin degradation products were analyzed by liquid chromatography (LC, L-6000, Hitachi Ltd.). The column of Waters Co., Ltd. (μ-bond sphere 5 μm, C_{18}-100 Å) was used. And the eluent mixed with 3% acetic acid and 30% methanol, 67% water was used. The absorbance measurement conditions were a column temperature of 50°C. and a flow rate of 0.5 mL/min, and a light wavelength of 234 nm.

4.1.4 Combustion Analysis of Dissolved Polymer Resin

The total calorific value during combustion of dissolved UP was determined according to No. 2279 of the JIS K standard. For qualitative and quantitative analyses of combustion gas, the combustion method in silica tubes specified in the JIS K 2541 standard was used to obtain combustion gas. CO, CO_2 and HC compounds were collected in separate bags according to the JIS K 0098 standard and analyzed by gas chromatography. SO_X was absorbed into H_2O_2 aqueous solution according to the JIS K 0103 standard, and NO_X has absorbed into an aqueous solution of N-(1-naphthyl) ethylene-diamine dihydrochloride, in accordance with the Saltzman method specified in the JIS K 0104 standard. Both SO_X and NO_X were subsequently analyzed by ion chromatography.

In these experiments, samples of UP dissolved under different conditions were used. The ratio of UP and bean oil was prepared to 1:2, and the degradation reaction was carried out for 30 minutes at 350°C (JIS Handbook 2003).

4.2 Results and Discussion

4.2.1 Degradation of Glass Fiber Reinforced Polymer (GFRP) Waste

Figure 7.15 shows the flow of FRP degradation under the atmosphere by the heated vegetable oil. The degradation product cooled below 100 °C could be separated into the organic degradation oil and the glass fiber by using the centrifuge. In our earlier research, the liquid degradation matter was recycled as fuel, and the glass fiber was recycled as the reinforcement in the asphalt sheet for vibration suppression of the car.

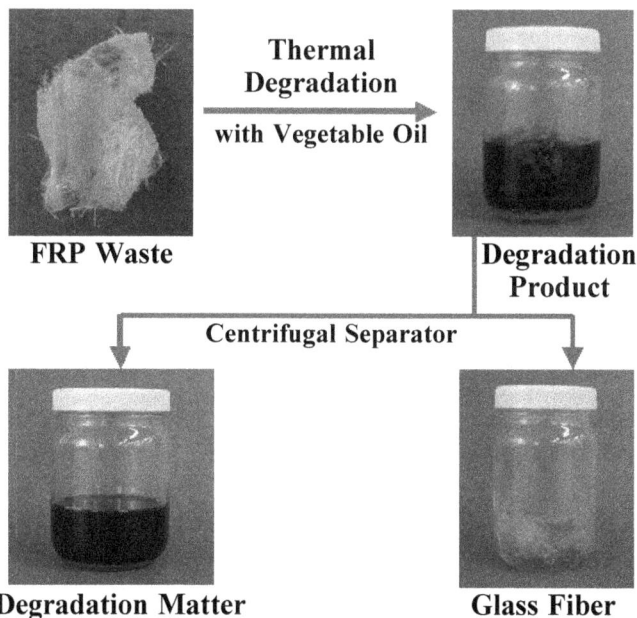

Figure 7.15 Flow of Glass Fiber Reinforced Polymer (GFRP) degradation.

4.2.2 Behavior of Unsaturated Polyester (UP) Dissolution

Figure 7.16 shows the Arrhenius plots of UP dissolution rate as a function of rape oil temperature. In both cases of autoclave under sealed pressurization and atmospheric reaction, UP dissolution rates have a similar activation energy of about 35 kcal/mol. Both dissolution rates increased with dissolved after 70 minutes. These results have suggested that the vegetable oil is very suitable for the solvent of GFRP degradation and the atmospheric reaction is very practical. The dissolution rate under sealed pressurization was higher than that of the reaction under atmosphere. This is the rising of temperature. The dissolution rate of UP in rape oil was dependent mostly upon the temperature. At the heating temperature of 320°C under sealed pressurization, the dissolution reaction of the solid UP is an activated state, and the amount of UP dissolution was reached to 100 wt.% when 60 minutes elapsed. Further, at the heating of 340°C, the dissolution rate increased, and UP dissolved completely after 20 minutes. In the case of 340°C under atmosphere, the UP considered due to the collision probability of molecules from rape oil and UP in the liquid phase since different data of frequency factor A. Further, it is considered due to the high penetration of vegetable oil into the interstice of solid UP under the pressurization condition. The following reaction steps are considered to have occurred as the process of UP dissolution in the rape oil. First, radicals generate from the rape oil by heating. In addition, UP is heated in rape oil, and radicals are generated from the surface and the interstice of solid UP. Next, these radical species react complicatedly, and UP is dissolved in the rape oil (Keiichiro Sano 2005).

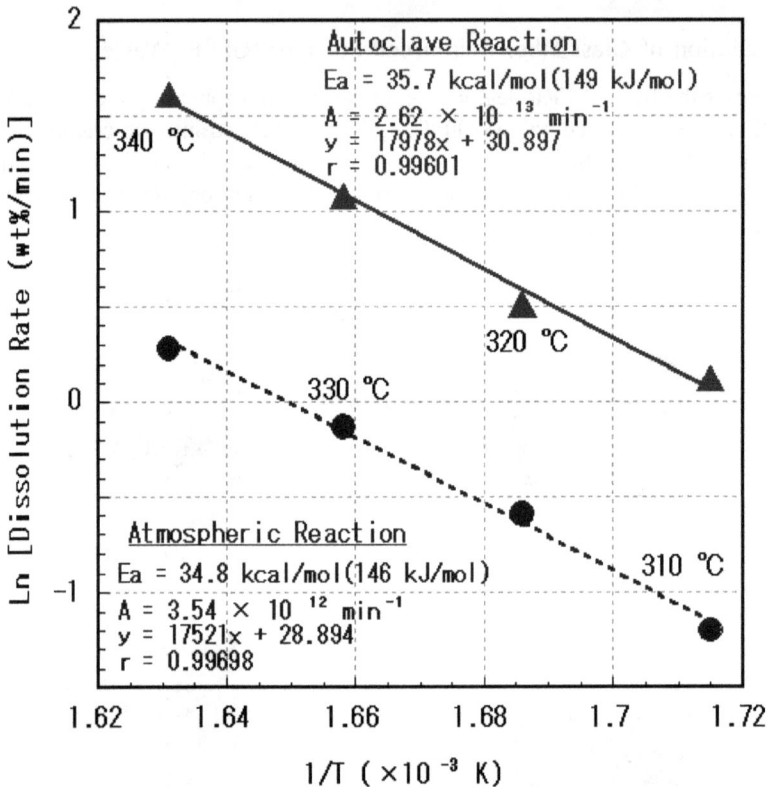

Figure 7.16   Arrhenius plots of Unsaturated Polyester (UP) dissolution rate.

Figure 7.17 shows the molecular weight distribution curves by Gel Permeation Chromatography (GPC) of UP dissolution products reacted at different temperatures in the bean oil for 3 hours. When the heating temperature increased, the molecular weight decreased. In the dissolution products obtained at 320°C, the distribution had a molecular weight of more than 10 thousand. On the other hand, at the temperature of 350°C, the molecular weight distribution was in the range below 10 thousand. Further, the distribution at 350°C became small with the passing of reaction time, the distribution of more than one thousand decreased, and the distribution of less than one thousand increased. It is considered that the difference in these molecular weight distributions was caused by the difference in the dissolution reaction of the UP resin. The dissolution of the UP in the bean oil depends upon the heating temperature and reaction time (Keiichiro Sano and Masaaki Takayanagi 2003).

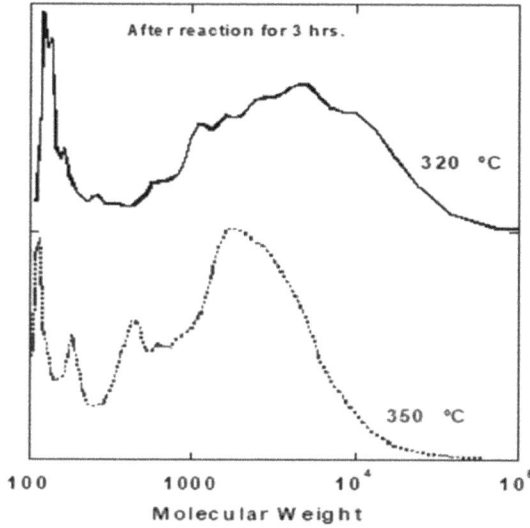

Figure 7.17 Gel Permeation Chromatography (GPC) curves of Unsaturated Polyester (UP) dissolved with bean oil.

4.2.3 Dissolution of Degraded Unsaturated Polyester (UP)

Figure 7.18 shows the FTIR spectra of the original UP resin and soybean oil. In the spectrum of UP, the absorption peaks of C_6H_6 were observed at 705 cm^{-1}, 725 cm^{-1}, and 1450–1600 cm^{-1} associated with aromatic groups of UP and cross-linked styrene. And the peaks of C=O at 1730 cm^{-1} and C–O at 1000–1340 cm^{-1} were related to aromatic ester groups in UP observed. Further, the peak related to CHn at 2920 cm^{-1} in UP cross-linked with styrene was shown. In the spectrum of bean oil, the peaks of (CH_2) n observed at 720 cm^{-1} and 1470 cm^{-1} are associated with triglycerides of bean oil. And the peaks of C=O at 1750 cm^{-1} and C–O at 1000–1330 cm^{-1} related ester group of tri-glyceride observed. Further, the peaks related to CHn at 2850 cm^{-1} and 2920 cm^{-1}, and the C=C double bond of fatty acids in triglyceride at 3010 cm^{-1} were shown (Keiichiro Sano and Masaaki Takayanagi 2003).

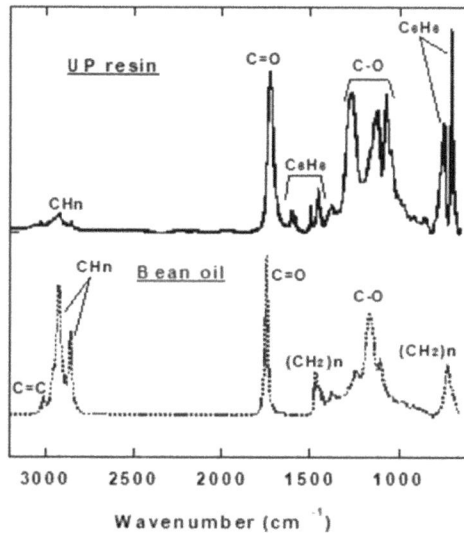

Figure 7.18 Fourier Transform Infrared Spectroscopy (FTIR) spectra of Unsaturated Polyester (UP) resin and bean oil.

Figure 7.19 shows FTIR spectra of UP dissolution oil compared with autoclave and atmospheric reaction. In both cases, the reaction temperature was controlled at 320°C, and the reaction was finished when the solid UP dissolved completely. Both spectra have some peaks from elements of UP and rape oil, the absorption peaks related to C_6H_6 of an aromatic compound from UP, CHn, (CH_2)n and C=C of fatty acids from rape oil and C=O and C–O from both of UP and rape oil were detected. However, in the case of atmospheric reaction, the absorption peaks for the aromatic compound were decreased. Many organic compounds such as ortho-phthalic acid, phthalic anhydride, and styrene were vaporized under the atmospheric reaction (Keiichiro Sano 2005).

4.2.4 Combustion Analysis of Dissolved Unsaturated Polyester (UP)

Table 7.3 shows the total calorific value of combustion of each specimen. The dissolved UP including bean oil was obtained with a higher total calorific value as compared with that of solid

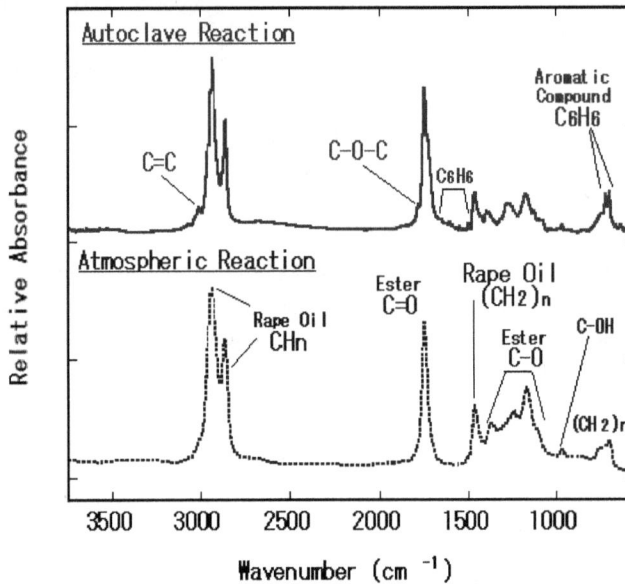

Figure 7.19 Fourier Transform Infrared Spectroscopy (FTIR) spectra of Unsaturated Polyester (UP) degradation oil.

UP because the combustion efficiency was higher than that of solid UP. Although the calorific value of dissolved UP is lower than that of heavy oil, it can be used as an effective heat-energy source because the calorific value is sufficiently high for combustion in boilers. In addition, the actual calorific value will be closer to that of heavy oil, as dissolved UP was diluted with heavy oil or kerosene.

Table 7.3 Combustion energy of sample (Keiichiro Sano and Masaaki Takayanagi 2003)

	Dissolved UP	UP	Heavy Oil	Soybean Oil
E (kJ/g)	38.5	30.0	44.0	39.5

UP: Unsaturated Polyester

Table 7.4 shows the results of the analysis of generation gas with the combustion of dissolved UP at 800°C. The CO_2 gas was almost 100 wt.%, and the amounts of CO, NO_x and HC compounds were not detection levels. These observations suggested that dissolved UP in the liquid state can readily become a gas at higher temperatures and that complete combustion occurred under this condition.

In the case of direct combustion of pulverized FRP, solid UP is first liquefied and then converts to a gas at high temperature. Hence, ready perfect combustion is difficult. Further, the glass fiber in FRP normally prevents combustion.

Table 7.4   Combustion gas of dissolved UP (Unsaturated Polyester) (Keiichiro Sano and Masaaki Takayanagi 2003)

Temperature	Amount of gas (mg)*				
	CO_2	CO	SOx	NOx	HC Comp.
800°C	2900	—**	0.06	—**	—**

*Conversion value of using UP dissolution sample of 1.0 g
**Non-detection

4.2.5   Phthalic Materials Extracted from Unsaturated Polyester (UP)

Figure 7.20 shows the precipitate in the UP degradation oil by autoclave reaction. In this sample, glass fiber was removed. Figure 7.21 shows the needle-shaped crystal which was reduced when the vaporized gases from the UP degradation reaction under atmospheric conditions were cooled. Most precipitates did not appear in the degradation oil.

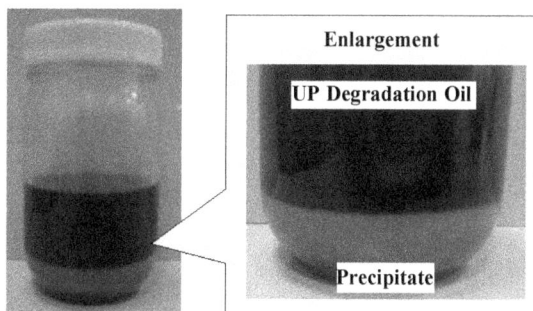

Figure 7.20   Precipitate in the Unsaturated Polyester (UP) degradation oil.

Figure 7.21   Phthalic acid of needle-shaped crystal from gases of degradated Unsaturated Polyester (UP).

Figure 7.22 shows the detection peaks of the precipitate and the needle-shaped crystal in the liquid chromatography analysis. Both samples which were adhered with rape oil were washed carefully with hexane before the analysis. In the precipitate, the ortho-phthalic acid of UP raw material was mainly observed, and some unknown substances were detected. On the other hand, in the needle-shaped crystal, two peaks due to ortho-phthalic acid and phthalic anhydride were observed. They are predicted that the crystal is a pure mixture of two substances. The generation reaction of phthalic materials through the UP degradation considered from these analysis data was shown in Fig. 7.23. First, the main chain scission of cross-linked UP is progressed in the heated oil, and the diester phthalate radicals generate. These radicals are accepted as hydrogen atoms from the rape oil and change to ortho-phthalic acid. Further, with the needle-shaped crystal under atmospheric reaction, the dehydration of ortho-phthalic acid under high temperature occurs and some phthalic anhydride are obtained. If the phthalic anhydride contacts with the moisture in the atmosphere, it may be rechanged to the ortho-phthalic acid by hydration.

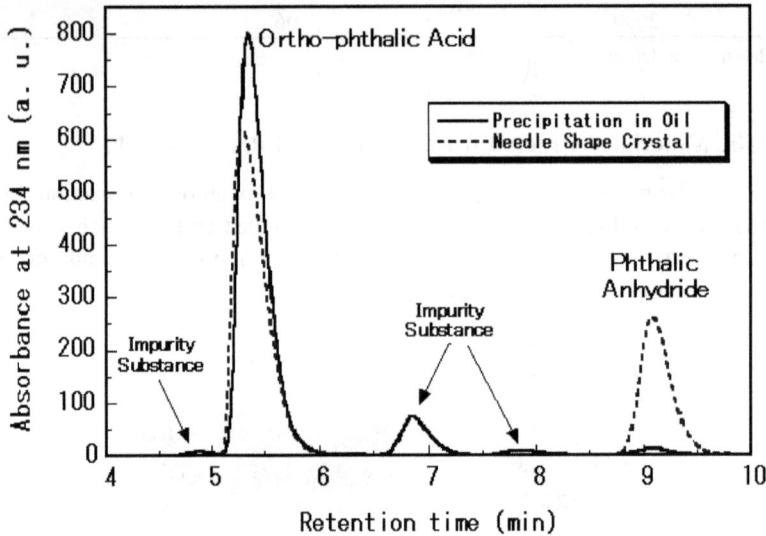

Figure 7.22   Liquid chromatograms of extract from gases degraded Unsaturated Polyester (UP).

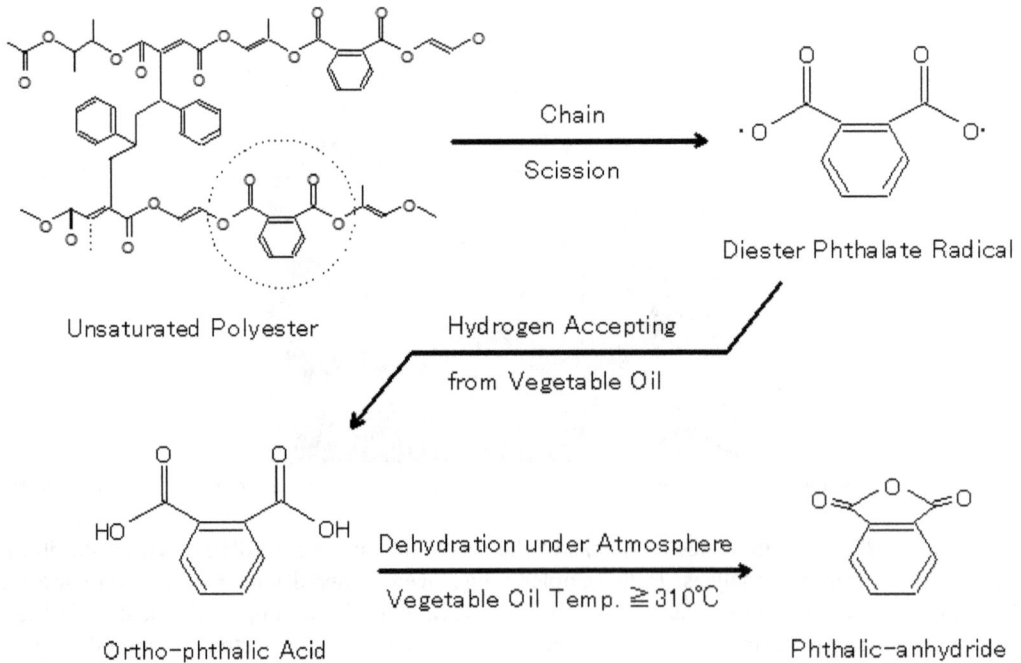

Figure 7.23   Generation reaction of phthalic materials from Unsaturated Polyester (UP).

4.2.6   Process Design for Glass Fiber Reinforced Polymer (GFRP) Recycling

Figure 7.24 shows the proposal for the GFRP recycling process. This process was designed from the above-mentioned experimental results. First, the granular GFRP waste is degraded in heated vegetable oil. In the case of reaction under the atmosphere, the sublimation substance is collected. Next, the degradation products are cooled and the products are separated into organic degradation matter and inorganic fiber. In the autoclave reaction, the precipitate of degradation matter is collected. The collected precipitate and sublimate are washed and extracted in the solvent. The reduced phthalic materials can be recycled chemically, for example, they may be able to be reused for the UP raw material. The remaining degradation matter can be recycled thermally or as another material. Glass fiber is used as a material.

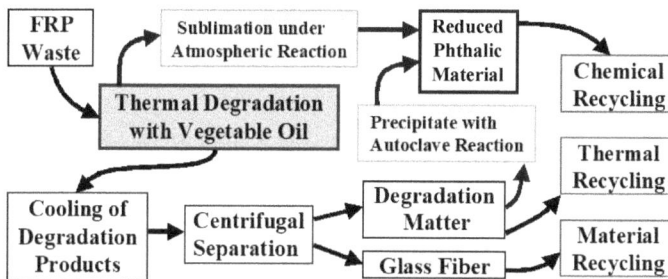

Figure 7.24   Process design of Glass Fiber Reinforced Polymer (GFRP) recycling.

4.2.7   Recycling Machine for Glass Fiber Reinforced Polymer (GFRP)

Author and Nisshin OiiliO group, Ltd. developed the recycling unit for thermal degradation of GFRP waste using vegetable oil. Figure 7.25 shows the GFRP recycling unit produced by the Nisshin OilliO group. In this method, the UP degradation and the separation of glass fiber from degraded UP were achieved very simply. Further, it has been found that the ortho-phthalic acid of UP raw material was precipitated in the degradation oil when the GFRP degraded under sealed pressurization, on the other hand, the needle-shaped crystal which mixed the ortho-phthalic acid and the phthalic anhydride was reduced when the vaporized gases were cooled at the GFRP degradation reaction under atmospheric pressure. These decomposition materials extracted from the UP degradation reaction have a possibility of being chemically recyclable.

Figure 7.25   Recycling Unit of Glass Fiber Reinforced Polymer (GFRP) waste.

5 CONCLUSION

In this study, it has been found that the UP in FRP waste can be degraded easily in heated vegetable oil, and the glass fiber in FRP can be? easily removed from the degraded substance by a centrifuge, and the ortho-phthalic acid and the phthalic anhydride of UP raw materials can be easily extracted by the UP degradation reaction. As a/the last result, the FRP waste degraded in vegetable oil could be separated as much as possible into each material, and the new FRP treatment process involving chemical recycling, thermal and material recycling was designed. Furthermore, analysis of these separated materials and development of the recycling methods will be needed. For the following purpose, the study of UP reproduction from the phthalic materials extracted from FRP waste has been planned.

ACKNOWLEDGMENTS

The author would like to thank Dr. Renate Luetzkendorf and Dr. Thomas Reussmann of TITK in Germany for lending their expertise on the application of natural fiber. I also thank Mr. Masaaki Takayanagi Nisshin OilliO Group, Ltd. in Japan for provision oil with the experiments.

■ References

Green Car Congress, Toyota to Show Plug-in Flex-Fuel Hybrid Concept with Double the Fuel Efficiency of the Prius, http://www.greencarcongress.com/2007/10/toyota-to-show-.html

Hitachi Chemical News Release 2007.

JIS Handbook 2003. Chemistry, JIS K, Test Method, Issued by the Japan Standards Association.

Keiichiro, S. and M. Takayanagi 2003. Proceedings of Polytronic 2003.

Keiichiro, S. and M. Nishimaki and M. Takayanagi. 2005. Proceedings of 3rd International Symposium on Feedstock Recycling (ISFR) of Plastics & Other Innovative Plastics Recycling Techniques. 207–216.

Keiichiro, S. and Y. Motomura, T. Seno and K. Yokota. 2010. Development trends of natural fiber reinforced plastics for automobile industries in Germany and Japan, Proceedings of JSAE Annual Congress. No. 59-10: 1–3.

Keiichiro, S. and K. Takeda. 2014. Proceedings of 3rd International Conference on Materials, Science and Environments (ICMEE). Honolulu, Hawaii, USA: Hawaii.

Reate Luzkendorf, bulletin, Institute of human environmental studies, Kanto Gakuin University (2009).

Takeru, N. 2014. Presentation slide for next-generation automobile regional industry-academia-government forum, p.24. https://docsplayer.net/140659217-%E3%82%B9%E3%83%A9%E3%82%A4%E3%83%89 -1.html

Panasonic Electric Works Corporate Global Environment Exhibition 2008.

Reussmann, T. and R. Luetzkendorf. 2008. Proceedings of 7th global WPC and natural fiber composites congress. A5: 1–13.

Tokyo Motor Show. 2007. http://www.tokyo-motorshow.com/en/history/40.html

Topics of NEC, NEC & UNITIKA Realize bioplastic reinforced with kenaf fiber for mobile phone use. http://www.nec.co.jp/press/en/0603/2001.html

White Paper on Land, Infrastructure and Transport in Japan. 2004. The Ministry of Land, Infrastructure, Transport and Tourism. p. 64. https://www.mlit.go.jp/english/white-paper/mlit04/p2c7.pdf

Chapter 8

Liquid Phase Deposition of Functional Nanocomposites

Yong X. Gan* and Andrew Izumi

Department of Mechanical Engineering, California State Polytechnic University Pomona,
3801 W. Temple Avenue, Pomona, CA 91768, USA

yxgan@cpp.edu

Jeremy B. Gan

Department of Chemical and Biomolecular Engineering,
University of California Los Angeles, 405 Hilgard Ave, Los Angeles, CA 90095, USA

Ahalapitiya H. Jayatissa

Department of Mechanical, Industrial and Manufacturing Engineering,
University of Toledo, 2801 W. Bancroft Street, Toledo, OH 43606, USA

1 INTRODUCTION

Liquid Phase Deposition (LPD) is a direct deposition technique using ligand-exchange hydrolysis of metal-fluoro complex for generating metal oxide thin films (Deki et al. 2002). The F^- consumption is due to the reaction with boric acid or aluminum metal. The feasibility of using this technique for depositing silica (SiO_2) coating was demonstrated (Nagayama et al. 1988). In their research, the silica (SiO_2) layer was obtained on the surface of soda lime silicate glass slides by immersing the glass in a hydrofluosilicic acid (H_2SiF_6) solution supersaturated with silica gel at low temperature. Later, it was reported that SiO_2 thin films can be made by immersing a substrate into the hexafluorosilisic acid (H_2SiF_6) solution with pre-added boric acid (Hishinuma et al. 1991). It was found that more stable SiO_2 thin films can be obtained in H_2SiF_6 solution by dissolving aluminum. The mechanism of the liquid phase deposition process was explored as well. For the silica film deposition, the following reaction mechanisms formulae (1) through (3) were proposed (Hishinuma et al. 1991).

---

*Corresponding Author

$$H_2SiF_6 + 2H_2O \leftrightarrow 6HF + SiO_2 \tag{1}$$

$$H_3BO_3 + 4HF \leftrightarrow BF_4^- + H_3O^+ + 2H_2O \tag{2}$$

$$2Al + 12HF \leftrightarrow 2H_3AlF_6 + 3H_2\uparrow \tag{3}$$

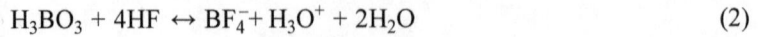

Equation (1) presents the formation of oxide. Equations (2) and (3) show the fluorine ion consumption through the usage of two different scavengers (the boric acid and the aluminum metal).

Preparation of vanadium oxide film by the LPD method was also investigated (Deki et al. 1996). During the liquid phase deposition process, Vanadium (V) oxide was dissolved in a 5% hydrofluoric acid (HF) to form a saturated solution with a vanadium ion concentration of 0.384 M. The solution was then diluted with distilled water to 0.15 M of vanadium ion concentration and used for the liquid phase deposition of vanadium oxide film. A cleaned glass substrate was inserted vertically into the solution. A pure aluminum metal plate as a free F^- ion scavenger was placed into the solution and set near the suspended substrate. The reaction cell was maintained at 30°C for 20–40 hours. After the deposition, the sample was rinsed in distilled water and air-dried at room temperature. Heat treatment on the sample at various temperatures for 1 hour in airflow was also performed using a muffle furnace. The ligand exchange reaction mechanism proposed can be given as reaction formula (4).

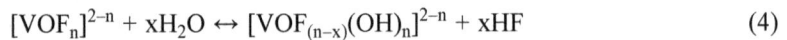

$$[VOF_n]^{2-n} + xH_2O \leftrightarrow [VOF_{(n-x)}(OH)_n]^{2-n} + xHF \tag{4}$$

Deposition of β-FeOOH thin films on borosilicate glass and Au wires by the LPD method was described (Deki et al. 2002). By adding NH_3 to adjust the pH value, FeOOH was precipitated by the hydrolysis of $Fe(NO_3)_2$ from an aqueous solution. After the separation, the precipitate was washed repeatedly with distilled water and air-dried at room temperature. Then the FeOOH precipitate was dissolved in $NH_4F \cdot HF$ to form the FeOOH–$NH_4F \cdot HF$ aqueous solution for β-FeOOH deposition. By adding H_3BO_3 into the solution, β-FeOOH films were uniformly deposited on the substrates. The deposition time was varied from 40 minutes to 20 hours and the temperature for the deposition was kept at 30°C. Stepwise hydrolysis was found in the LPD process (Deki et al. 2002). First, the iron (III) ions, Fe^{3+}, were coordinated by fluorine ions to generate $[FeF_6]^{3-}$ species in the solution. The release of F^- ions from $[FeF_6]^{3-}$ in the liquid phase resulted in the transitions as described by the following equation (Deki et al. 2002):

$$[FeF_6]^{3-} \rightarrow [FeF_{6-n}(OH)_n]^{3-} \rightarrow [Fe(OH)_6]^{3-} \tag{5}$$

The ligand exchange of $[FeF_6]^{3-}$ and $[FeF_{6-n}(OH)_n]^{3-}$ can be accelerated by boric acid-promoted consumption reaction of F^- ions in the solution as shown earlier by Equation (2). The final hydrolysis product is attached to the surface of the substrates to form the iron oxyhydroxide thin films. Typically, the presence of Cl^- or F^- ions during hydrolysis of Fe^{3+} led to the formation of β-FeOOH instead of α-FeOOH (Flynn Jr. 1984).

The liquid phase deposition (LPD) method is not just suitable for preparing the above-mentioned metal oxide thin films, but also applicable for making various ordered nanoparticles, for example, titanium dioxide and tin dioxide nanoparticles (Deki et al. 2004b). During the LPD process, Reverse Micelles (RMs) with inner water pools were served as both nanoreactors and templates of the nanoparticles to be synthesized. As known, reverse micelles are nanoscopic aggregates consisting of three components: a water pool, a surfactant and a nonpolar solvent. To synthesize titanium dioxide and tin dioxide nanoparticles, a nonionic surfactant containing Triton X-100 (TX-100) and 1-hexanol was made (Deki et al. 2004b). Cyclohexane was used as the oil phase for RMs formation. The relative amount of TX-100 to 1-hexanol was set as 0.2. The ratio of water to TX-100, W_0, was kept as 10. As a water phase in RM the LPD reaction solution contains $(NH_4)_2TiF_6$ and H_3BO_3 for titanium dioxide nanoparticle deposition. For tin dioxide nanoparticle deposition, the LPD reaction solution consists of tin fluorocomplex (SnO_n HF) and H_3BO_3 The LPD solutions were injected into the initially made T-100/1-hexanol/cyclohexane/ water RM solution under vigorous stirring to get transparent RM solutions. The transparent RM solutions were placed into a warm bath at 30°C for 20 hours. After the LPD reactions, the RM sample solutions were removed by centrifugation. The generated precipitates were ultrasonically cleaned in distilled water three times. To remove the residual surfactant in nanoparticles, the LPD samples were re-dispersed in cyclohexane for purification. The purified samples were dried and the final products of titanium dioxide and tin dioxide nanoparticles were obtained. The average diameter and standard deviation for the titanium dioxide nanoparticles were 59 nm and 0.84 nm, respectively. For SnO_2 nanoparticles, the average size was found to be 3.0 nm. Its standard deviation was 0.71 nm (Deki et al. 2004b).

In general, the liquid deposition method depends on the chemical equilibrium between a metal-fluoro complex and a metal oxide which can be expressed as (Deki et al. 2002).

$$MF_n^{(2-n)} + H_2O \leftrightarrow MO_x + HF \tag{6}$$

As can be seen from Equation (6), no specific substrate material is involved in the reaction. Thus, homogeneous thin films can be deposited on various kinds of substrates. The substrates include those with large surface areas and complex morphologies. Next the discussion will concentrate on ordered nanostructure deposition.

2   LIQUID PHASE DEPOSITION OF ORDERED NANOSTRUCTURES

Ordered nanostructures have found various applications in surface engineering, for example, superhydrophobic surface design and metamaterial preparation. Recently, the work on making ultra-low light reflection surfaces consisting of aluminum oxyhydroxide nanolens and nanopillar arrays has been reported (Jung et al. 2019). Such nanostructures were generated by depositing aluminum oxyhydroxide on anodic oxidized aluminum via hydrothermal synthesis followed by platinum sputtering. The anti-reflective properties were evaluated on several types of nanostructures such as nanolens and nanopillar arrays with different aspect ratios. It was found that the nanopillar arrays with an aspect ratio of 1:14 showed reflectance as low as 0.18% to the light of 550 nm wavelength at all measured incident angles.

Nanorod and nanopore arrays can also be prepared by the liquid phase deposition (Deki et al. 2004a). First, electron lithography was used to make silicon templates with regularly aligned nanopores. Then, the Liquid-Phase Infiltration (LPI) method (as an extended one of the LPD) was used for preparing nanopillars and nanopores with two-dimensional (2D) periodicity. As shown in Fig. 8.1, a one-step approach resulted in the deposition of titanium dioxide into the pores of the silicon template. The peeling-off or etching-off the template left nanopillars on a layer of the solid oxide coating.

Also shown in Fig. 8.1 is the two-step process in which a polymer was cast into the nanopore of the silicon template. The polymer replica was lifted off from the silicon template and used as the substrate for liquid phase deposition of TiO_2. After that, the polymer replica was removed by acetone and the porous TiO_2 was produced.

The micrographs in Fig. 8.2 are the scanning electron microscopic images of the LPD TiO_2 nanopillar array and the porous silicon template. At lower magnification (Fig. 8.1(a)), the 2D periodicity of the LPD TiO_2 nanopillar array can be seen. The rod diameter is about 440 nm. Figure 8.2(b) reveals the nanoscale pores in the silicon template. Figure 8.2(c) is the SEM image of the TiO_2 nanopillar array with a rod diameter of 660 nm at a higher magnification. The 3D morphology of the thicker nanorods can be seen clearly from this SEM image. Most of the nanorods are aligned vertically to the base coating.

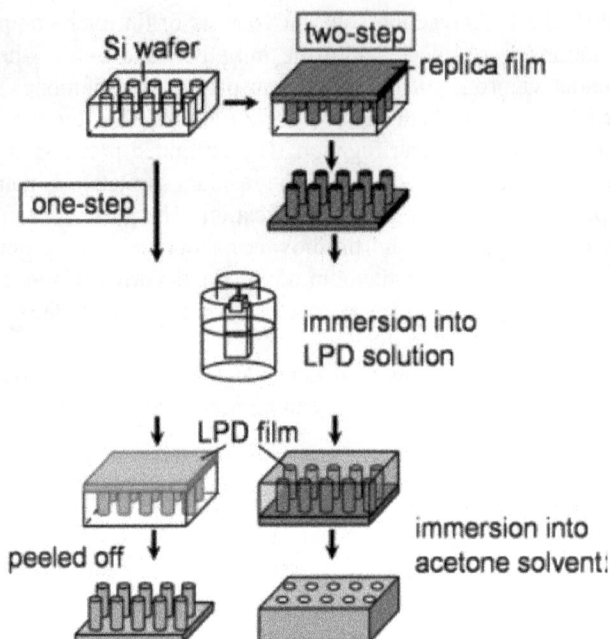

Figure 8.1 Schematic showing the procedures for nanopillar and nanopore arrays preparation by liquid phase deposition. Reproduced with permission from literature (Deki et al. 2004a); Liquid-phase infiltration (LPI) process for the fabrication of highly nano-ordered materials. *Chemistry of Materials* 16(9): 1747–1750, © 2004 American Chemical Society.

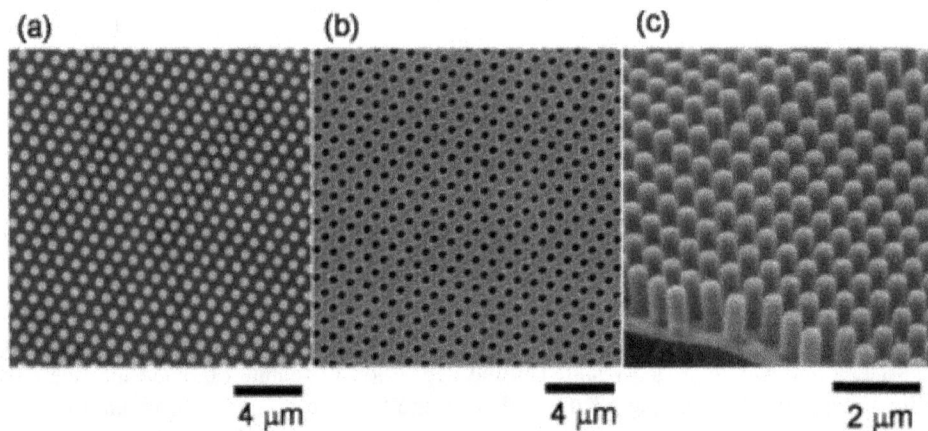

Figure 8.2 Scanning electron microscopic images showing: (a) the LPD TiO$_2$ nanopillar array with a rod diameter of 440 nm, (b) the porous silicon template, and (c) the TiO$_2$ nanopillar array with a rod diameter of 660 nm at a higher magnification. Reproduced with permission from literature (Deki et al. 2004a); Liquid-phase infiltration (LPI) process for the fabrication of highly nano-ordered materials. *Chemistry of Materials* 16(9): 1747–1750, © 2004 American Chemical Society.

The morphology of nanohole arrays made by Deki et al. (2004a) was observed by SEM. Figure 8.3(a) shows the porous TiO$_2$ with a hole diameter of 310 nm. The magnified nanopores are revealed in Fig. 8.3(b). In Fig. 8.3(c), the silicon template with a pore diameter of 44 nm is shown. From the SEM images in Fig. 8.3, it is clear that the liquid phase deposited porous TiO$_2$ has a hole diameter about 30% smaller than that of the pores in the silicon template.

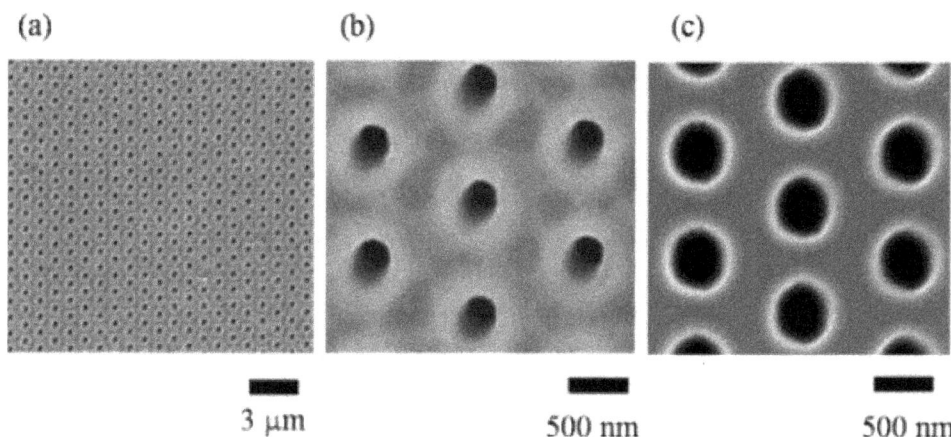

Figure 8.3   Scanning electron microscopic images showing: (a) the LPD TiO$_2$ nanopore array, (b) the TiO$_2$ nanopore array at a higher magnification, and (c) the porous silicon template. Reproduced with permission from literature (Deki et al. 2004a); Liquid-phase infiltration (LPI) process for the fabrication of highly nano-ordered materials. *Chemistry of Materials* 16(9): 1747–1750, © 2004 American Chemical Society.

Since liquid phase deposition reactions can be controlled by adjusting the concentration of chemicals in the liquid phase, the dimensions of the deposited particles or films can be changed easily. In addition, the reaction time is another important parameter for tailoring the structure and size of the deposited nanostructures. The two examples as shown in Figs. 8.2(a) and 8.2(c) support this statement. More examples will be shown later on nanocomposite deposition using the LPD method.

3   LIQUID PHASE DEPOSITION OF NANOCOMPOSITES

Since liquid phase deposition reactions occur in a single phase of liquid, uniform mixing of various chemicals is relatively easy to achieve. This allows the LPD technique to be suitable for making multi-component materials including nanoscale composite materials. Next emphasis will be placed on nanocomposite deposition.

3.1   Photosensitive Cadmium Sulfide/Titanium Dioxide Nanocomposite

3.1.1   Liquid Phase Deposition of Core/Shell CdS@TiO$_2$

Here a heterogeneous nanocomposite containing liquid phase deposited titanium dioxide nanotube sheath and cadmium sulfide nanorod core will be introduced. The titanium dioxide nanotube was made by template-based deposition. Electrochemical oxidation of pure aluminum foil in an organic or inorganic acid-containing solution resulted in the anodic aluminum oxide template formation. Then the cadmium sulfide nanorod core was deposited onto the regularly aligned titanium dioxide nanotube array via chemical bath deposition which can be considered as another extended form of liquid phase deposition. As known, both cadmium sulfide and titanium oxide nanotube are n-type semiconducting materials. The enhanced visible light absorption by such a hybrid composite structure was observed.

As shown by Hsu et al. (2005), CdS@TiO$_2$ hybrid coaxial nanocables with the wall thickness controlled precisely can be made using the liquid-phase deposition (LPD) method with porous Anodic Aluminum Oxide (AAO) as the template. During the LPD process, the thickness of the

TiO$_2$ sheaths was controlled through the variation of the reaction conditions. The continuous and polycrystalline CdS nanotube or nanorod core was deposited onto the titanium oxide nanotube sheath by chemical bath deposition. The obtained coaxial CdS@TiO$_2$ nanocables were characterized in view of the structure and UV-visible light absorption property. To make the CdS/TiO$_2$ composite, well-ordered TiO$_2$ nanotubes were prepared first by liquid phase deposition (LPD) (Hsu et al. 2005). The solution for LPD was made by mixing 0.05 M ammonium hexafluorotitanate, (NH$_4$)$_2$TiF$_6$ with 0.15 M boric acid, H$_3$BO$_3$ in the volume ratio of 1:3. Electrochemically processed anodic alumina porous membranes with a pore diameter of 200 nm were used as the template. Before deposition to the TiO$_2$ shell, both faces of the membrane were deposited with OTS-SAMs (octadecyltetrachlorosilane-self-assembled monolayers) by microcontact printing technique (Steps 1 and 2 in Fig. 8.4). The OTS-SAMs can prevent the deposition of TiO$_2$ on the faces of the membrane. As a result, the TiO$_2$ thin coating can only be deposited along the pore walls but not on the faces of the membrane as shown in Step 3 in Fig. 8.4. Without OTS-SAMs modification, the faces become preferentially coated and the TiO$_2$ nanotubes are not obtained. These modified AAO templates were put vertically into a vessel with an appropriate amount of the solution containing hexafluorotitanate and boric acid. At the end of the reaction, the AAO template was taken out and washed by deionized (DI) water several times and dried in nitrogen flow. Then the templates containing the precursor were subsequently heated at 500°C for 2 hours. This thermal annealing allows the titanium dioxide to be crystallized.

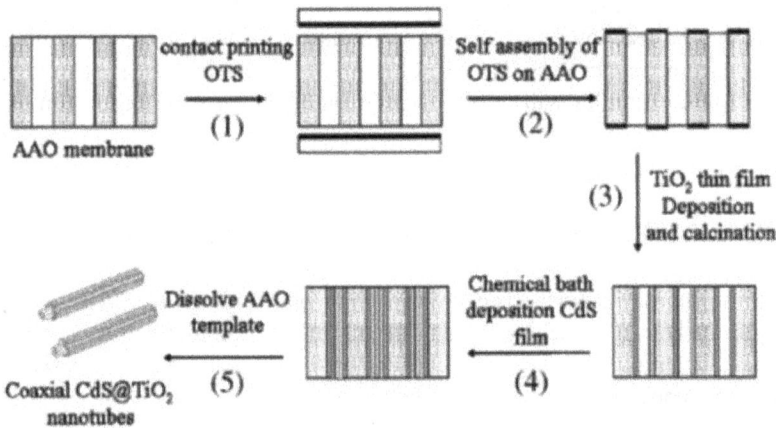

Figure 8.4   Schematic showing the procedures for liquid phase deposition of CdS/TiO$_2$ nanocomposite cable. Reproduced with permission from reference (Hsu et al. 2005); Fabrication of CdS@TiO$_2$ coaxial composite nanocables arrays by liquid-phase deposition. *Journal of Crystal Growth* 285(4): 642–648, © 2005 Elsevier B.V.

The CdS nanotube or nanorod core was deposited in the inner wall of the TiO$_2$ nanotube by soaking a solution of 0.2 M cadmium perchlorate first. After the TiO$_2$ nanotubes were impregnated with Cd^{2+} solution by the capillary effect, the sample was washed with DI water and dried in N$_2$ flow. After that, the sample was inserted into a solution containing thioacetamide (TAA) with the formula of C$_2$H$_5$NS. This procedure was repeated three to five times so that CdS sufficient nanocrystals grew inside the TiO$_2$ nanotube. The AAO template was eventually removed by selective etching in a 6.0 M NaOH solution. The final product of CdS@TiO$_2$ coaxial nanotubes or nanocables was obtained. The morphology of the harvested CdS@TiO$_2$ coaxial nanotubes was examined and can be found in Fig. 8.5. Figure 8.5(a) is a low magnification image of the CdS@TiO$_2$ composite. The aligned composite nanocables are shown. In Fig. 8.5(b), at a higher magnification, the inner CdS core and the outer TiO$_2$ layer of the composite nanocables can be seen clearly.

Figure 8.5 Scanning electron microscopic images of liquid phase deposited CdS/TiO$_2$ nanocomposite: (a) at low magnification, (b) at high magnification. Reproduced with permission from reference (Hsu et al. 2005); Fabrication of CdS@TiO$_2$ coaxial composite nanocables arrays by liquid-phase deposition. *Journal of Crystal Growth* 285(4): 642–648, © 2005 Elsevier B.V.

3.1.2   UV-Visible Light Responses

The photosensitivity of the CdS/TiO$_2$ composite was measured and compared with that of the pure TiO$_2$ nanotube (Hsu et al. 2005). Absorption of the two materials to both ultraviolet (UV) and visible light was tested and their UV-visible absorption spectra are presented in Fig. 8.6 as curve (a) and curve (b). The characteristic absorption band of TiO$_2$ as shown by curve (a) only lies in the UV light spectrum range. The other curve, (b) in Fig. 8.6 shows the absorption in both UV and visible light bonds. This indicates that after the CdS layer was deposited into the TiO$_2$ nanotubes, the obtained composite CdS@TiO$_2$ core-shell nanocomposite shows a stronger light absorption capability. It is meaningful for the composite to be used as a more effective solar energy harvest material. Typical examples of the applications of CdS@TiO$_2$ on photocatalytic degradation of organics can be found in the papers published by (Alizadeh et al. 2020a, b) and Al-Fandi et al. (2019). For hydrogen generation through water splitting under sunlight, the CdS@TiO$_2$ core-shell nanocomposite has played an important role as shown by Wu et al. (Wu et al. 2019).

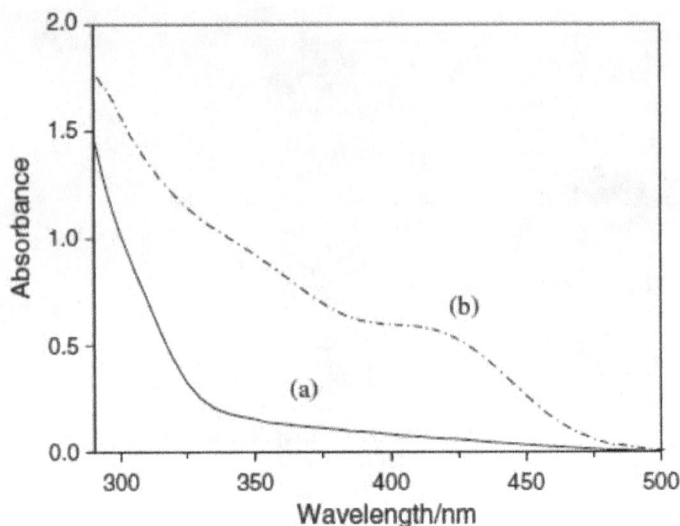

3.2 Thermoelectric Cobalt Oxide/Titanium Dioxide Nanocomposite

Liquid phase deposition of CoO@TiO$_2$ coaxial nanocables was deposited into anodic aluminum oxide (AAO) porous templates for making nanocomposites with thermoelectric properties (Su et al. 2011). The morphology and structure of the nanocomposites were revealed by electron microscopic analysis. Seebeck coefficients of the composites were measured as well. The highest absolute value of Seebeck coefficient reached 393 μV/K for the TiO$_2$ nanotube-filled AAO. The TiO$_2$-CoO coaxial nanocable-filled AAO showed a lower absolute Seebeck coefficient value of 300 μV/K. Both composites were n-type. The effect of Ag nanoparticles addition on the thermoelectric behavior of the nanocomposites was also studied.

3.2.1 Liquid Phase Deposition of Core/Shell CoO@TiO$_2$

The following procedures for nanocomposite preparation are cited from previous work (Su et al. 2011). First, the TiO$_2$ nanotubes were generated on the porous walls of an AAO membrane with a pore diameter of 20 nm. Before the synthesis of the TiO$_2$ nanotubes via liquid phase deposition (LPD), both sides of the AAO membrane have to be covered with a self-assembled monolayer to inhibit the deposition of the TiO$_2$ particles on the top and bottom surfaces of the AAO membrane. Consequently, TiO$_2$ nanotubes are only generated on the inner walls of the nanopores rather than on the two flat surfaces of the AAO membrane. The self-assembled monolayer was prepared by mixing 10 mM octadecyltetrachlorosilane in 50 mL of hexane solution. After that, TiO$_2$ nanotubes were generated by immersing the treated AAO membrane with the monolayer into 10 mL of a mixed aqueous solution containing 0.05 M (NH$_4$)$_2$TiF$_6$ and 0.1 M H$_3$BO$_3$ for about 30 minutes to guarantee that the TiO$_2$ nanotubes within the nanopores were of a suitable thickness. The addition of the H$_3$BO$_3$ acid was found to accelerate the TiO$_2$ deposition. The deposited AAO membrane was washed with deionized (DI) water several times and dried completely. The acquired AAO membrane was then heated at 550°C for 2 hours, when the crystallized TiO$_2$ nanotube arrays were eventually obtained. Figure 8.7a gives the top view of the crystallized TiO$_2$ nanotubes.

To promote the hydrolysis of the $(NH_4)_2TiF_6$ compound, the fluorine ions in the solution should be consumed to ensure the formation of the TiO_2. The added boric acid exactly served the role of fluorine ion scavenger. In order to enhance the thermoelectric performance of the composite material, Ag precipitations were generated and placed into the TiO_2 nanotubes to decrease the electrical resistivity. To generate fine Ag nanoparticles and make sure they are dispersed uniformly on the inner wall of the TiO_2 nanotubes, TiO_2-AAO was dipped into 10 mL of 0.05 M $AgNO_3$ solution. After being washed with DI water and dried, it was heated at 500°C for 1 hour. $AgNO_3$ decomposed and produced Ag nanoparticles (NPs). These particles are distributed uniformly into the inner wall of the TiO_2 nanotubes. Figure 8.7b provides the top view of the silver nanoparticle-containing TiO_2 nanotubes (NTs). It must be noted that the Ag particles may not be easily seen due to their tiny sizes.

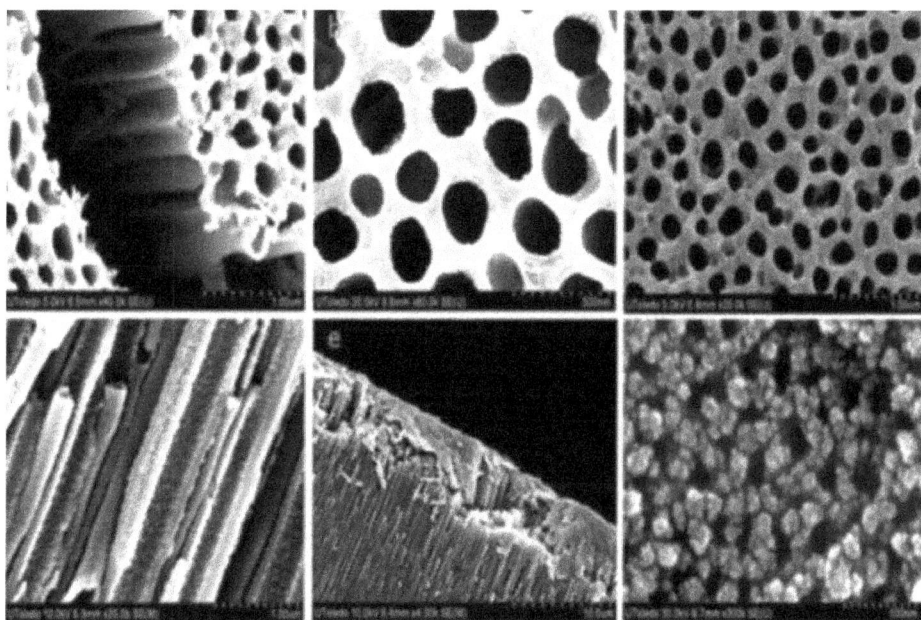

Figure 8.7 SEM images of (a) the top view of the TiO_2 nanotubes; (b) the surface morphology of the TiO_2 nanotubes and Ag particles; (c) the top view of the CoO nanotubes; (d) the cross-section of the TiO_2-CoO nanocables; (e) the cross-section of the TiO_2-CoO nanocables after sputtering coating of Au and (f) the top view of the TiO_2-CoO nanocables after sputtering coating of Au. Reproduced with permission from literature (Su et al. 2011); Thermoelectricity of nanocomposites containing TiO_2-CoO coaxial nanocables. *Scripta Materialia* 64(8): 745–748, © 2010 Elsevier B.V.

The deposition of core/shell CoO@TiO_2 followed the steps as described below. A self-assembled monolayer was applied again to prevent the uneven deposition of CoO on the top and bottom surfaces of the AAO template decorated with TiO_2 (NTs) and/or Ag (NPs). The self-assembled monolayer was the same as mentioned above. The AAO template with TiO_2 nanotubes and/or Ag (NPs) was immersed into 10 mL solutions of 0.05 M Co $(NO_3)_2$ for 20 minutes to obtain TiO_2-Ag-CoO nanocables. After the Co^{2+} ions and NO_3^- ions were impregnated into the TiO_2 nanotubes, the template was air-dried. Then, it was heated at 450°C for 30 minutes to obtain CoO@TiO_2 bilayer nanotubes. This method can be used to deposit a pore CoO thin layer into the inner wall of pares in AAO. The CoO nanotubes within the AAO was observed and an SEM image is shown in Fig. 8.7(c). The top view of the specimen presents the structure with a fairly smooth surface. The cross-section view of the CoO@TiO_2 bilayer nanotubes can be found in Fig. 8.7(d). These tubes have a core/shell nanocable structure with some precipitates inside the wall of the pores. After the CoO@TiO$_2$ bilayer nanotube sample was sputtered with Au coating, the side view was shown in Fig. 8.7(e) and the top surface of the Au-coated CoO@TiO$_2$ bilayer nanotube specimen was revealed in Fig. 8.7(f).

It must be noted that, by dissolving the AAO template into the solution with H$^+$ and F$^-$ ions, the TiO$_2$-CoO nanocables can be released. The dissolution of the AAO substrate in the solution follows the reaction described by Su et al. (2011). Figure 8.8 shows the high-resolution Transmission Electron Microscopic (TEM) images. In Fig. 8.8(a), the TiO$_2$ nanotubes without the high-temperature heat treatment are shown. Figure 8.8(b) reveals the diffraction pattern of the undertreated TiO$_2$ nanotubes. As can be seen from the diffraction pattern that the untreated tubes are amorphous. In Fig. 8.8(c), the TEM image of the heat-treated TiO$_2$-CoO nanocables is shown, and Fig. 8.8(d) represents the diffraction pattern of the heat-treated TiO$_2$-CoO nanocables. Obviously, crystallization during the high-temperature treatment on the TiO$_2$ and CoO occurred because the selected X-ray diffraction pattern came from the crystalline phase or phases (Su et al. 2011).

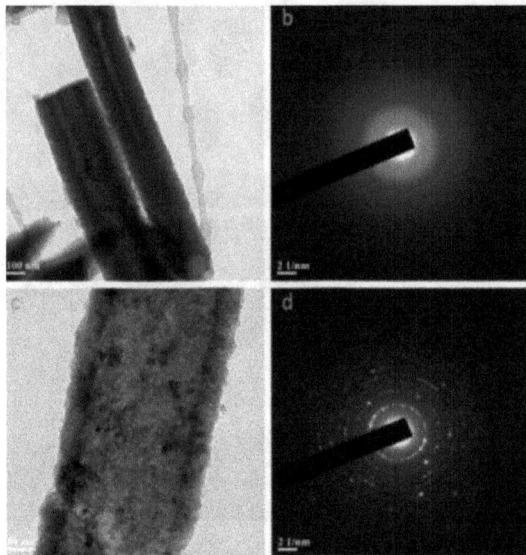

Figure 8.8 TEM images showing (a) TiO$_2$ nanotubes, (b) the diffraction pattern of the TiO$_2$ nanotubes, (c) TiO$_2$-CoO nanocables, and (d) diffraction pattern of the TiO$_2$-CoO nanocables. Reproduced with permission from literature (Su et al. 2011); Thermoelectricity of nanocomposites containing TiO$_2$-CoO coaxial nanocables. *Scripta Materialia* 64(8): 745–748, © 2010 Elsevier B.V.

3.2.2 Thermoelectric Behavior Characterization

To measure the Seebeck coefficients of the TiO$_2$-CoO nanocomposite samples, the two ends of each sample were bonded to strips of Al foils using a silver-based conductive adhesive. The aluminum foil strips can provide good conductive property at the composite/electrode interfaces. One end of the Al foil as the hot end was heated up to a required temperature. The hot end temperature ranged from 40°C to 130°C in the experiments. The other end (cold end) was kept at the ambient temperature of 25 °C (Su et al. 2011).

The Seebeck coefficients at different temperatures were calculated for the four AAO-based nanocomposite specimens (named TiO$_2$, CoO, TiO$_2$/Ag, and TiO$_2$/Ag/CoO) (Su et al. 2011). The calculated results show that the AAO-based TiO$_2$ + Ag has the highest absolute value of Seebeck coefficient, which may be due to the enhanced electrical property due to the incorporation of the Ag nanoparticles. These highly conductively metallic nanoparticles possess a high density of electrons under thermal excitation. In addition, electron tunneling exists among the fine Ag nanoparticles within the nanotubes, which could improve the electrical conductivity of the nanocomposite significantly. The AAO-based CoO nanotube composite showed the lowest absolute value of Seebeck coefficient. This could be due to the semiconductor type change of CoO. As known, CoO shows n- to p-type transition behavior, while the TiO_2 nanotube is n-type. Another reason could be the reduced electrical conductivity of the $CoO@TiO_2$ nanocable as compared with the TiO_2 nanotube. The positive influence of Ag nanoparticles was found to be more intensive than the negative effect of the CoO for the Seebeck coefficient. This is why the AAO-based TiO_2/Ag/CoO nanocomposite has a higher absolute Seebeck coefficient value than the AAO-based TiO_2 nanocomposite. In view of the absolute values of the Seebeck coefficients, the nanocomposites are ranked from high to low as follows: TiO_2/Ag > TiO_2/Ag/CoO > TiO_2 > CoO. It is concluded that the oxide-based nanocomposites containing TiO_2 nanotubes and $CoO@TiO_2$ coaxial nanocables possess a strong Seebeck effect. The absolute value of the Seebeck coefficient for the TiO_2 nanotube-filled AAO is 393 µV/K, while the $CoO@TiO_2$ coaxial nanocable-filled AAO has a slightly lower absolute value of 300 µV/K. Both composites are n-type. The thermoelectric figure of merit of such nanocomposites could potentially be very high due to the low value of the thermal conductivity of the AAO matrix (Su et al., 2011).

4 INTEGRATING LIQUID PHASE DEPOSITION WITH OTHER TECHNIQUES

4.1 Integrating Liquid-Phase Deposition and Hydrothermal Synthesis

Sadeghzadeh-Attar (2020) reported the work on combining hydrothermal synthesis with liquid phase deposition for making the SnO_2/V_2O_5 nanocomposite doped with Nb. The photocatalytic function of the nanocomposite was illustrated. As compared with the titanium dioxide nanoparticle (P25), pristine SnO_2 nanotube and SnO_2/V_2O_5 heterogeneous nanostructure, the Nb-doped SnO_2/V_2O_5 nanocomposite exhibited higher photocatalytic efficiency in decomposing Basic Red 46 (BR46) and producing H_2 production under visible-light with the wavelength λ greater than 420 nm.

The synthesis process is detailed in literature (Sadeghzadeh-Attar 2020). First, hydrous-SnO_2 was deposited in-situ into the nanopores of alumina membranes using the liquid-phase deposition method. Both a hydrolysis equilibrium reaction of $[SnF_6]^{2-}$, and an F^- consuming reaction were involved in the process. The nominal pore size, average thickness and diameter of the alumina membranes were 100 nm, 60 µm and 21 mm, respectively. The alumina membranes served as both the scavenger for F^- and the template for SnO_2 nanotube formation. Hydrous-SnO_2 nanotubes were subsequently converted to crystalline SnO_2 nanotubes after annealing.

Second, Nb-doped SnO_2 nanotubes were synthesized through the liquid phase co-deposition from the mixed solutions consisting of 0.1 M ammonium hexafluorostannate, $(NH_4)_2SnF_6$ and several Nb-containing solutions with different concentrations. The Nb-containing solutions were made by dissolving niobium (V) oxide, Nb_2O_5 and ammonium fluoride, NH_4F in deionized water. The solutions containing Nb were added to the aqueous solution of $(NH_4)_2SnF_6$. The nominal concentration ratios of Nb dopant in the solutions were 0, 0.5, 1, 2 and 4 mol%. Porous alumina membranes were then inserted vertically into the treatment solutions. The liquid phase deposition was sustained at room temperature for 4 hours. After deposition, the specimens were taken out and rinsed with distilled water and ethanol before air-drying. Then, the samples were placed in 1 M NaOH solution for the removal of alumina membranes.

Next, the hydrothermal synthesis was used to add V_2O_5 onto the Nb-doped SnO_2 to obtain the Nb-doped SnO_2/V_2O_5 nanocomposite. Briefly, a solution of 0.1 M ammonium metavanadate, NH_4VO_3, was prepared in distilled water as the source for V_2O_5. The synthesized Nb-doped SnO_2 nanotubes were added to the solution. The mixture was then transferred to a Teflon-lined stainless-steel autoclave. The autoclave was sealed and heated up to 120°C and for 10 hours. Then the hydrothermal synthesis installation was cooled down naturally to room temperature. The reaction products were centrifuged with ethanol and distilled water and naturally dried at ambient temperature. The amount of added vanadium was controlled at 6 wt.% approximately. The samples were placed into a muffle furnace and calcinated at 600 °C for 2 hours in air. Finally, the Nb-doped SnO_2/V_2O_5 heterogeneous nanostructured composites with 0, 0.5, 1, 2 and 4 mol% Nb were obtained and these samples were named SV, 0.5 Nb-SV, 1.0 Nb-SV, 2.0 Nb-SV and 4.0 Nb-SV, respectively.

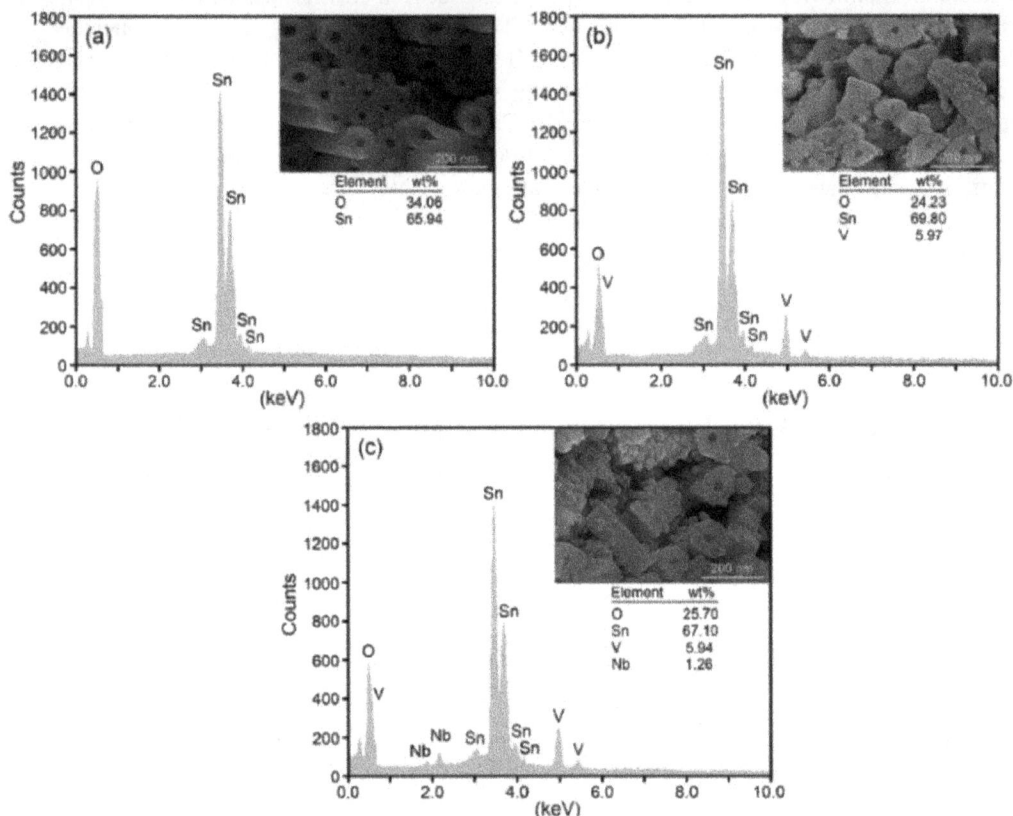

Figure 8.9 Energy dispersive spectra (EDS) and field emission scanning electron microscopic images: (a) SnO_2 nanotubes, (b) SnO_2/V_2O_5 heterogeneous nanostructure, and (c) 2.0 Nb-doped SnO_2/V_2O_5 nanocomposite. Reproduced from literature (Sadeghzadeh-Attar 2020); Enhanced photocatalytic hydrogen evolution by novel Nb-doped SnO_2/V_2O_5 hetero nanostructures under visible light with simultaneous basic red 46 dye degradation. *Journal of Asian Ceramic Societies* 8(3): 662–676, © 2020 The author.

Figure 8.9 presents the Energy Dispersive Spectra (EDS) and Field Emission Scanning Electron Microscopic (FE-SEM) images for various materials studied (Sadeghzadeh-Attar 2020). The inset of each subfigure shows the corresponding FE-SEM image of each material. From Fig. 8.9(a), it can be seen that the nanotubes contain the elements Sn and O without any impurity on the surfaces of the nanotubes, confirming the high purity of synthesized SnO_2 nanotubes. Further, the semiquantitative analysis shows that the atomic percentages of Sn and O in the nanotubes are 65.94% and 34.06%, respectively (Sadeghzadeh-Attar 2020). The calculated theoretical stoichiometric atomic proportion of O(66.7%) contained in SnO_2 is close to the actual EDS result. The FE-SEM image of SnO_2 nanotubes is presented in the inset of Fig. 8.9a. This inset image reveals that a highly-ordered tubular structure, uniform and parallel to each other with an outer diameter of about 100 nm and a wall thickness of about 35 nm is formed in the alumina membrane pores by the LPD process. Figure 8.9(b) presents the EDS of SnO_2/V_2O_5 heterogeneous nanostructure, and the SEM image is shown by the inset. Obviously, the V_2O_5 were deposited in the form of nanoparticles by hydrothermal synthesis. Figure 8.9(c) provides the electron microscopic analysis results of the 2.0 Nb-doped SnO_2/V_2O_5 nanocomposite. The inset image shows Nb- and V-containing oxide nanoparticles adhered to the tin oxide nanotubes.

The optoelectrical properties of the nanocomposites were studied and presented (Sadeghzadeh-Attar 2020). Comparison of the light emission from various nanomaterials can be found in Fig. 8.10(a). The SnO_2 nanotubes exhibit an obvious PL emission signal around 568 nm. Introducing dopants to form nanocomposites resulted in photon quenching effect. This allows the enhanced solar light absorption due to the incorporation of the Nb and V containing oxides. This behavior can be further revealed by the photocurrent measurement experiments. The photo responses of selected nanocomposites were measured and the results were plotted in Fig. 8.10(b). The results clearly show that the light absorption of the liquid-phase deposited SnO_2 nanotubes was enhanced by the hydrothermal synthesized V_2O_5 nanoparticles. The doping of Nb can further increase the light absorption of the SnO_2/V_2O_5 nanocomposite. This is the main reason for the significant improvement in the photocatalysis performances of the Nb-doped SnO_2/V_2O_5 nanocomposite in hydrogen generation and organic dye decomposition as shown in the paper (Sadeghzadeh-Attar 2020).

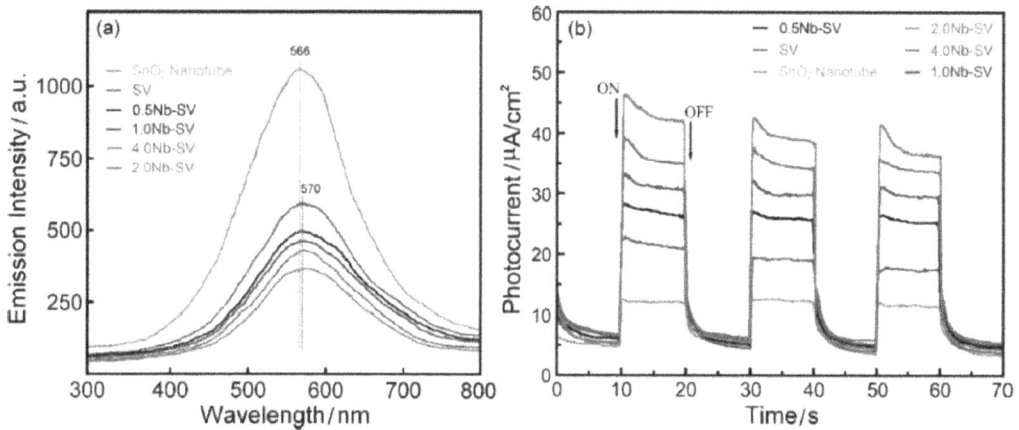

Figure 8.10 Results from comparative studies on the photoelectronic properties of SnO_2 nanotubes, SnO_2/V_2O_5 heterogeneous nanostructure, 0.5, 1.0, 2.0 and 4.0 Nb-doped SnO_2/V_2O_5 nanocomposite: (a) photoluminescence spectra, (b) photocurrent changes under visible light excitation. Reproduced from literature (Sadeghzadeh-Attar 2020); Enhanced photocatalytic hydrogen evolution by novel Nb-doped SnO_2/V_2O_5 hetero nanostructures under visible light with simultaneous basic red 46 dye degradation. *Journal of Asian Ceramic Societies* 8(3): 662–676, © 2020 The author.

4.2   Integrating Liquid-Phase Deposition and Carbonization

Gou et al. (2019) prepared a porous ternary $MnTiO_3/TiO_2/C$ composite from Mn_2O_3/polyaniline precursor through liquid phase deposition (LPD) and subsequent carbonization. The generated $MnTiO_3/TiO_2/C$ composite was used as the anode material for Lithium-Ion Batteries (LIBs). It has been found that the composite has good electrochemical performance. The cycling stability is high (418 mAh g^{-1} after 200 cycles at 500 mA g^{-1}). The rate capability reached 270 mAh g^{-1} at 4 A g^{-1}. The coulombic efficiency was greater than 98% after 200 cycles. As compared with pristine Mn_2O_3 or Mn_2O_3/TiO_2 composite, the $MnTiO_3/TiO_2/C$ composite has better electrochemical performance. This is due to the unique porous structure and the synergistic effect of the three constituents. It is believed that this porous ternary $MnTiO_3/TiO_2/C$ composite may be used as a new generation of anode material for LIBs.

4.3   Integrating Liquid-Phase Deposition and *in-situ* Chemical Polymerization

In the work performed by Yuan et al. (2018), combining liquid-phase deposition (LPD) and *in-situ* chemical polymerization was used to make a carbon nanotube (CNT)@SnO_2@polypyrrole (PPy) nanocomposite for the anode of sodium-ion batteries. The CNT@SnO_2@PPy nanocomposite has the one-dimensional structure of the CNT. The diameter of the carbon nanotube is around 40 nm. The thickness of the PPy coating is about 7 nm. The sandwich formed CNT@SnO_2@ PPy electrode showed excellent rate capability. It maintained a high capacity of 226 mAh/g after 100 cycles at the current density of 100 mA/g. It was believed that the synergistic effects among CNT, SnO_2 and PPy enhanced the electrochemical performance of the anode. The good electrical conductivity could reduce the aggregation tendency of Sn during the charge/discharge cycles; thus, the pulverization of the electrode was prevented.

The CNT@SnO_2@PPy was prepared using a similar route as for making the SnO_2/polypyrrole (SnO_2/PPy) hollow nanospheres as reported in (Yuan et al. 2017). During the liquid-phase deposition, carbon was used as the template. Subsequent *in-situ* chemical-polymerization was carried out for PPy coating formation. Figure 8.11 illustrates the two steps for synthesizing the CNT@SnO_2@PPy nanocomposite (Yuan et al. 2018). The core/shell CNT@SnO_2 nanostructure was made first through the LPD approach. CNTs were ultrasonically dispersed in distilled water. Then, SnF_2 was added under stirring. The product from the reaction was centrifuged, cleaned and air-dried. Calcination was conducted at 500°C for 2 hours in an Ar atmosphere to crystalize the tin oxide (Yuan et al. 2018). The second step for preparing the CNT@SnO_2@PPy composite was the PPy coating deposition on the CNT@SnO_2 nanostructure through a chemical polymerization process. Such a chemical polymerization process was also demonstrated for making CuO/PPy composite in (Yin et al. 2012). Briefly, the CNT@SnO_2 and surfactant sodium dodecylsulfate (SDS) were added into the water by ultrasonication. Then, pyrrole monomer, HCl (1.0 M) and $(NH_4)_2S_2O_8$ (0.1 M) were added. The polymerization of pyrrole was maintained at 0°C for 3 hours under magnetic stirring.

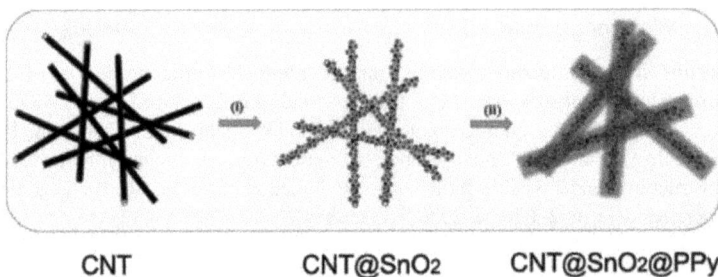

CNT   CNT@SnO2   CNT@SnO2@PPy

Figure 8.11 Schematic showing synthesis of the CNT@SnO_2@PPy nanocomposite. Reproduced with permission from reference (Yuan et al. 2018); Sandwiched CNT@SnO_2@PPy nanocomposites enhancing sodium storage. *Colloids and Surfaces A: Physicochemical and Engineering Aspects* 555: 795–801, © 2018 Elsevier B.V.

The CNT/SnO_2/PPy nanocomposite was examined by transmission electron microscopy (TEM). The microscopic analysis results can be found in Fig. 8.12 (Yuan et al. 2018). Figure 8.12(a) is a TEM image showing the locations of CNT, SnO_2, and PPy in the nanocomposite. Figure 8.12(b) presents an image with the Selected Area Electron Diffraction (SAED) pattern as an inset. Such a pattern indicates the existence of the SnO_2 rutile phase. Figure 8.12(c), a High-Resolution

Transmission Electron Microscopic (HRTEM) image of the CNT@SnO$_2$@PPy nanocomposite, reveals an interplanar spacing of 0.33 nm which matches that of the (110) plane of the tetragonal SnO$_2$ crystal.

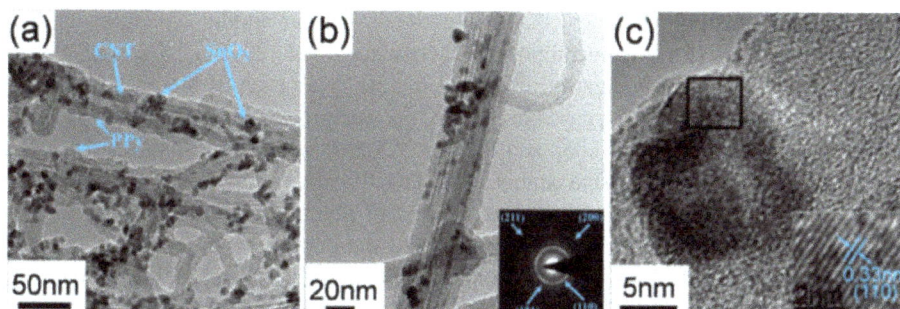

Figure 8.12 TEM analysis results of the CNT@SnO$_2$@PPy nanocomposite: (a) image showing the CNT, SnO$_2$ and PPy in the nanocomposite, (b) image with selected area electron diffraction (SAED) pattern as the inset, and (c) a high-resolution transmission electron microscopic (HRTEM) image of the CNT@SnO$_2$@PPy nanocomposite. Reproduced with permission from reference (Yuan et al. 2018); Sandwiched CNT@SnO$_2$@PPy nanocomposites enhancing sodium storage. *Colloids and Surfaces A: Physicochemical and Engineering Aspects* 555: 795–801, © 2018 Elsevier B.V.

Figures 8.13(a) and 8.13(b) present the Electrochemical Impedance Spectra (EIS) of the electrodes made from CNT@SnO$_2$ and CNT@SnO$_2$@PPy, respectively (Yuan et al. 2018). The Nyquist plots for both the electrodes present a similar shape with a depressed semicircle in the high-frequency region and a straight line in the low-frequency domain. The semicircle is responsible for charge transfer resistance at the electrode/electrolyte interface. The semicircle's diameter of CNT@SnO$_2$@PPy electrode after 10 cycles is much smaller than that of a fresh cell, demonstrating the enhanced electron transport after cycling. Compared with CNT@SnO$_2$ electrode (inset of Fig. 8.13(a)), CNT@SnO$_2$@PPy electrode (inset of Fig. 8.13(b)) shows a smaller semicircle diameter, which indicates that the PPy coating on the surface of CNT@SnO$_2$ promotes the electron transport. The EIS curves of CNT@SnO$_2$ and CNT@SnO$_2$@PPy electrodes were also simulated by the equivalent circuit in inset of Fig. 8.13(b). R_0 is the solution resistance, and R_{ct} stands for the charge transfer resistance. CPE2 is the constant phase element. It can be seen that the simulated lines fit the EIS data nicely.

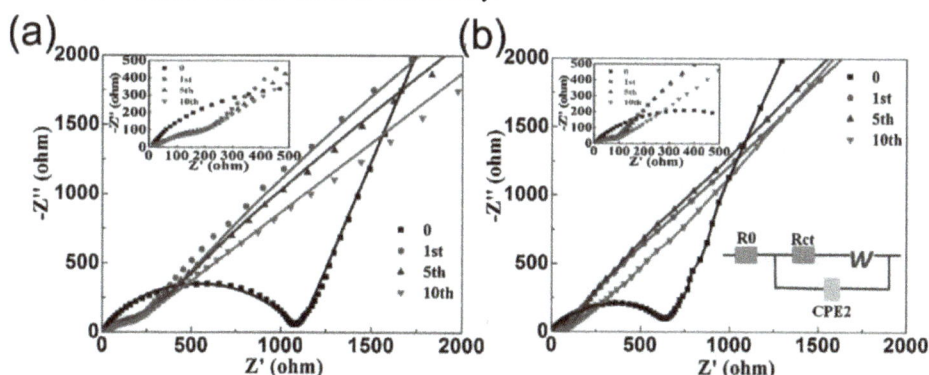

Figure 8.13 Nyquist plots of the two synthesized nanocomposites before and after 1, 5 and 10 cycles: (a) CNT@SnO$_2$, and (b) CNT@SnO$_2$@PPy. The insets in both subfigures represent the magnified regions close to the origins. The equivalent electrical circuit is shown by the inset of (b) which was used to fit the experimental impedance data. The dots stand for experimental results and the lines were generated by simulations. Reproduced with permission from reference (Yuan et al. 2018); Sandwiched CNT@SnO$_2$@PPy nanocomposites enhancing sodium storage. *Colloids and Surfaces A: Physicochemical and Engineering Aspects* 555: 795–801, © 2018 Elsevier B.V.

4.4 INTEGRATING LIQUID PHASE DEPOSITION WITH ANODIC OXIDATION

The liquid phase deposition method can be integrated with an anodic oxidation approach for making TiO_2/porous Si (PSi) nanocomposites (Mizuhata et al. 2015). TiO_2 was obtained by liquid-phase deposition. The anodization was used for the generation of the porous silicon (PSi). Scanning electron microscopy-energy-dispersive X-ray spectroscopy showed that the TiO_2 was only deposited in the fine pores of anodized PSi (ca. 7.4 nm) when the PSi surface was anodized in the presence of Ti ions as F^- scavengers. The TiO_2/PSi nanocomposites were fabricated by anodization of PSi in a H_2TiF_6 electrolyte at a constant potential. The amount of Ti deposited was maximum at 300 mV vs. Ag/AgCl electrode, and the deposition process was controlled by varying the applied potential for PSi anodization. The charge/discharge capacities of the fabricated TiO_2/PSi nanocomposites as Li-ion battery anodes were determined. Improvements in the charge/discharge capacity were achieved.

During the processing, the Si wafer or PSi working electrode, platinum mesh counter electrode, Ag/AgCl reference electrode and a salt bridge were fixed in a three-electrode electrochemical cell; H_2TiF_6/H_2O/EtOH solution was used as the electrolyte. The Si working electrode was anodized in the electrolyte at a constant potential from 0 to 550 mV vs. the Ag/AgCl electrode. Anodization was performed for 30 minutes at room temperature. After anodization, the working electrode was removed from the electrolyte solution and rinsed with deionized distilled water and methanol, and dried in an Ar atmosphere at room temperature. The TiO_2 thin film on the Si wafer was characterized.

4.5 INTEGRATING LIQUID PHASE DEPOSITION WITH PLASMA SYNTHESIS

Liquid-phase plasma synthesis integrates the plasma synthesis technique with the liquid phase deposition. Wei et al. (2013) showed that silicon quantum dot/carbon (SiQD/C) can be successfully prepared by the liquid-phase plasma synthesis. The synthesis procedures for SiQDs can also be found in more detail in an earlier published paper by Kang et al. (Kang et al. 2007a, b; 2009). Absolute ethanol was used as the medium for preparing silicon quantum dot/carbon composites. The prepared SiQDs were mixed with ethanol via ultrasonication. The mixed solution was then placed in the chamber with the Radio Frequency (RF) generator setup as shown in (Yuan et al. 2008). Modulating the RF power allowed the plasma generation from the tungsten electrode tip of the RF microelectronic device. The SiQD/C nanocomposites were obtained after the sparking for a time period of 20 minutes (Yuan et al. 2008). To remove the SiQDs from the nanocomposite, the SiQDs/C composite sample was added into a mixed solution of HF (50 mL, 10%) and H_2O_2 (5 mL, 30%) and stirring for 5 hours. This allowed the Si to dissolve into the solution. The released carbon was separated from the solution by centrifugation. After that, the carbon material was washed with deionized water and absolute ethanol and dried in a vacuum oven at 60°C. Figure 8.14 shows the liquid phase deposition and plasma synthesis scheme and the TEM analysis results of the silicon quantum dot/carbon nanocomposite. For comparison, pure carbon was made by this liquid-phase plasma synthesis technique in ethanol (Wei et al. 2013).

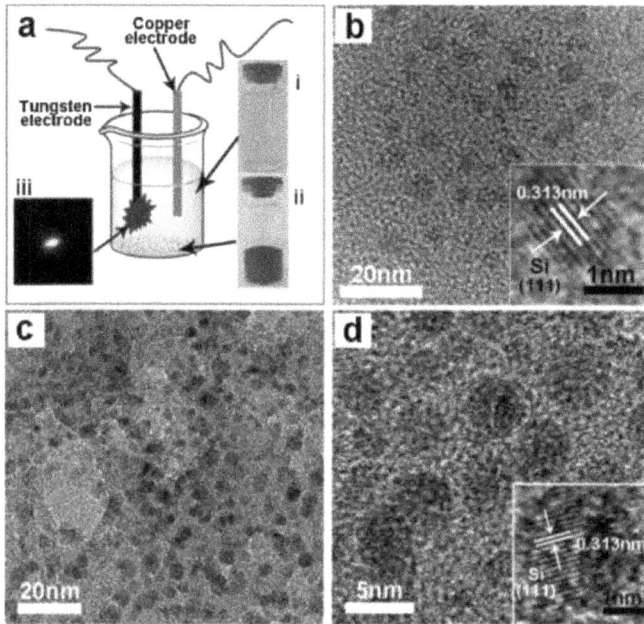

Figure 8.14 Liquid phase deposition and plasma synthesis of silicon quantum dot/carbon nanocomposite: (a) schematic showing the integrated liquid phase deposition and plasma synthesis technique, (b) TEM and HRTEM images of SiQDs, (c) TEM image of the SiQDs/C nanocomposite. (d) HRTEM image of the SiQDs/C nanocomposite. Reproduced with permission from reference (Wei et al. 2013); Liquid-phase plasma synthesis of silicon quantum dots embedded in carbon matrix for lithium battery anodes. *Materials Research Bulletin* 48(10): 4072–4077, © 2013 Elsevier B.V.

5 CONCLUSION

Liquid phase deposition (LPD) is a low-cost, scalable processing technique for various nanocomposites including metal oxide thin films, nanoparticles, ordered nanofibers, nanowires or nanorods. LPD reactions generally occur in uniform solutions. No specific substrate material is involved into the reaction. Thus, homogeneous structures can be deposited on various kinds of substrates. The substrates include those with large surface areas and complex morphologies. Consequently, the LPD technique is especially suitable for ordered nanostructure deposition. Various nanocomposites can be made by the liquid phase deposition (LPD) technique. These nanocomposites have been used for photocatalysis, Li- and Na-ion storage, photovoltaics and thermoelectric energy conversion.

■ References

Al-Fandi, T., F. Al Marzouqi, A.T. Kuvarega, B.B. Mamba, S.M.Z. Al Kindy, Y. Kim, et al. 2019. Visible light active CdS@TiO$_2$ core-shell nanostructures for the photodegradation of chlorophenols. Journal of Photochemistry and Photobiology A: Chemistry 374: 75–83.

Alizadeh, S., N. Fallah and M. Nikazar. 2020a. Synthesis and characterization of direct Z-scheme CdS/TiO$_2$ nanocatalyst and evaluate its photodegradation efficiency in wastewater treatment systems. Chemical Papers 74(1): 133–143.

Alizadeh, S., N. Fallah and M. Nikazar. 2020b. Photocatalytic degradation of dimethyl sulphoxide by CdS/TiO$_2$ core/shell catalyst: A novel measurement method. The Canadian Journal of Chemical Engineering 98 (2): 491–502.

Deki, S., Y. Aoi, Y. Miyake, A. Gotoh and A. Kajinami. 1996. Novel wet process for preparation of vanadium oxide thin film. Materials Research Bulletin 31(11): 1399–1406.

Deki, S., N. Yoshida, Y. Hiroe, K. Akamatsu, M. Mizuhata and A. Kajinami. 2002. Growth of metal oxide thin films from aqueous solution by liquid phase deposition method. Solid State Ionics 151(1–4): 1–9.

Deki, S., S. Iizuka, A. Horie, M. Mizuhata and A. Kajinami. 2004a. Liquid-phase infiltration (LPI) process for the fabrication of highly nano-ordered materials. Chemistry of Materials 16(9): 1747–1750.

Deki, S., A. Nakata and M. Mizuhata. 2004b. Fabrication of metal-oxide nanoparticles by the liquid phase deposition method in the heterogeneous system. Electrochemistry 72(6): 452–454.

Flynn Jr., C.M. 1984. Hydrolysis of inorganic iron(III) salts. Chemical Reviews 84(1): 31–41.

Gou, Q.Z., C. Li, X.Q. Zhang, B. Zhang, D.Y. Huang and C.X. Lei. 2019. Facile synthesis of porous ternary $MnTiO_3/TiO_2/C$ composite with enhanced electrochemical performance as anode materials for lithium ion batteries. Energy Technology 7(5): 1800761.

Hishinuma, A., T. Goda, M. Kitaoka, S. Hayashi and H. Kawahara. 1991. Formation of silicon dioxide films in acidic solutions. Applied Surface Science 48–49: 405–408.

Hsu, M.C., I.C. Leu, Y.M. Sun and M.H. Hon. 2005. Fabrication of $CdS@TiO_2$ coaxial composite nanocables arrays by liquid-phase deposition. Journal of Crystal Growth 285(4): 642–648.

Jung, J.H., E.D. Han, B.H. Kim, Y.H. Seo and Y.M. Park. 2019. Ultra-low light reflection surface using metal-coated high-aspect-ratio nanopillars. Micro & Nano Letters 14(3): 313–316.

Kang, Z.H., C.H.A. Tsang, N.B. Wong, Z.D. Zhang and S.T. Lee. 2007a. Silicon quantum dots: A general photocatalyst for reduction, decomposition, and selective oxidation reactions. Journal of the American Chemical Society 129(40): 12090–12091.

Kang, Z.H., C.H.A. Tsang, Z.D. Zhang, M.L. Zhang, N.B. Wong, J.A. Zapien, et al. 2007b. A polyoxometalate-assisted electrochemical method for silicon nanostructures preparation: From quantum dots to nanowires. Journal of the American Chemical Society 129(17): 5326–5327.

Kang, Z.H., Y. Liu, C.H.A. Tsang, D.D.D., Ma, X. Fan, N.B. Wong, et al. 2009. Water-soluble silicon quantum dots with wavelength-tunable photoluminescence. Advanced Materials 21(6): 661–664.

Mizuhata, M., A. Katayama and H. Maki. 2015. On-site fabrication and charge-discharge property of TiO_2 coated porous silicon electrode by the liquid phase deposition with anodic oxidation. Journal of Fluorine Chemistry 174(SI): 62–69.

Nagayama, H., H. Honda and H. Kawahara. 1988. A new process for silica coating. Journal of the Electrochemical Society 135(8): 2013–2016.

Sadeghzadeh-Attar, A. 2020. Enhanced photocatalytic hydrogen evolution by novel Nb-doped SnO_2/V_2O_5 heteronanostructures under visible light with simultaneous basic red 46 dye degradation. Journal of Asian Ceramic Societies 8(3): 662–676.

Su, L., Y.X. Gan and L. Zhang, L. 2011. Thermoelectricity of nanocomposites containing TiO_2-CoO coaxial nanocables. Scripta Materialia 64(8): 745–748.

Wei, Y., H. Yu, H.T. Li, H. Ming, K.M. Pan, H. Huang, et al. 2013. Liquid-phase plasma synthesis of silicon quantum dots embedded in carbon matrix for lithium battery anodes. Materials Research Bulletin 48(10): 4072–4077.

Wu, K.L., P.C. Wu, J.F. Zhu, C. Liu, X.J. Dong, J.N. Wu, et al. 2019. Synthesis of hollow core-shell $CdS@TiO_2/Ni_2P$ photocatalyst for enhancing hydrogen evolution and degradation of MB. Chemical Engineering Journal 360: 221–230.

Yin, Z.G., Y.H. Ding, Q.D. Zheng and L.H. Guan. 2012. CuO/polypyrrole core–shell nanocomposites as anode materials for lithium-ion batteries. Electrochemistry Communications 20: 40–43.

Yuan, Q.H., Y. Xin, G.Q. Yin, X.J. Huang, K. Sun and Z.Y. Ning 2008. Effect of low-frequency power on dual-frequency capacitively coupled plasmas. Journal of Physics D: Applied Physics 41(20): 205209.

Yuan, J.J., C.H. Chen, Y. Hao, X.K. Zhang, B. Zou, R. Agrawal, et al. 2017. SnO_2/polypyrrole hollow spheres with improved cycle stability as lithium-ion battery anodes. Journal of Alloys and Compounds 691: 34–39.

Yuan, J.J., Y.C. Hao, X.K. Zhang and X.F. Li. 2018. Sandwiched $CNT@SnO_2@PPy$ nanocomposites enhancing sodium storage. Colloids and Surfaces A: Physicochemical and Engineering Aspects 555: 795–801.

Chapter 9

Epoxy Nanocomposites Containing Hybrid Montmorillonite/Geopolymer Filler for Piping Application

Yusrina Mat Daud*[1,3], Azlin Fazlina Osman[1,3] and Mohammad Firdaus Abu Hashim[1,2,3]

[1]Faculty of Chemical Engineering Technology, Universiti Malaysia Perlis (UniMAP), Kompleks Pusat Pengajian Jejawi 2, 02600 Arau Perlis, Malaysia

[2]Faculty of Mechanical Engineering Technology, Universiti Malaysia Perlis (UniMAP), Pauh Putra Campus, 02600 Arau, Perlis

[3]Center of Excellence Geopolymer and Green Technology (CEGeoGTech), Universiti Malaysia Perlis (UniMAP), 02600, Arau, Perlis, Malaysia

[1]*yusrina@unimap.edu.my*; [2]*azlin@unimap.edu.my*; [3]*firdaushashim@unimap.edu.my*

1 INTRODUCTION

Polymer nanocomposites having layered nanosilicates (clay) as filler have been extensively studied in recent years involving the use of various types of polymer matrices such as polyamide, polystyrene, polypropylene, polyurethane, ethylene-vinyl acetate, phenolic resin, unsaturated polyester and epoxy resin (Fu and Qutubuddin 2001; Leszczyńska et al. 2007; Pavlidou et al. 2008; Fengge Gao 2012; Shiravand et al. 2016; Taghaddosi et al. 2017). The incorporation of layered nanosilicates in polymer matrices provides several benefits to the produced nanocomposites such as enhancement in mechanical, physical and barrier properties. For instance, several works proved that the layered nanosilicates can significantly improve the creep and fatigue resistance, toughness, thermal stability, hardness, abrasion resistance, permeability and biostability of the host polymer (Alexandre and Dubois 2000; Sinha Ray and Okamoto 2003; Pavlidou et al. 2008; Azeez et al. 2013). Additionally, the advantages of layered nanosilicates include a high aspect ratio nanolayer structure, unique intercalation and exfoliation behavior, which lead to efficient stiffening and strengthening effects. Natural and synthetic layered nanosilicates are abundantly available and low in cost. There are many types of layered nanosilicates existing in clay mineral sphere that can be classified by their characteristics, such as sheet structure, chemical structure and mineral group

---

*Corresponding Author

such as kaolinite, montmorillonite or smectite, illite and chlorite (Alexandre and Dubois 2000; Okamoto 2003; Uddin 2008; Uddin 2008). However, among the different types of clay minerals, montmorillonite (MMT) is the most commonly used for the preparation of composites. MMT receives special attention among the smectite group by the ability to show extensive interlayer expansion or swelling because of the structure of the montmorillonite consists of an octahedral sheet of alumina or magnesia that is surrounded by two tetrahedral sheets of silica. Besides, MMT has great potential in terms of a specific area and high aspect ratio with a thickness of each individual layer is about 1 nm and width around 500 nm (Zhu et al. 2011). Literature suggests that MMT has been used as a reinforcing filler in various types of thermoplastic, thermoset, thermoplastic elastomer and copolymer matrices (Sinha Ray and Okamoto 2003; Mittal 2009; Osman et al. 2014; Taghaddosi et al. 2017).

Thermosetting polymers have received greater interest for structural and high-performance applications as compared to other types of polymers due to their strength and thermal stability, which originate from their three-dimensional network of bonds (crosslinking). Among thermoset materials, epoxy resin is largely highlighted by many researchers since the material shows versatility with excellent chemical, heat resistance, high adhesive strength, good impact resistance, hardness and good adhesion to many substrates (Ramesh and Velmurugan 2006; Phonthammachai et al. 2011). Moreover, the properties of the epoxy can be tailored and improved by incorporating various types of nanoparticle such as montmorillonite, carbon nanotubes and nano-titania, producing a new form of material called epoxy nanocomposite. Nowadays, epoxy nanocomposites are demanded by many industrial applications such as adhesive, construction material, piping, composites, laminates and coating (Isik et al. 2003; Jumahat et al. 2012a). In the growth of nanotechnology research area, the incorporation of nanoparticles such as nanoclay was observed as an effective way to toughen epoxy resin. The researches on epoxy/layered silicate nanocomposites are widely explored and commonly reported. However, there is a limitation on the use of a single filler or nanofiller for epoxy reinforcement. It cannot guarantee the matrix to achieve high mechanical performance when subjected to various mechanical loads (for example in compression and flexural) (Miyagawa et al. 2006; Aziz 2010; Jumahat et al. 2012b). The enhancement in mechanical properties was sometimes accompanied by a reduction in thermal properties (Li et al. 2012). Therefore, researchers have started to employ hybrid fillers in order to obtain epoxy nanocomposite materials with greater combination properties.

Hybridization on the epoxy matrix is a method that allows a third phase material to integrate into the epoxy composite/nanocomposite system whereby the incorporation of two reinforcement materials into epoxy resin significantly demonstrated beneficial improvement in mechanical properties compared to single filler (Wang et al. 2011). Currently, there are three types of reinforcement materials for epoxy hybrid that have been studied; soft rubber particles, rigid inorganic fillers and semi-rigid organic fillers where substantial differences in mechanical properties have been reported (Liang and Pearson 2010). There are many other publications reported on the production of epoxy composites/nanocomposites incorporating two or more particulate reinforcement (hybrid fillers) (Chen and Morgan 2009; Liang and Pearson 2010; Wang et al. 2011). However, only a small number of published researches focused on the use of two physically/chemically different silicate/clay materials as hybrid fillers in the epoxy matrix (Osman et al. 2016a). Specifically, there is no published work on the use of layered clay and clay-based geopolymer as hybrid fillers in an epoxy matrix.

Geopolymer is a new family type of thermosetting inorganic polymer resin (Song et al. 2013). It is a cementitious material formed by aluminosilicates produced predominantly silica (Si) and aluminum (Al) materials. Generally, the process of geopolymerization involves chemical reactions under highly alkaline conditions. The discovery of geopolymer-based materials contributes to excellent mechanical properties, high early strength, abrasion resistance and thermal stability that lead to many application fields such as ceramics, binder, matrices for hazardous waste stabilization, fire resistance and high tech materials. Currently, geopolymer-based composites have become popular among researchers in order to improve several properties of geopolymer materials such as brittleness and their low flexural strength in order to extend the limit of application in the structural field. The well-known method is by combining the geopolymer with organic polymers such as polyvinyl acetate, polypropylene, polyvinyl alcohol or water-soluble organic polymers (Colangelo et al. 2013). However, the organic polymer usually acts as an additive in geopolymer systems.

2 EPOXY/LAYERED NANOSILICATES NANOCOMPOSITES

Epoxy is thermosetting matrix material used often in the composites industry. However, it is highly susceptible to internal damage caused by the low-velocity impact due to the inherent brittleness of the cured resin. Therefore, the deficiency of epoxy resin can be resolved with a combination of layered nanosilicates.

Generally, incorporating layered nanosilicates in the epoxy resin offers synergistic improvement in the properties and performance of the resultant nanocomposites. The main attraction of adding layered nanosilicates is the low cost and well-developed intercalation chemistry to achieve a nanostructure from a micron size filler. The layered silicate minerals can also easily disperse in liquid thermosetting resin such as epoxy by simple mechanical mixing, shear mixing or sonication (Ratna et al. 2005). Several works are reporting on layered silicates as fillers in the epoxy. Most of the works focused on the synthesis, characterization and properties of composites made by adding a nanofiller to epoxy matrix system as well as the effect of processing technique on the composites. The following paragraphs summarize the findings.

Alsagayar et al. (2015) investigated the tensile and flexural properties of the epoxy/nanoclay nanocomposites prepared by exfoliation method (solvent-based method). The tensile and flexural strength of the epoxy has been successfully increased with the addition of the nanoclay, especially when the content of the nanoclay was greater than 2 phr. However, it was observed that the addition of high loading of nanoclay (more than 2 phr) decreased both strains at break and toughness of the epoxy.

Wang et al. (2015) investigated the use of the modified copolymer which has been *in-situ* synthesized for modifying the nanoclay during the epoxy curing process. The incorporation of this copolymer resulted in the formation of the interpenetrating network (IPN) structure within the epoxy network, leading to a double IPN structure. The main purpose of this strategy was to enhance the tensile and fracture toughness of the epoxy/nanoclay nanocomposites without compromising the strength of the material. It was observed that a topological interlocking between the nanofiller-modifier-matrix has made the matrix strong and tough. At optimum nanoclay loading, the tensile strength, modulus and elongation at break of the epoxy/nanaoclay nanocomposite were enhanced, indicative of a remarkable improvement of the tensile and fracture toughness while maintaining reasonable strength. Additionally, due to the small amounts of the modified clay incorporated, the epoxy/clay prepolymer remains as a non-Newtonian liquid and its shear thinning behavior at higher frequencies renders the system processable by simple mechanical stirring.

In another research, Wang and Qin (2007) investigated the effect of ultrasonic stirring time on the mechanical and thermal properties of the epoxy–nanoclay nanocomposites. The ultrasonic stirring was suggested to promote the diffusion of the epoxy and curing agent into the nanoclay layer and fasten the curing reaction of epoxy with an amine. With the increase in ultrasonic stirring time, the nanoclay layer could be separated further; thus curing reaction was promoted, and the cross-linking system with branched alkyl chain was obtained.

Mohan and Kanny (2011) studied the catalytic effect of the organically modified montmorillonite (organo-MMT) in the epoxy matrix by monitoring the curing parameters, which are exothermic temperature and curing time. The data were compared with those of the epoxy nanocomposite with the untreated MMT. It was observed that the change in temperature versus curing time provided more insight into the role of the organoclay in the nanocomposite system such as its concentration and structure in the nanocomposites system. For instance, they found that the addition of a greater amount of the organo-MMT in the epoxy matrix reduces the catalytic effect and leads to the formation of the intercalated nanocomposites structure due to improper curing at the inter-gallery and extra-gallery region of the polymer. Meanwhile, the untreated clay shows no catalytic effect but acts as a micron-scale filler in the epoxy matrix. Improvement in tensile properties was observed in the epoxy nanocomposite with the organo-MMT, but not with the untreated MMT.

Gârea et al. (2010) investigated the new epoxy-amine adducts as modifier agents for the nanoclay filler to enhance the compatibility of the organic and organic phases of the epoxy matrix. The adducts were synthesized by reacting different epoxy resin types with a stoichiometric amount of polyetheramine which includes hydrophobic and hydrophilic units. They found that these epoxy-amine adducts can be intercalated within the layers of montmorillonite and thus lead to an increase in the basal distance. Their findings suggest that the final structure of the epoxy–montmorillonite nanocomposite strongly depends on the agents used for the montmorillonite modification.

Zaman et al. (2011) used three types of modifiers to investigate the interface strength of the epoxy/clay nanocomposites: ethanolamine (denoted ETH), Jeffamine M2070 (M27) and Jeffamine XTJ502 (XTJ). Among these three types of the modifier, the XTJ led to a good interface between the clay layers and the matrix as it bridged the layers with matrix by a chemical reaction, proving through the Fourier transform infrared spectroscopy analysis. On the other hand, the presence of the M27 resulted in an intermediate interface strength between the epoxy and clay due to the molecular entanglement between the grafted M27 chains and the matrix molecules. Meanwhile, the ETH resulted in weak interface bonding between the matrix and nanofiller because neither chemical bridging nor molecular entanglement was involved. The studies of mechanical and thermal properties and morphology at a wide range of magnification show that the strong interface promoted the highest level of exfoliation and dispersion of clay layers, and achieved the most increment in Young's modulus, fracture toughness and glass transition temperature (Tg) of the matrix.

Nuhiji et al. (2011) investigated the effect of both the mixing technique and heating rate of the epoxy curing process on the dispersion of montmorillonite (MMT) clay in the epoxy matrix. A two-part epoxy thermoset 'EPON 828' based on diglycidyl ether of bisphenol A/epichlorohydrin (DGBEA) and 'Ardur 2954' along with I.30E MMT clay was used to synthesize the specimen. MMT clay corresponding to 5 wt.% of epoxy resin was combined with a given amount of 'EPON 828' resin and heated to 70°C. This mixture was stirred for 2 hours using a magnetic stirrer. The sonication technique was used for 30 minutes to facilitate the dispersion of clay. Epoxy/clay nanocomposites were heated to 130°C at either $3°C \pm 1°C$/min or $10°C \pm 2°C$/min and held isothermally for 30 minutes. The vacuum had been maintained at -70 KPa during cure to prevent the mixture from boiling. The increase in heating rate from 3°C/min to 10°C/min during cure reduced the viscosity by 60% facilitating the penetration of polymer chains into clay galleries. The use of a high amplitude sonication dispersion technique that combined with a rapid heating rate resulted in a two-fold increase in clay gallery distances. As the degree of dispersion was enhanced, the flexural modulus and strength properties were found to increase by 15 and 40% respectively.

3 THE COMBINATION OF GEOPOLYMER AND POLYMER TO ENHANCE PROPERTIES AND PROCESSABILITY

Geopolymer is inexpensive and an eco-friendly material obtained through synthetic procedures with a natural silica-aluminates source and it is well known in the civil engineering field for construction purposes and other engineering applications. These materials show excellent mechanical properties, low shrinkage, thermal stability, freeze-thaw, acid and fire resistance, long-term durability and recyclability, therefore the geopolymer-based materials can be applied in various applications (Ferone et al. 2013). Despite being used in so many construction industries, geopolymer materials have been widely investigated as composite materials which are comprised of several types of filler such as basalt fibers, carbon fibers, glass fibers, wood fibers, cotton or woven fabric (Yuana et al. 2004; Alomayri et al. 2014; Omar et al. 2015). Moreover, the polymeric materials in fiber form were frequently added as reinforcement in geopolymer matrix systems (Yunsheng et al. 2008). On the contrary, the use of geopolymer as a filler material in the polymer is a new and interesting field to explore. However, very few researches were found to deal with this exciting field (Mustafa Al Bakri et al. 2011). The related works will be emphasized next. To date, the geopolymer was used either as a matrix or filler in the composite system. The following paragraphs discuss the use of geopolymer as a matrix and polymeric materials as filler.

Literature studies on geopolymer functionalities with polymeric materials have been reported in some of the organic polymers such as polypropylene (Liu and Liang 2011). The polypropylene (PP) fiber was added to reinforce the geopolymer in order to improve the tensile failure, shrinkage and brittleness. In other research, polyvinyl alcohol (PVA) short fibers were used to improve the ductility of the hardened geopolymeric cement (Yunsheng et al. 2008). The addition of PVA fiber has changed the impact failure mode of the geopolymer from brittle to ductile resulting in great impact toughness. Whereas, Du et al. (2016) reported the use of epoxy resin in geopolymer to enhance the thickening properties of the epoxy resin. The addition of epoxy has resulted in the enhancement of the compressive strength of the hardened geopolymer paste. The studies also reveal the potential of epoxy resin as a compatibilizer when diluted and dispersed in the geopolymer suspension. Table 9.1 summarizes the use of polymer filler in geopolymer composites.

Table 9.1   Geopolymer composites with polymer fillers

Geopolymer materials	Polymer fillers	Mechanical properties	References
Low calcium fly ash	Woven cotton fabric	Compressive and flexural strength decreased after being conditioned at the temperature of 200°C to 1000°C.	Alomayri et al. 2014
Fly ash and Metakaolin	Polyvinyl alcohol (PVA) fibers	Impact strength decreased but impact stiffness and toughness increased.	Yunsheng et al. 2008
Fly ash	Cotton fibers	Compressive strength improved from 24.78 MPa (geopolymer matrix) to 28.42 MPa.	Korniejenko et al. 2016

4   GEOPOLYMER FILLER IN POLYMERIC MATERIALS

Geopolymer is also used as filler to reinforce polymer in the form of polymer/geopolymer composite/nanocomposite. For instance, the combination of geopolymer and epoxy matrix was conducted by Hussain et al. (2005) for automobile and marine applications. The geopolymer in viscous paste was added into a mixture of epoxy resin and curing agent and then allowed to cure. This combination has enabled the improvement in thermal stability and fire resistance of the epoxy. However, there was no mechanical testing data reported to describe the effect of geopolymer filler on the mechanical properties of the epoxy composite. Table 9.2 summarizes the use of geopolymer filler in polymer composites.

Table 9.2   Polymer composites with geopolymer filler

Polymer materials	Geopolymer fillers	Mechanical properties	References
Low-density polyethylene	Wollastonite and metakaolin	Tensile modulus increased with a higher proportion of geopolymer, tensile strength decreased when the mass of the geopolymer exceeds 20%.	Yuana et al. 2004
Natural rubber	Fly ash	Tensile and modulus 100% increased with 10 phr geopolymer filler addition.	Yangthong et al. 2018

The addition of geopolymer filler in natural rubber also has been investigated using fly ash geopolymer filler to observe the effect of geopolymer on cure characteristic, mechanical, dynamic, thermal and morphology of the matrix (Korniejenko et al. 2016). The findings show that the geopolymer filler can fasten the curing time and increase the tensile strength and thermal properties of the natural rubber.

Another investigation on developing composites via geopolymer and thermoplastic mixing was performed by Yuana et al. (2004) using low-density polyethylene (LDPE) for glass-ceramic application. The LDPE powder was added into a viscous paste of metakaolin-based geopolymer and mixed in the mechanical processor. In order to improve the adhesion between the inorganic powders and polymeric matrix, a coupling agent was used besides the preheating process in the oven. The preliminary results obtained from the study showed the potential of geopolymer for enhancing the mechanical properties of LDPE. However, the research only involved the application of geopolymer filler produced from one source of geopolymer material, which is metakaolin thereby limiting the knowledge on geopolymer composition and subsequent effects to the polymer composite properties.

The above literature studies suggest that very few researches involved with combining geopolymer and polymer, especially in studying the geopolymer as filler material in the polymer matrix. Moreover, the published works only reported the application of geopolymer filler in viscous paste form rather than dry powder form.

5   FABRICATION OF EPOXY/LAYERED NANOSILICATES/GEOPOLYMER NANOCOMPOSITES FOR PIPING APPLICATION

Epoxy is a commonly used polymer to replace heavier non-polymeric materials for structural applications such as ceramic and metal. However, the mechanical and physical properties of epoxy can be severely reduced when exposed to harsh and corrosive environments such as in the oil and gas pipeline. The structure of epoxy can chemically degrade and leads to deterioration in its mechanical and physical properties. These may hamper the long-term use of epoxy in piping applications, where mechanical and physical integrity are required to allow stability in various environmental conditions. In that case, epoxy has been restored by adding the nanofiller such as organo-montmorillonite (organo-MMT). Organo-MMT is a promising reinforcing nanofiller for epoxy matrix due to its high aspect ratio, hydrophobicity and capability to enhance the mechanical properties of the matrix when added in small quantities. However, the morphology and structure properties found in the nanoclay type filler affect the stability of the mechanical properties of the composite. Therefore, due to this limitation of single nanofiller, researchers have initiated the idea of using hybrid filler in the epoxy matrix to obtain composite with greater combination properties. However, only a small number of published researches focused on the use of two physically/ chemically different silicate/clay materials as hybrid fillers in the epoxy matrix (Wang et al. 2011, Osman et al. 2016b).

Geopolymer is a new revolution of existing concrete that formed by alkaline activation from aluminosilicates source. These inorganic materials have been studies over the past decade by many researchers and have superior characteristics such as high compressive strength, low shrinkage, acid resistance and thermal resistance. Therefore, the use of organo-MMT and geopolymer filler (geo-filler) in producing the epoxy hybrid nanocomposite engenders better strength and durability as compared to neat epoxy, thereby, a novel and exciting field of exploration. It was anticipated that the incorporation of these hybrid fillers would improve the physical and mechanical properties of the epoxy without compromising its lightweight characteristic. Our research team has produced the epoxy/MMT/geopolymer hybrid nanocomposite for piping application. The product of pipes (with epoxy and polypropylene as matrix) are shown in Fig. 9.1. We have prepared the epoxy/MMT/geopolymer hybrid nanocomposites using two different types of geopolymer source material, which are fly ash and kaolin. The mechanical properties of the resultant hybrid nanocomposites were investigated and the results are discussed next.

Figure 9.1   Pipes made of polymer/MMT/geopolymer nanocomposites.

6 THE ROLE OF GEOPOLYMER FILLER IN ENHANCING THE MECHANICAL PROPERTIES OF THE EPOXY/LAYERED NANOSILICATES NANOCOMPOSITES FOR PIPING APPLICATION

Recently, the application of geopolymer in fine dry form has been developed with a combination of epoxy layered nanosilicates. This combination has enabled the excellent improvement in mechanical properties compared to single-layered silicates filler in the epoxy system.

The effect of fly ash-based geopolymer in epoxy-layered nanosilicates nanocomposites over flexural properties and morphological characterization were investigated (Mat Daud et al. 2015). Nanocomposite materials were prepared using an epoxy resin, digilcydyl ether of bisphenol A (DGEBA) as a polymer matrix, and MMT surface modified with 35 wt.% to 45 wt.% dimethyl dialkyl (C14–C18) amine as nanofiller. The curing agent, isophorondiamine (IPDA), was used to harden the nanocomposites. Fly ash with particle sizes of range 50 μm to 100 μm was used as source material to form the geopolymer for use as a hybrid filler with the MMT. The hybrid nanocomposites were prepared with fly ash-based geopolymer filler in 1 wt.% to 7 wt.% loadings. As presented in Fig. 9.2, the addition of low content of geopolymer filler (1 wt.%) resulted in the decrease of the flexural strength of the epoxy/MMT nanocomposite. The flexural strength of the epoxy/MMT and epoxy/MMT nanocomposite with 1 wt.% geopolymer filler was 9.19 MPa and 6.09 MPa, respectively. However, flexural strength has been significantly increased to 14.98 MPa when 3 wt.% of fly ash geopolymer filler was used to form the hybrid nanocomposite. This work proved that the fly ash-based-geopolymer filler has brought a positive impact to the enhancement of the flexural properties of the epoxy/MMT nanocomposite system. Therefore, it is a potent co-filler for the MMT in the development of the epoxy nanocomposite with high flexural strength.

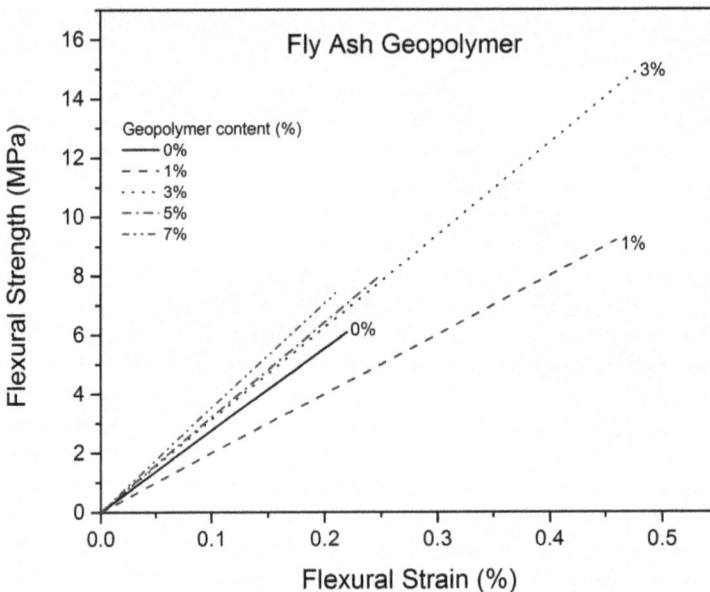

Figure 9.2   The effect of fly ash geopolymer content on the flexural strength of the epoxy/MMT nanocomposite.

We also investigated the effect of fly ash-based geopolymer filler addition in the mechanical properties of the epoxy layered nanosilicates nanocomposites using a compressive test. A series of nanocomposites with fly ash-based geopolymer content of 1 phr to 7 phr was prepared. Apparently, the addition of low content of fly ash (1 phr) as the hybrid filler in the epoxy/MMT composite resulted in the decrease of compressive strength of the material, from 24.6 MPa to 12.69 MPa. However, the compressive strength has successfully increased to 31.06 MPa, when 3 phr fly ash geopolymer filler was added as hybrid filler in the epoxy/MMT nanocomposite.

In a more recent publication, we reported the effects of kaolin-based geopolymer filler in the compressive strength of the epoxy/MMT nanocomposites. A series of nanocomposites with kaolin-based geopolymer filler varies from 1 phr to 7 phr were prepared and the compressive strength results are illustrated in Fig. 9.3. It was discovered that the addition of low content of kaolin-based geopolymer filler (1 phr) caused a reduction in the compressive strength of the epoxy/MMT nanocomposite. However, the compressive strength was markedly increased when 3 phr of kaolin-based geopolymer filler was employed to produce the nanocomposite. This finding validates the capability of the kaolin-based-geopolymer filler in enhancing the compressive strength of the epoxy nanocomposite when being used as a co-filler with the MMT.

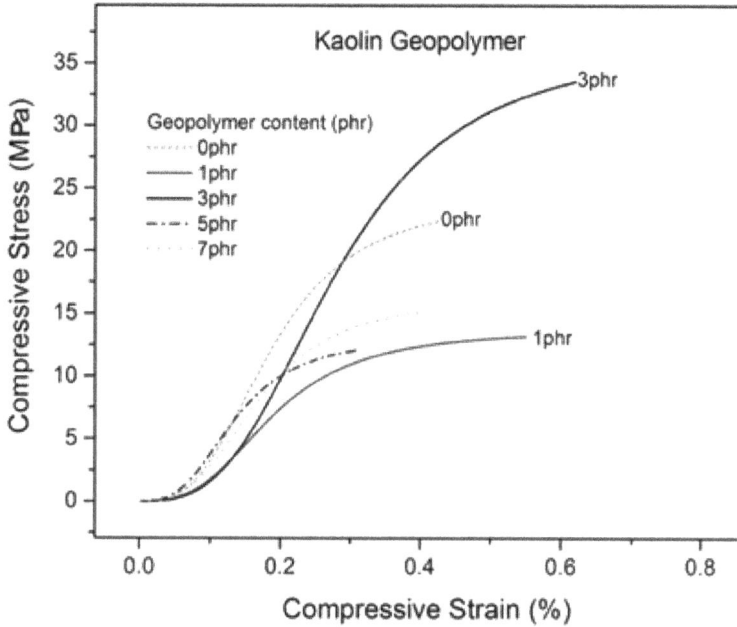

Figure 9.3   The effect of fly ash geopolymer content on the compressive strength of the epoxy/MMT nanocomposite.

The morphological study was conducted to analyze the fractured surface of the epoxy-MMT hybrid nanocomposite with fly ash-based geopolymer filler and epoxy-MMT hybrid nanocomposite with kaolin-based geopolymer filler using Field Emission Scanning Electron Microscope (FESEM). The images are shown in Fig. 9.4. The fractured surface morphology of these materials was compared with the neat epoxy and epoxy-MMT nanocomposite. It can be seen in Fig. 9.4(a) that the neat epoxy exhibits a relatively smooth fractured surface with single crack propagation. However, at high magnification FESEM image in Fig. 9.4(b) neat epoxy illustrates a typical fracture feature, indicating poor crack resistance and brittle failure behavior. This finding is in line with that the one reported by Zainuddin et al. 2010. It is believed that this feature contributes to the production of epoxy with possessing low strength in bending.

On contrary, the fractured surface of the epoxy-MMT nanocomposite shows considerably different fractographic features as shown in Fig. 9.4(c) and magnifies the image in (d). Generally, a rougher surface is seen upon adding organo-MMT into the epoxy matrix (Subramaniyan and Sun 2007; Zainuddin et al. 2010; Castrillo et al. 2015). As the epoxy-MMT is incorporated with optimum geo-fillers (kaolin) and fly ash Fig. 9.4(e, f, g, and h) a much rougher surface is seen in the hybrid nanocomposite, with the presence of few voids. According to Lin et al. (2006), the surface roughness of the epoxy matrix may contribute to the toughening and strengthening effects. This is because more energy is needed to allow the breaking of the nanocomposite structure. This factor supports the mechanical data obtained, where the epoxy-MMT hybrid nanocomposite with geo-filters possesses higher compressive and flexural strength than the epoxy-MMT nanocomposite without geo-fillers.

The XRD patterns of the best epoxy-MMT nanocomposite and hybrid epoxy-MMT nanocomposites containing different geo-fillers; kaolin and fly ash (in 1 phr to 7 phr) are shown in Fig. 9.5(a) and (b). The broad diffraction peak observed around $2\theta = 19°$ for the neat epoxy is related to the amorphous structure of the epoxy (Motahari et al. 2013). This amorphous structure is attributed to the reaction between the epoxy monomer and amine hardener when the crosslinking process initiates.

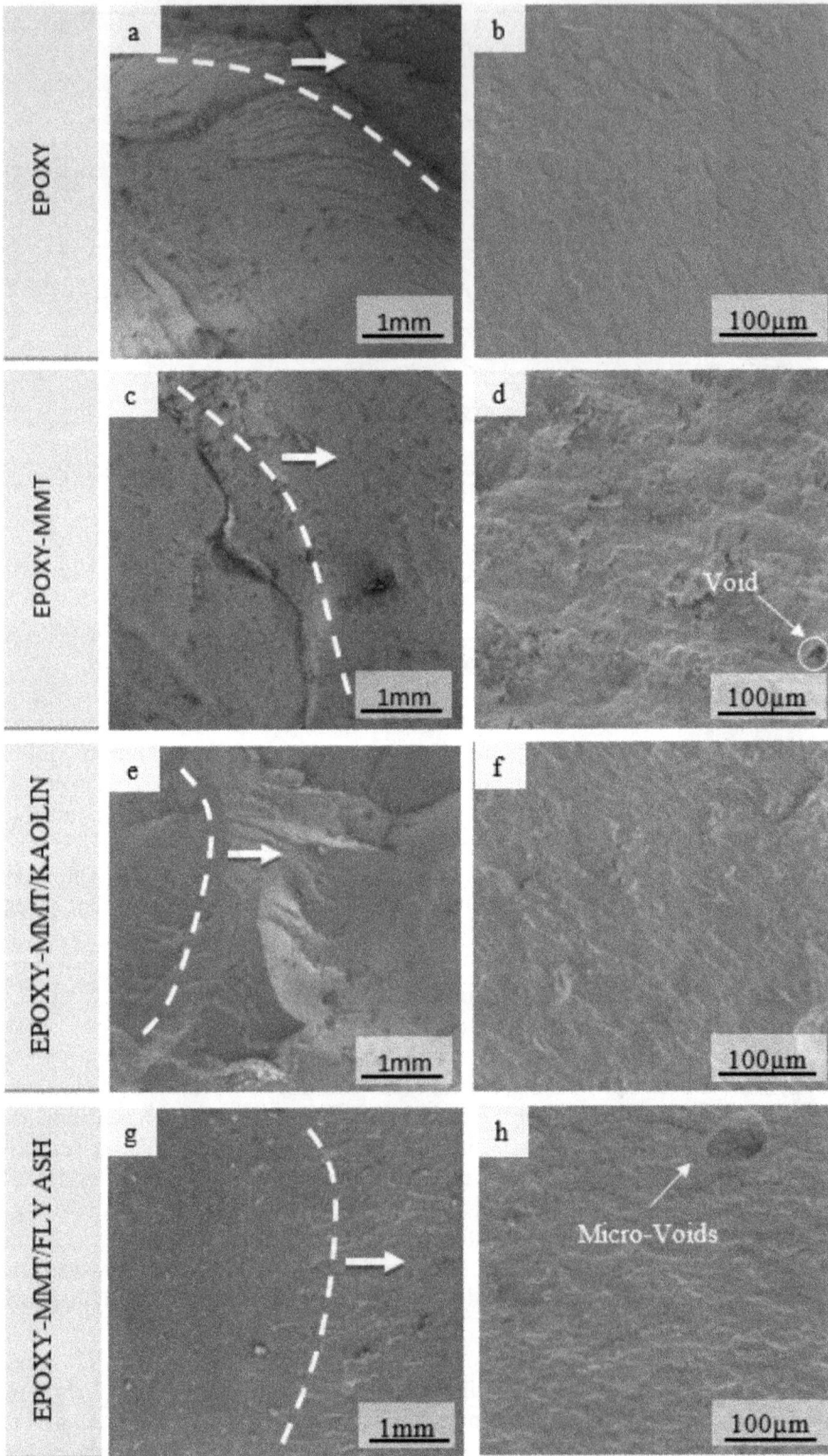

Figure 9.4 Fractured surface by FESEM analysis for (a) and (b) neat epoxy, (c) and (d) Epoxy-MMT, (e) and (f) Epoxy-MMT/Kaolin and (g) and (h) Epoxy-MMT/Fly Ash nanocomposites.

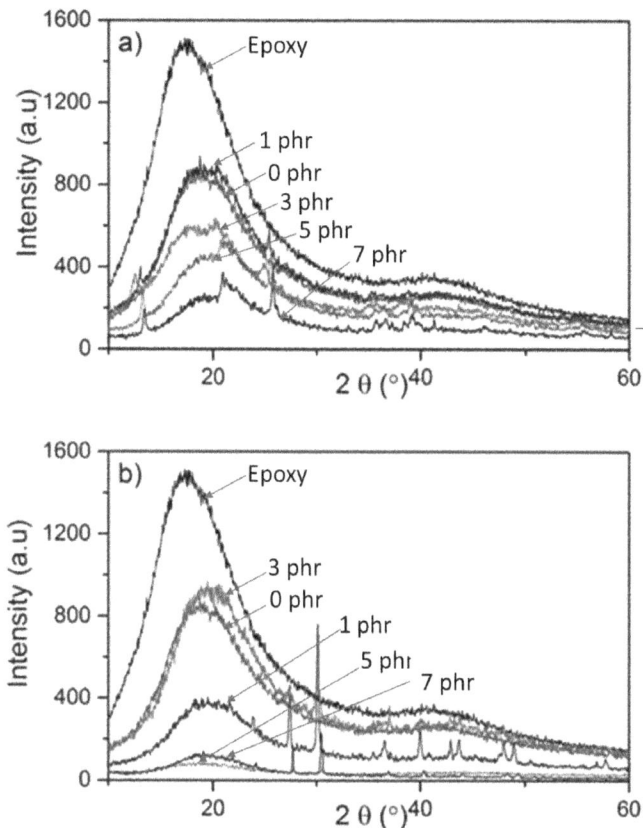

Figure 9.5 XRD patterns of epoxy-MMT and its hybrid composites and different geopolymer filler (a) kaolin geopolymer filler and (b) fly ash geopolymer filler.

With the addition of an optimum (3 phr) MMT and geo-filler (kaolin and fly ash) to the epoxy matrix, the diffraction peaks become broader and less intense. The peak intensity at 2θ of around 20° (20° 2θ) was slightly decreased as the MMT incorporated into the epoxy system. However, when 1 phr kaolin geo-filler was added into the epoxy-MMT system, the intensity increased slightly with those peaks related to epoxy-MMT and epoxy-MMT/kaolin 1phr were almost identical, suggesting the geo-fillers have good interaction with the organo-MMT in the epoxy system, thereby resulted in better performance in the mechanical properties. Since the peak at 20° 2θ can be related to the crosslinking structure of the epoxy, the addition of MMT and MMT/kaolin geopolymer fillers was found to less significantly reduce the crosslink density of the host epoxy as compared to other hybrid composites containing a higher amount of kaolin geo-filler. The addition of a high amount of geopolymer filler (>1 phr) could result in the agglomeration and poorly dispersed particles inside the epoxy matrix. As a result, the bulky and agglomerated geo-filler might disrupt the epoxy crosslinking structure, and this can be observed through the lowering of the 20° 2θ peak intensity. Such prominent interference in the epoxy structure could reduce its molecular bonding forces and subsequently lowered the mechanical and physical properties of the epoxy. This could be one of the reasons why the epoxy hybrid nanocomposites containing a high amount of geopolymer filler (>1 phr) exhibit deteriorating mechanical and thermal properties.

It can also be seen that the addition of 3 phr fly ash geopolymer filler into the epoxy-MMT system resulted in the least intensity reduction of the 20° 2θ peak. This suggests that this particular amount of fly ash geopolymer filler brings only insignificant modification in the epoxy structure and therefore, could result in enhancement of the mechanical and thermal properties of the epoxy through optimal dispersion of both MMT and fly ash geopolymer particles.

SUMMARY

This chapter highlights the promising properties of epoxy-based composites and nanocomposites incorporating single and hybrid fillers. The use of layered silicate (nanoclay) filler may provide several benefits that can improve the mechanical and thermal performance of the polymer, reducing the cost and widening the application of the epoxy as it turned into polymer composites/ nanocomposites. Based on the research outcomes, it can be highlighted that the use of hybrid MMT/geopolymer fillers can more efficiently improve the mechanical properties of the epoxy as compared to the single MMT filler, given that the optimum content of the geopolymer filler was employed. The findings indicate that the hybrid epoxy nanocomposites with layered silicates (MMT) and geopolymer filler have great potential to be developed for structural application, particularly to produce pipes.

■ References

Alexandre, M. and P. Dubois. 2000. Polymer-layered silicate nanocomposites: preparation, properties and uses of a new class of materials. Materials Science and Engineering: R: Reports 28: 1–63.

Alomayri, T., F.U.A. Shaikh and I.M. Low. 2014. Effect of fabric orientation on mechanical properties of cotton fabric reinforced geopolymer composites. Materials and Design 57: 360–365.

Alsagayar, Z.S., A.R. Rahmat, A. Arsad and S.N.H. binti Mustaph. 2015. Tensile and flexural properties of montmorillonite nanoclay reinforced epoxy resin composites. Advanced Materials Research 1112: 373–376.

Azeez, A.A., K.Y. Rhee, S.J. Park and D. Hui. 2013. Epoxy clay nanocomposites—Processing, properties and applications: A review. Composites Part B: Engineering 45: 308–320.

Aziz, M. 2010. A Study on the effect of hardener on the mechanical properties of epoxy resin.

Castrillo, P.D., D. Olmos, H.J. Sue and J. González-Benito. 2015. Mechanical characterization and fractographic study of epoxy-kaolin polymer nanocomposites. Composite Structures 133: 70–76.

Chen, C. and A.B. Morgan. 2009. Mild processing and characterization of silica epoxy hybrid nanocomposite. Polymer 50: 6265–6273.

Colangelo, F., G. Roviello, L. Ricciotti, C. Ferone and R. Cioffi. 2013. Preparation and characterization of new geopolymer-epoxy resin hybrid mortars. Materials 6: 2989–3006.

Du, J., Y. Bu, Z. Shen, X. Hou and C. Huang. 2016. Effects of epoxy resin on the mechanical performance and thickening properties of geopolymer cured at low temperature. Materials & Design 109: 133–145.

Fengge, G. 2012. Advances in Polymer Nanocomposites. Woodhead Publishing.

Ferone, C., G. Roviello, F. Colangelo, R. Cioffi and O. Tarallo. 2013. Novel hybrid organic-geopolymer materials. Applied Clay Science 73: 42–50.

Fu, X. and S. Qutubuddin. 2001. Polymer – clay nanocomposites: Exfoliation of organophilic montmorillonite nanolayers in polystyrene. Polymer 42: 807–813.

Gârea, S.A., H. Iovu and G. Voicu. 2010. The influence of some new montmorillonite modifier agents on the epoxy-montmorillonite nanocomposites structure. Applied Clay Science 50: 469–475.

Hussain, M., R. Varely, Y.B. Cheng, Z. Mathys and G.P. Simon. 2005. Synthesis and thermal behavior of inorganic-organic hybrid geopolymer composites. Journal of Applied Polymer Science 96: 112–121.

Isik, I., U. Yilmazer and G. Bayram. 2003. Impact modified epoxy/montmorillonite nanocomposites: Synthesis and characterization. Polymer 44: 6371–6377.

Jumahat, A., C. Soutis, J. Mahmud and N. Ahmad. 2012a. Compressive properties of nanoclay/epoxy nanocomposites. Procedia Engineering 41: 1607–1613.

Jumahat, A., C. Soutis, S.A. Abdullah and S. Kasolang. 2012b. Tensile properties of nanosilica/epoxy nanocomposites. Procedia Engineering 41: 1634–1640.

Korniejenko, K., E. Frączek, E. Pytlak and M. Adamski. 2016. Mechanical properties of geopolymer composites reinforced with natural fibers. Procedia Engineering 151: 388–393.

Leszczyńska, A., J. Njuguna, K. Pielichowski, J.R. Banerjee, M. Science, M. Sciences, et al. 2007. Polymer/montmorillonite nanocomposites with improved thermal properties. Part I. Factors influencing thermal stability and mechanisms of thermal stability improvement. Thermochimica Acta 453: 75–96.

Li, X., Z.J. Zhan, G.R. Peng and W.K. Wang. 2012. Nano-disassembling method-A new method for preparing completely exfoliated epoxy/clay nanocomposites. Applied Clay Science 55: 168–172.

Liang, Y.L. and R.A. Pearson. 2010. The toughening mechanism in hybrid epoxy-silica-rubber nanocomposites (HESRNs). Polymer 51: 4880–4890.

Lin, L.Y., J.H. Lee, C.E. Hong, G.H. Yoo and S.G. Advani. 2006. Preparation and characterization of layered silicate/glass fiber/epoxy hybrid nanocomposites via vacuum-assisted resin transfer molding (VARTM). Composites Science and Technology 66: 2116–2125.

Liu, S.P. and C.W. Liang. 2011. Preparation and mechanical properties of polypropylene/montmorillonite nanocomposites – After grafted with hard/soft grafting agent. International Communications in Heat and Mass Transfer 38: 434–441.

Mat Daud, Y., K. Hussin, C.M. Ruzaidi, A.F. Osman, M.M. Al Bakri Abdullah and M. Binhussain. 2015. Characterization of epoxy-layered silicates filled with fly ash-based geopolymer. Applied Mechanics and Materials 754–755: 225–229.

Mittal, V. 2009. Polymer layered silicate nanocomposites: A review. Materials 2: 992–1057.

Miyagawa, H., R.J. Jurek, A.K. Mohanty, M. Misra and L.T. Drzal. 2006. Biobased epoxy/clay nanocomposites as a new matrix for CFRP. Composites Part A: Applied Science and Manufacturing 37: 54–62.

Mohan, T.P. and K. Kanny. 2011. Study of catalytic effect of nanolayered montmorillonite organoclays in epoxy polymer. ISRN Nanotechnology 2011: 1–7.

Motahari, A., A. Omrani, A.A. Rostami and M. Ehsani. 2013. Preparation and characterization of a novel epoxy based nanocomposite using tryptophan as an eco-friendly curing agent. Thermochimica Acta 574: 38–46.

Mustafa Al Bakri, A.M., H. Kamarudin, M. Bnhussain, I.K. Nizar and W. Mastura. 2011. Mechanism and chemical reaction of fly ash geopolymer cement- A review. Journal of Asian Scientific Research. 1: 247–253.

Nuhiji, B., D. Attard, G. Thorogood, T. Hanley, K. Magniez and B. Fox. 2011. The effect of alternate heating rates during cure on the structure–property relationships of epoxy/MMT clay nanocomposites. Composites Science and Technology 71: 1761–1768.

Okamoto, M. 2003. Polymer Layered Silicates Nanocomposites, Vol. 14. Rapra Technology Limited.

Omar, M.F., L.Y. Wei, M.M. Al Bakri Abdullah and K. Hussin. 2015. Effect of Hybrid Fillers on the Thermal Properties of UHMWPE/Chitosan-ZnO Composites. Applied Mechanics and Materials 754–755: 71–76.

Osman, A.F., G. Edwards and D. Martin. 2014. Effects of processing method and nanofiller size on mechanical properties of biomedical thermoplastic polyurethane (TPU) nanocomposites. Advanced Materials Research 911: 115–119.

Osman, A.F., A.R. Abdul Hamid, M. Rakibuddin, G. Khung Weng, R. Ananthakrishnan, S.A. Ghani, et al. 2016a. Hybrid silicate nanofillers: Impact on morphology and performance of EVA copolymer upon in vitro physiological fluid exposure. Journal of Applied Polymer Science 44640.

Osman, A.F., T.F.M. Fitri, M. Rakibuddin, F. Hashim, S.A.T. Tuan Johari, R. Ananthakrishnan, et al. 2016b. Pre-dispersed organo-montmorillonite (organo-MMT) nanofiller: Morphology, cytocompatibility and impact on flexibility, toughness and biostability of biomedical ethyl vinyl acetate (EVA) copolymer. Materials Science and Engineering: C 74: 194–206.

Pavlidou, S. and C.D. Papaspyrides. 2008. A review on polymer–layered silicate nanocomposites. Progress in Polymer Science 33: 1119–1198.

Phonthammachai, N., X. Li, S. Wong, H. Chia, W.W. Tjiu and C. He. 2011. Fabrication of CFRP from high performance clay/epoxy nanocomposite: Preparation conditions, thermal–mechanical properties and interlaminar fracture characteristics. Composites Part A: Applied Science and Manufacturing 42: 881–887.

Ramesh, T.P.M.Æ.M. and K.Æ.R. Velmurugan. 2006. Thermal, mechanical and vibration characteristics of epoxy-clay nanocomposites. Journal of Materials Science 41: 5915–5925.

Ratna, D., P. Joong, Y. Chang and D. Han. 2005. Epoxy Composites: Impact Resistance and Flame Retardancy. Macromolecules 16: 2171–83.

Shiravand, F., J.M. Hutchinson and Y. Calventus. 2016. A novel comparative study of different layered silicate clay types on exfoliation process and final nanostructure of trifunctional epoxy nanocomposites. Polymer Testing 56: 148–155.

Sinha Ray, S. and M. Okamoto. 2003. Polymer/layered silicate nanocomposites: A review from preparation to processing. Progress in Polymer Science 28: 1539–1641.

Song, X.L., X.M. Cui, K.S. Lin, G.J. Zheng and Y. He. 2013. Reprint of hot-pressure forming process of PVC/geopolymer composite materials. Applied Clay Science 73: 51–55.

Subramaniyan, A.K. and C.T. Sun. 2007. Toughening polymeric composites using nanoclay: Crack tip scale effects on fracture toughness. Composites Part A: Applied Science and Manufacturing 38: 34–43.

Taghaddosi, S., A. Akbari and R. Yegani. 2017. Preparation, characterization and anti-fouling properties of nanoclays embedded polypropylene mixed matrix membranes. Chemical Engineering Research and Design 125: 35–45.

Uddin, F. 2008. Clays, nanoclays, and montmorillonite minerals. Metallurgical and Materials Transactions A 39: 2804–2814.

Wang, J. and S. Qin. 2007. Study on the thermal and mechanical properties of epoxy-nanoclay composites: The effect of ultrasonic stirring time. Materials Letters 61: 4222–4224.

Wang, J. and G. Hou. 2011. Preparation and properties of PS/PSEG composite materials. Advanced Materials Research 284–286: 1842–1845.

Wang, Y., B. Zhang and J. Ye. 2011. Microstructures and toughening mechanisms of organoclay/polyethersulphone/epoxy hybrid nanocomposites. Materials Science & Engineering A 528: 7999–8005.

Wang, M., X. Fan, W. Thitsartarn and C. He. 2015. Rheological and mechanical properties of epoxy/clay nanocomposites with enhanced tensile and fracture toughnesses. Polymer 58: 43–52.

Yuana, X.W., A.J. Easteal and D. Bhattacharyyaa. 2004. Geopolymer reinforced polyethylene nanocomposites. pp. 796–802. In: L. Ye, Y.-W. Mai and Z. Su (eds). Composite Technologies for 2020: Proceedings of the Fourth Asian–Australasian Conference on Composite Materials (ACCM 4). Woodhead Publishing Limited, Cambridge CB 1 6AH: England.

Yangthong, H., S. Pichaiyut, S. Jumrat, S. Wisunthorn and C. Nakason. 2018. Novel natural rubber composites with geopolymer filler. Advances in Polymer Technology 37: 2651–2662.

Yunsheng, Z., S. Wei, L. Zongjin, Z. Xiangming, Eddie and C. Chungkong. 2008. Impact properties of geopolymer based extrudates incorporated with fly ash and PVA short fiber. Construction and Building Materials 22: 370–383.

Zainuddin, S., M.V. Hosur, Y. Zhou, A.T. Narteh, A. Kumar and S. Jeelani. 2010. Experimental and numerical investigations on flexural and thermal properties of nanoclay-epoxy nanocomposites. Materials Science and Engineering A 527: 7920–7926.

Zaman, I., Q.-H. Le, H.-C. Kuan, N. Kawashima, L. Luong, A. Gerson, et al. 2011. Interface-tuned epoxy/clay nanocomposites. Polymer 52: 497–504.

Zhu, S., J. Chen, Y. Zuo, H. Li and Y. Cao. 2011. Montmorillonite/polypropylene nanocomposites: Mechanical properties, crystallization and rheological behaviors. Applied Clay Science 52: 171–178.

Chapter 10

The Prospect of Microcrystalline Cellulose and Nanocrystalline Cellulose Composites from Agricultural Residues in Biomedical/Pharmaceutical

Nurul Huda Abd Kadir, Tan Ching Mig and Aima Ramli
Faculty Science and Marine Environment,
Universiti Malaysia Terengganu, 21030 Kuala Nerus, Terengganu, Malaysia
nurulhuda@umt.edu.my; tanching@umt.edu.my; aima.ramli@umt.edu.my

Marshahida Mat Yashim and Masita Mohammad*
Solar Energy Research Institute,
Universiti Kebangsaan Malaysia, 43600, Bangi, Selangor, Malaysia
marsh5166@uitm.edu.my; masita@ukm.edu.my

1 INTRODUCTION

Global demand for biodegradable and biobased goods has increased in recent years. As a result, biodegradable materials such as polymers, plastics and other organic materials are being used and produced on a wide scale in various sectors. Any solid or liquid, including polymers and agricultural wastes, can be used to create biodegradable products. These are some of the primary reasons why biodegradable items, such as medical and pharmaceutical products, textiles and plastics, have become required throughout the supply chain and numerous sectors.

Agricultural residues are the remainders from daily human life, including animals' dung and crops residue (Liu et al 2018). They are also organic compounds that are eliminated when agricultural products are manufactured. Based on the Organization for Economic Cooperation and Development and the Food and Agricultural Organization (OECD-FAO) data, 39.35 million tons of natural plant fibers were assembled per year from farmers worldwide. Approximately 27 million tons of agro-wastes were produced on 3.35 million hectares of planted lands per year

---

in Egypt. While in China, 56.2 million tons of agro-waste has remained from the 332.83 million hectares of land (Nagalakshmaiah et al. 2016).

In 2019, the Malaysian agriculture industry generated 7.1% of GDP, with plantation crops accounting for 72.9% (DOSM 2020). This significant contribution has been growing despite the needs and challenges of land, water resources and power sources. As a result, agricultural residues have been regarded as an innovative source of renewable raw materials for alternative markets.

Over the past 10 few years, nanomaterials development in the areas of measures and knowledge has made considerable headway, especially in the medicinal field. To establish a new area of advanced nanomaterials, the cooperation of technologies and sustainable resources are essential elements so that such planning will be able to succeed (Kamel et al. 2020a). Based on the study from Brinchi et al. (2013), cellulose $(C_6H_{10}O_5)$n is a biocompatible, biodegradable and renewable biopolymer that exists in a vast quantity in our mother earth. A polymer made up of carbohydrate linear contains repeating units of β-D-glucopyranose by a linkage called $\beta(1 \rightarrow 4)$ and three hydroxyls (OH) groups. In 1959, Richard Feynman, a famous physicist, introduced nanotechnology, contributing well-off impacts to the sciences and technologies industry. Thus, the unique characteristics of nanomaterials, such as the scale of 10–9 m nano and large surface area to volume ratio, have been used in different professions, such as the medical and electronic industry. In addition, nanomaterials can be classified into different classes, an example like metal nanoparticles (MNPs) and fullerene (Alavi 2019), while nanocrystalline cellulose (NCC) is a new class from nanomaterials (Sacui et al. 2014).

NCC is produced by the acidic extraction method of lignocellulosic; thus, it possesses the crystallites structure, a kind of nano-dimensional cellulose rods (Habibi et al. 2010). Due to its negative charges and large surface area, NCC acts as an excipient in drug delivery. It has been proved that it allows the binding of large quantities of drugs to its surface and thus, it owns the high potential in payloads and dosage control (Shaikh et al. 2007) compared to microcrystalline cellulose (MCC). Although MCC has also been applied in a drug-delivery system due to its inert and biocompatible properties, it does not participate in the molecular level of drug release control directly through the binding interactions of drugs. The limitation is caused by the small amount of hydroxyl residues which lead to the limited amounts of negative charges on its large surface, and it shows that the adsorption of major drugs is not easy (Guo et al. 2009). From the toxicology perspective, NCC is considered of low toxicity in both animal and plant cells due to its natural biopolymers source, cellulose (Domingues et al. 2014; Zoppe et al. 2014).

Several studies on cytotoxicity assessment have been conducted to evaluate the safety of nanomaterials as a support of its significant usage in pharmaceutical and biomedical applications (Xu et al. 2018; O'Brien et al. 2006). This assessment is growingly acknowledged as a productive way to test the potential human toxicity to optimize the percentage of developing successful compounds, including the small size of particles such as MCC and NCC (Jones and Grainger 2009; Pereira et al. 2013). In biomedical components production, biocompatibility is the precondition that defines an external component's capability to coexist in harmony when being inserted into the body instead of bringing out damaging effects (Dugan et al. 2013). Cellulose is a biocompatible material, but it will lead to moderate foreign body response *in vivo* (if any). Moreover, it will also cause minor incompatibility as our body does not contain cellulolytic enzymes to break them down. However, there is still a lack of studies about the biocompatibility of crystalline cellulose (Lin and Dufresne 2014). Thus, it has the potential risks of NCC and MCC that need to be evaluated (Jones and Grainger 2009).

Additionally, Domingues et al. (2014) also stated that NCC becomes cytotoxic when it appears in high concentration, which is low compared to the toxic concentration in carbon nanotubes and crocidolite asbestos. This particular benefit has brought a huge benefit to NCC in various applications. So, it is important to conduct more analysis on both MCC and NCC with cytotoxicity and toxicity evaluation points. Therefore, research will focus on the review of MCC and NCC from agricultural waste as a new alternative approach to the pharmaceutical industry.

MCC and NCC are the potential excipients in formulating tablets due to their low reactions and rejection properties when binding with API. Moreover, NCC obtained from different sources possess different properties, which may also affect their functions when used as an excipient (Azubuike et al. 2012). There are several reports stated that some excipients expressed negative effects on drug reactions. Such as they will affect the concentration of the drug when they bind with the API, and some unknown elements and biological activities would occur when the excipient reacts with the active pharmaceutical substances (Giorgio and Patrizia 2003). Thus, the review of cytotoxicity evaluation MCC and NCC through *in vivo* and *in vitro* should be used to reference future use.

This chapter highlights new alternative approaches of both MCC and NCC from agricultural waste in the pharmaceutical industry that may provide more information and understanding on the agricultural waste sources and the new applications of both MCC and NCC in the pharmaceutical area. Both MCC and NCC from agricultural waste possess bio-composite characterizations that offer significant sustainability and eco-efficiency in developing products, especially biomedical applications. There are a lot of different types of crystalline cellulose that possess different functions and properties, such as Micro-Cellulose Crystal (MCC), cellulose nanocrystal (CNC), cellulose nanofibrils (CNF), and Bacterial Cellulose (BC) (Klemm et al. 2011; Dufresne 2013). However, there is still a lack of review and analysis on the data of agricultural waste and review mentioning the new usage of MCC and NCC from agricultural waste in the pharmaceutical industry. The safety evaluation of the agricultural waste-made crystalline cellulose is still insufficient. Thus, cytotoxicity evaluation is required to conduct and ensure the consumption safety of tablets to patients. This study on cytotoxicity is a part of toxicology. So, the concept of redox biology and medicine will be used in this study.

Consequently, more studies and research on the cytotoxicity of MCC and NCC were obtained from fibrous plants related to pharmaceuticals. Furthermore, it may become a reference in sustaining the development of agricultural waste in the pharmaceutical industry in the future. Despite its potential, the safety evaluation of the pharmaceutical products from agriculture waste-made crystalline cellulose is still inadequate. Thus, cytotoxicity evaluation and toxicological tests are essential to ensure the consumption safety of patients.

In addition, this chapter will emphasize the properties and characterizations of both MCC and NCC from different sources of agricultural wastes in pharmaceutical product development, including binding and coating agents.

2 PROPERTIES AND CHARACTERISTICS OF MCC AND NCC DERIVED FROM AGRICULTURAL RESIDUES

Agricultural residues are a byproduct of food production and/or crop deposits in the form of solids or liquids. In an earlier journal, it was proposed (Habibi et al. 2010) that cellulose accounted for the largest proportion of sustainable organic material produced in mother earth, with the manufacture of approximately up to 7.5 to 1010 tons per year. The vast quantity of agricultural waste left over from the crops manufacture has picked up the steam on society that purposed to minimize non-renewable energy usage (oil and gas). However, the compliance of energy transformation from biomass is still low because of the weak characteristics of biomass, for example like high moisture and low energy content.

Several studies have focused on obtaining more materials from agricultural waste that can be used for various industries using more environmentally friendly methods. Rice husks, banana peel and sugarcane bagasse are examples of Cellulose Fibers (CFs), which are agricultural wastes that contain cellulose. They were used to produce essential items, such as biomedical products or automotive materials (Dungani et al. 2016). The example of sugarcane bagasse, is the pre-potency element contained in cellulose. It is also the major agro-biomass produced in the sugarcane industry (Antunes et al. 2018). It comprises 45% of crystalline structure form of cellulose, 26% hemicellulose and 24% of lignin with other minor components such as extractives (Sluiter et al. 2016).

Due to its biocompatibility, biodegradability, non-toxicity and excellent renewability properties, cellulose has been known as a potential alternative material used in the production industry (Du et al. 2019). According to (Siqueira et al. 2011) and (Fortunati et al. 2013), Cellulose Nanomaterials (CNs) can be obtained from the sources of animals, plants or mineral plants. There are two general classes of CNs, cellulose nanocrystals (CNCs) and Cellulose Nanofibrils (CNFs). NCC can also be called cellulose nanowhiskers (CNW), one of the examples from cellulose nanocrystals, while nanofibrillated cellulose (NFC) is classified under CNFs. NCC is utilized as reinforcement in hydrophilic polymers due to the existence of a vast quantity of hydroxyl group (–OH) on its surface which forms strong hydrogen bonding (Klemm et al. 2011). Moreover, few journals stipulated the way of extractions and the source of cellulose will affect the performance of CNC as a nano-reinforcing agent indirectly due to the differences in characteristics and morphologies of the nanoparticles (Moon et al. 2011; Peng et al. 2011). However, there is a lack of study regarding the different properties of NCC cellulose that can be contributed to various fields of application.

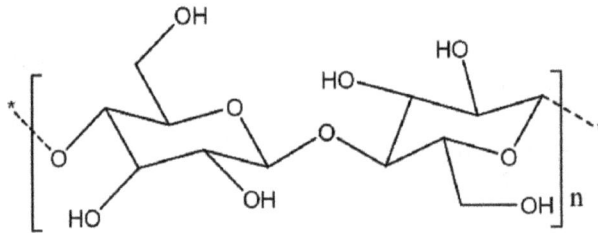

Figure 10.1   Chemical Structure of NCC.

MCC with the chemical structure of $C_{14}H_{26}O_{11}$ is one of the cellulose derivatives defined as a purified, odorless and partially depolymerized crystalline powder prepared by dilute mineral acids treatment (Thoorens et al. 2014). In an earlier article, it was proposed by (Jia et al. 2015) MCC was the only crystalline cellulose commercialized in a powdered and colloidal form with hydrophilic properties. Moreover, it can also be manufactured by various corporations with different trade names, such as Avicel and Emcocel. Due to its high mechanical strength and non-toxicity characteristics, MCC was utilized as a binder and stabilizer in the areas of medical tablets and the food industry (Singh et al. 2015).

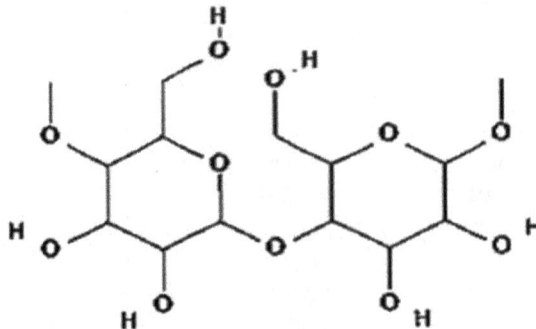

Figure 10.2   Chemical Structure of MCC.

A high content of lignocellulosic constituents in agricultural waste such as cellulose and lignin content makes them a potential raw material to produce micro and nanocrystals materials with distinctive properties. Various promising characteristics of cellulose microcrystal and nanocrystals from oil palm biomass have recently gained the interest of many researchers. In addition, some outstanding properties of these two components have widened their applications in various industries including the biomedical field. The interest in micro and nanosized cellulose arise from its fascinating properties, which vary depending on the source, purification procedures and isolation process conditions. Agricultural waste originated from wood and non-wood resources contain mainly cellulose, hemicelluloses, lignin and extractives. The chemical composition of some cellulose-containing agricultural waste is shown in Table 10.1. Cotton, hemp and jute have the highest cellulose content, whereas coir has the highest lignin content. This composition will facilitate the selection of appropriate treatment for various feedstock sources.

Table 10.1   Chemical composition of some agricultural waste cellulose-containing fiber

Source	Composition (%)			
	Cellulose	**Hemicellullose**	**Lignin**	**Extractives**
Bagasse	40	30	20	10
Coir	32–43	10–20	43–49	4
Corn cobs	45	35	15	5
Cotton	95	2	1	0.4
Empty fruit bunch	50	30	17	3
Hemp	70	22	6	
Jute	71	14	13	2
Kenaf	36	21	18	2
Ramie	76	17	1	6
Sisal	73	14	11	1
Wheat straw	30	50	15	5

Adapted from (Abdul Khalil et al. 2012)

Different treatment conditions can influence the chemical properties of MCC and NCC from agricultural waste. The chemical structures of MCC and NCC are mainly investigated using Fourier-transform infrared spectroscopy (FTIR), which identifies chemical bonds in a molecule that produces an infrared absorption spectrum. Extracted MCC and NCC possess chemical structures that are quite similar. These are due to the fact that they contain the same element, however, in varying sizes to suit their functions.

The absorptions bands allocated approximately 3400 cm^{-1} are attributed to –OH stretching in the hydroxyls groups, abundant in cellulosic compounds, and appear in MCC and NCC and are responsible for inter- and intramolecular bonds cellulose components. This large bundle of hydroxyl groups and their ability to form hydrogen bonds play a major role in creating a highly ordered crystalline structure and governing the physical properties of cellulose. As these crystalline structures are isolated, they are called MCC and NCC. Another important absorption band key for MCC and NCC is the absorption bands centered at 890 cm^{-1} attributed to the β-glycosidic linkages in the cellulose components, responsible for polymerization (DP) in cellulose. Regardless of its source, cellulose consists of a linear homopolysaccharide composed of glucopyranose units linked together by β-1-4-linkages (Brinchi et al. 2013). The number of DP in cellulose varies but can go up to 20000, and the number of glycosidic linkages and length of the chain depends on the source of cellulose (Abdul Khalil et al. 2012).

During the extraction of MCC or NCC from its raw sources, the employment of the applied chemical treatments effectively removed most of all non-cellulosic components such as hemicellulose and lignin molecules (Kassab et al. 2020a). The reduction of the band intensity assigned to the C=O stretching vibration of carbonyl and acetyl groups in the xylan component of hemicelluloses (at 730–1720 cm^{-1}) and the disappearance of characteristic peaks assigned for lignin (at approximately 1520–1510 cm^{-1} in the absorption spectrum) would indicate the removal of both hemicellulose and lignin before the isolation of NCC and MCC (Liu et al. 2018).

The application of strong acid treatments would result in NCC production with different surface functionalities. The use of phosphoric acid during hydrolysis results in the formation of phosphorylated CNC with phosphate groups inserted on the surface. Yu et al. 2016, reported the bands relative to sulfate and phosphate negatively charged groups were detected in the spectra upon the extraction of NCC using sulfuric and phosphoric acid hydrolysis. For the NCC that was isolated using acid hydrolysis treatment with citric acid/hydrochloric acid mixture, a small peak in the FTIR spectra (1729 cm^{-1}) associated with carboxylate groups (COO–) was recorded (Kassab et al. 2020b), resulting from the esterification reaction between the hydroxyl groups of cellulose and the carboxyl groups of citric acid. However, acid hydrolysis using hydrochloric acid only has shown unmodified structure onto hydroxyl of NCC.

The morphology of MCC and NCC were significantly affected by the source of agricultural biomass and isolation techniques (Tang et al. 2014). The fundamental method of MCC and NCC extraction is to digest the amorphous portions while retaining the crystalline parts. For example, isolation of cellulose from its sources via acid hydrolysis has resulted in the formation of rod-shaped NCC with the diameter ranging between 5 and 30 nm and length between 100 and 500 nm (Araki et al. 1999). It is due to the acid hydrolysis treatment destruction of the amorphous domains of MCC, leaving the crystalline regions unaltered as this region persists hydrolysis. Further amorphous domains are removed when a significantly longer hydrolysis period is applied (Kassab et al. 2020b), leading to isolation of smaller CNC in terms of diameter and length. Eventually, the size of the cellulose fibers was reduced from the micron to the nanometer scale.

Apart from length and diameter, the morphology of MCC and NCC are also characterized based on aspect ratio, the ratio of the said dimension. It is well understood that the aspect ratios of MCC and NCC play a critical influence in their ability to reinforce to produce a composite. A filler with a high aspect ratio causes a complex entanglement between the matrix composite with enhanced reinforcement ability (Nagalakshmaiah et al. 2016). A high aspect ratio facilitates forming an anisotropic phase inside the polymeric matrix, resulting in superior attributes for nanocomposite materials. It is worth noting that, in most cases (under strictly controlled conditions of temperature and agitation), hydrolysis duration at around 40 and 45 minutes appears to be the optimum hydrolysis time to produce CNC with a high aspect ratio. The NCC extracted from kenaf fiber using sulfuric acid hydrolysis produced length and diameter approximately 158 nm and 12 nm, respectively, giving an aspect ratio of around 13 (Kargarzadeh et al. 2012). According to literature, an aspect ratio of roughly 50 is indicated for an effective strengthening effect of the CNC in nanocomposite materials (Eichhorn et al. 2010; Kassab et al. 2020c). The morphological dimension of various types of nanocellulose is tabulated in Table 10.2.

Table 10.2 Nanocellulose dimensions

Cellulose Structure	Diameter, d (nm)	Length, L (nm)	Aspect Ratio (L/d)
Microfibrillated cellulose	10–40	> 1000	100–150
Microcrystalline (MCC)	> 1000	> 1000	~1
Nanofibrils	2–10	> 10,000	> 1000
Nanocrystalline (NCC)	2–20	100–600	10–100

Slightly different observations can be obtained when employing different microscopy techniques; Field-Emission Scanning Electron Microscopy (FE-SEM), Transmission Electron Microscopy (TEM) or Atomic Force Microscopy (AFM). The most common procedure is TEM, which may directly offer high-resolution images; however, this technique frequently exhibits particle aggregation, owing to the drying step used to prepare the samples before the analysis. On the other hand, AFM has been widely used to provide valuable surface topography analysis of MCC and NCC under ambient temperature. A TEM image of MCC isolated from oil palm frond using acid hydrolysis is shown in Fig. 10.2 exhibiting irregular shape and size. Individual NCCs are completely suspended as the individual crystals in an aqueous solution contributed by electrostatic repulsion from negatively charged sulfate groups on NCC surfaces.

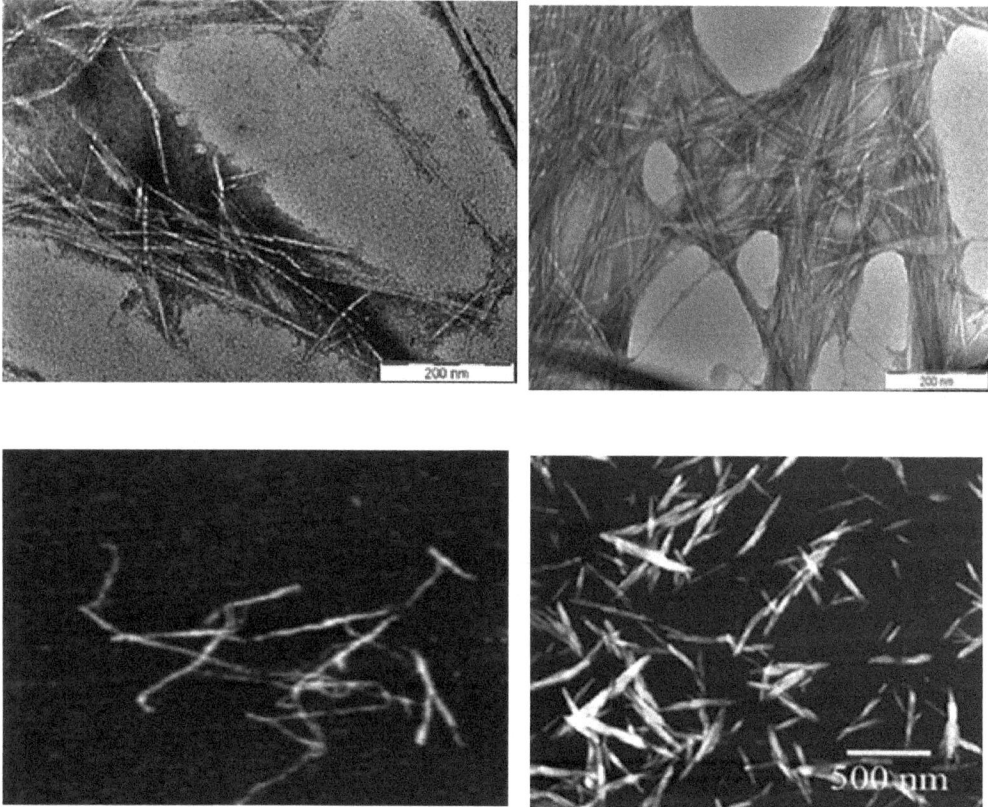

Figure 10.3 TEM image of MCC and NCC extracted from oil palm frond waste (top) and AFM image of NCC extracted from the sunflower-cake waste and pineapple crown waste (bottom). Reprinted from references (Nordin et al. 2017; Owolabi et al. 2017; Kassab et al. 2019, Prado and Spinacé 2019). Copyright (2017 and 2019) with permission from Elsevier.

Cellulose has been observed to have both crystalline and amorphous domains in its structure, in contrast to hemicellulose and lignin, which are amorphous in nature. The ratio of crystalline to the amorphous region assessed via X-Ray Diffraction (XRD) is used to determine the degree of crystallinity of the cellulose fiber. Major diffraction peaks observed in x-ray diffractogram for cellulosic compounds are around 15.1°, 16.6° and 22.5°, which correspond to crystallographic planes $1\bar{1}0$, 110 and 200, respectively. The crystallinity index of nanocellulose is calculated using the following equation:

$$CrI(\%) = \left(\frac{I_{002} - I_{Amorph}}{I_{002}} \right)$$

where I_{200} is the peak intensity of the reflection and I_{Amorph} is the minimum intensity between the (110) and (200) peaks (Prado and Spinacé 2019).

The presence of amorphous non-cellulosic components such as hemicellulose and lignin contribute to the amorphous region in the agricultural biomass fiber. As these components are largely removed during the cellulose extraction (from the alkalinization and bleaching treatments), the crystallinity index of microfiber fiber is much higher than its raw fiber (Bahloul et al. 2021) due to a better cellulose chain packing. An increment of the degree of crystallinity has been reported for the cellulose microfibrils (CMF) of sunflower oil cake and other sourced materials (Kassab et al. 2019; Kassab et al. 2020c) after further hydrolyzed during acid hydrolysis, confirming the removal of amorphous domains in cellulose chains without destroying the crystalline domains. There was a study that reported an increase of crystallinity index up to 38% as a result of the conversion of cellulose to nanocrystal (Prado and Spinacé 2019) from its raw fiber. The XRD patterns for sunflower oil-cake fiber and kenaf fiber at different stages of purification of NCC prepared using sulfuric acid hydrolysis treatment are shown in Fig. 10.4. Cellulose fiber with high crystallinity index is desirable for improving the mechanical properties of biocomposite fabrications, which is very useful for many applications. A comparison of morphological and structural properties of a few MCC and NCC isolated from various agricultural waste is shown in Table 10.3.

Table 10.3 Comparison of morphological and structural qualities of MCC and NCC extracted from agricultural waste using the same acid hydrolysis treatment.

Source	SEM/AFM			CrI(%)	References
	Diameter, D (nm)	Length, L (nm)	(L/D)		
Cotton, NCC	51	—	—	84.5	Raza et al. 2019
Sunflower, NCC	5	329	66	87	Kassab et al. 2019
Kenaf, NCC	12.4	158	13.2	81.8	Kargarzadeh et al. 2012
Coconut husk, MCC	150000	—	—	71.8	Abdullah et al. 2021b
Sisal, MCC	6000–13000	—	—	94	Sosiati et al. 2017
Corncob, MCC	~83 000	—	~1	52.8	Shao et al. 2020

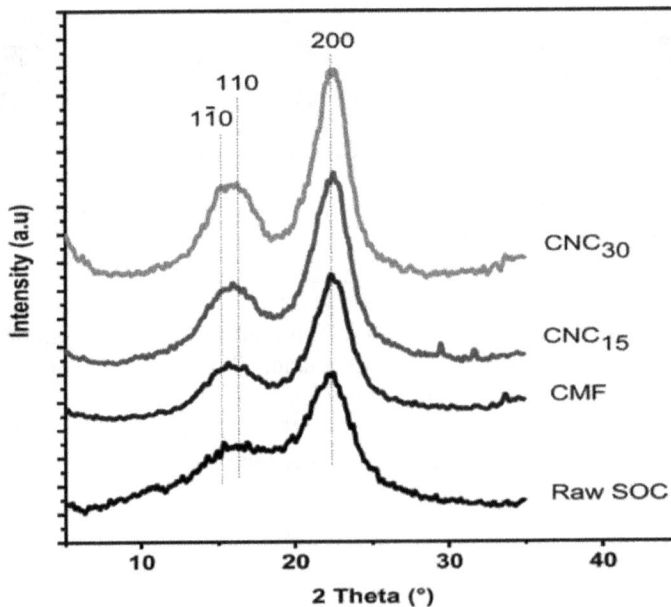

Figure 10.4 XRD spectra are comparing the crystallinity index of raw sunflower oil-cake fiber to its NCC. Reprinted from (Kassab et al. 2019). Copyright (2019) with permission from Elsevier.

Table 10.4 shows the summary of agricultural wastes, which focused on (i) the source of agricultural waste, (ii) the percentage of the cellulose yielded, and (iii) the crystallinity index (I_c) of the cellulose. Cellulose is a sustainable organic material that is able to be manufactured into MCC and NCC. The sources of agricultural waste that were able to formulate MCC and NCC came from rice husk (Ahmad et al. 2016; Hanani et al. 2017), oil palm fibers

Table 10.4   Agricultural waste sources, crystallinity index, and percentage yield of MCC and NCC

Agricultural Wastes	Percentage of Crystalline Index (I_c)	Treatment	Percentage of yield (Cellulose/MCC/NCC)	References
Rice husk	–63.9% (2M HNO_3 treatment) –52.2% (2M HCl) –42.0% (1M HNO_3) –37.0% (1M HCl) –39.3% (control group)	HNO_3 acid hydrolysis, HCl acid hydrolysis	–83.26% MCC (2M HNO_3) –80.64% MCC (1M HNO_3) –69.54% MCC (2M HCl) –60.58% MCC (1M HCl)	Ahmad et al. 2016
	–54.2% (HCl treatment) –52.4% (HNO_3 treatment) –49.7% (H_2SO_4 treatment) –44.1% (control group)	0.5M of HNO_3, HCl and H_2SO_4 acid hydrolysis	–83.5% MCC (HNO_3 treatment) –81.8% MCC (HCl treatment) –80.6% MCC (H_2SO_4 treatment)	Hanani et al. 2017
Banana pseudo-stem fibers	72.1%	HCl acid hydrolysis	99.0% MCC	Shanmugam et al. 2015
Bamboo	82.6%		80% of MCC	Rasheed et al. 2020
Coconut coir	N/A		18% MCC (fair)	Evangelista et al. 2017
Palm kernel	N/A		98%	Ndika et al. 2019
Water Hyacinth	73.28%		N/A	Semachai et al. 2018
Banana peel	66.4%	Sulfuric acid (H_2SO_4) hydrolysis	71.51% NCC (0.1% H_2SO_4) 62.06% NCC (1% H_2SO_4) 60.36% NCC (10% H_2SO_4)	Tibolla et al. 2018
Distilled waste of *Cymbopogon winterianus*	86%		83±1.6% crystalline cellulose	Mishra et al. 2018
Oil palm fibers	82.5%		42.43% cellulose	Xiang et al. 2016
Pineapple leaf fiber	N/A		81.27% NCC	Abdullah et al. 2021a
Corn stover	N/A	Glacial Acetic acid hydrolysis	–44.4% cellulose (Control group) –93.1% cellulose (after alkali and delignification treatment)	Xu et al. 2018
Walnut shell	29.5% (control group) 40.1% (After NaOH treatment) 42.9% (After bleach treatment)	Acetate buffer and NaClO treatment	27.4% NCC (Control group) 56.6% NCC (after NaOH treatment) 87.9% NCC (after bleach treatment)	Zheng et al. 2019

(Xiang et al. 2016), coconut fibers (Evangelista et al. 2017), banana pseudo-stem fibers (Semachai et al. 2018), banana peels (Tibolla et al. 2018), distilled waste of *Cymbopogon winterianus* (Mishra et al. 2018), corn stover (Xu et al. 2018), water hyacinth (Semachai et al. 2018), palm kernel (Ndika et al. 2019), walnut shell (Zheng et al. 2019), bamboo (Rasheed et al. 2020) and pineapple leaf fiber (Abdullah et al. 2021a; Abdullah et al. 2021b).

In order to strengthen the crystallinity and minimize the particle size of cellulose, acidic hydrolyzation is a popular method to conduct it (Sharma et al. 2019; Kusmono et al. 2020). Based on the journals from Ahmad et al. (2016), treatment of cellulose using acid hydrolysis method with different molarity of acid will affect the percentage yield of crystalline cellulose. 83.26% of MCC and 69.54% MCC were treated with 2M of HNO_3 and HCl, respectively, which were higher percentage yields than 80.64 and 60.58% MCC, respectively with 1 M of HNO_3 and HCl. For this reason, the solubility of the amorphous region of the cellulose will increase when the concentration of the treated acid increased. This correlation will lead to an increase in the MCC cellulose yield.

In the case of hydrolyzing crystalline cellulose using the same molarity (0.5 M) but different types of acid (Hanani et al. 2017), 83.5% of MCC is yielded by HNO_3 treatment, which is the highest compared to HCl (81.8%) and H_2SO_4 (80.6%). This can be explained by the presence of hydronium ions produced by the acids that was able to eliminate the excessive amorphous region in cellulose. Thus, the reduction of the amorphous region will contribute to the increase of percentage yield of crystalline cellulose, while HNO_3 has a higher capability to carry out such a reaction (Johar et al. 2012).

In HCl acid hydrolysis, the percentage yield of MCC and crystalline cellulose showed a high percentage with a range between 80–99%. Sharma et al. (2019) and Ventura-Cruz and Tecante (2019) stipulated that HCl is the most common type of acid applied in the acid hydrolyzation process. The insoluble crystalline cellulose will be left in the acid solution after been hydrolyzed as the acid removed the amorphous region (Nagalakshmaiah et al. 2016). In contrast, sulfuric acid (H_2SO_4) was also popular to be used in formulating cellulose with the aid of various conditions. The sulfate ions in the acid will esterify the hydroxyl groups in cellulose by attaching negative charges on the crystalline cellulose surface. As a result, the stability of the particular cellulose will be affected in the site of colloidal suspension due to the double attractive or repulsive forces which contributed from the stabilizing on anionic ions (Kargarzadeh et al. 2012). Based on Table 10.3, the percentage of NCC yield decreased when the percentage of H_2SO_4 applied increased, this can be explained by the probability of losing more samples if more extra steps were applied in the treatment. The acid hydrolysis method which is aided by 'extra' mechanical treatment possesses different percentage yields compared to that cellulose which is treated by just only acid hydrolysis. (Tibolla et al. 2018). To put it simpler, when the higher percentage of H_2SO_4 is used, there will be a higher possibility of extra mechanical treatment been added into the process simultaneously, caused the 'pureness' of the acid to change and affecting the percentage of the cellulose yield indirectly.

The percentage of the yield of crystalline cellulose has also been affected by alkali treatment (bleaching). A high percentage of cellulose (93.1%) (Xu et al. 2018) and NCC (87.9%) (Zheng et al. 2019) were produced after being treated by alkali and delignification process. The non-cellulosic materials and hemicellulose were eliminated when cellulose was subjected to alkali treatment. In the delignification process, the composition of lignin and hemicellulose will drop greatly which contributed to the production of highly purified cellulose (Karthik and Murugan 2013; Zheng et al. 2019).

X-ray diffraction was used to determine the plant origin of the particular cellulose based on its structure and investigate the isolation route effect (Robles et al. 2015). In the quantitative method, Crystallinity Index (I_c) was purposed to determine the strength and stiffness of fibers in powdered form (Shanmugam et al. 2015). In the process of acid hydrolysis, the crystallites of cellulose were released due to the hydrolytic break down of glycosidic bonds through the penetration of hydronium ions from acid into the amorphous region of cellulose (Johar et al. 2012). Based on Table 10.3, the high crystallinity index (I_c) was 66.4% (Tibolla et al. 2018), 73.28% (Semachai et al. 2018), 72.1% (Shanmugam et al. 2015), 82.5% (Xiang et al. 2016), 82.6% (Rasheed et al. 2020) and 86% (Mishra et al. 2018) respectively. In the site of explanation, higher I_C indicated there was higher efficiency in removing the amorphous region from the cellulose and left the crystallized cellulose behind (Deepa et al. 2011). In addition, the crystallinity index of cellulose will also be influenced by the treatment of chemicals. The journal from (Zheng et al. 2019) stipulated that the elimination of lignin, as well as amorphous hemicellulose, caused the I_c of NCC increased from 29.5% (untreated) to 40.1% after NaOH treatment was applied. In a nutshell, crystalline cellulose that possesses a higher crystallinity index owns a compact structure as well as denser particles (Rojas et al. 2011). Thus, MCC and NCC were popular to be used as an excipient or filler in pharmaceutical applications due to their compact structure and tensile properties.

3   MCC AND NCC COMPOSITE IN BIOMEDICAL PERSPECTIVE

Owing to its safety and efficacy, NCC has gained increasing attention to be used in biomedical applications. The use of cellulose nanocrystals has been broadly explored for targeted delivery of therapeutics since it is always possible to transform the surface hydroxyl groups of CNCs into other functional groups, such as amino or carboxyl groups, and make them useful for covalent or non-covalent binding of targeted drug moieties to the NCC's surface. It is also worth noting that the high surface hydrophilicity of nanocellulose inhibits the adsorption of proteins resulting in the delayed passage of the drug carrier from the bloodstream (Patra et al. 2018). The outstanding properties of MCC and NCC are beneficial for drug delivery applications. Significant amounts of drugs might be bound to the large surface of nanocellulose. And its negative charge might allow for high payloads and optimal dosage control. A great deal of literature reported the application of the composite froms that is nanocellulose extracted from bacteria (bacterial nanocellulose, BCN) for biomedical purposes such as wound dressing (Fu et al. 2013), implants, including cardio-vascular graft (Petersen and Gatenholm 2011). However, this chapter focuses on the application of MCC and NCC cellulose composite fabricated from agricultural waste.

3.1   Cytotoxicity

Researchers conducted several studies on the cytotoxicity of NCC during the past few years. Cytotoxicity refers to the detrimental effects of a substance on a cell's viability, and it is measured based on cell damage based on cell morphology. Cytotoxicity of any substance could also be evaluated by measuring specific metabolic activities. The hydroxyl groups in the cellulosic surface of MCC and NCC becomes partially esterified during acid hydrolysis and imparted some acidic properties to the fibers. On heating the sulfate group attached on the NCC's surface, known to cause desulfation and the release of sulfuric acid from the CNC surface, which might cause cytotoxicity to the cell. However, a few studies have shown only minimal impact as cytotoxicity study was done using NCC from agricultural waste isolated using sulfuric acid hydrolysis.

NCC isolated from eucalyptus pulp using sulfuric acid hydrolysis was used as a co-stabilizer for controlled release tablets induced negligible cytotoxicity when contact with the human gingival fibroblast cells in culture (Villanova et al. 2011). The same observations were recorded for NCC isolated from softwood pulp when tested on human brain microvascular endothelial cells (Roman et al. 2009). It is a necessary condition to assess the viability of NCC for the treatment of neurodegenerative diseases. Furthermore, the cytotoxicity study was conducted on a few more mammalian cell lines by Dong et al. 2012 quantified no cytotoxic effects against any cell lines in NCC concentration in the range of 0–50 µg/mL and exposure of 48 hours. A substantial cytotoxicity effect was only shown when the cytotoxicity study was done at a much higher concentration of 500 µg/mL and 1000 µg/mL (Hanif et al. 2014). In all these studies, the hydroxyl groups were fluorescently labeled by protruding hydroxyl groups of the outermost cellulose molecules with the aid of NCC ability to scatter the emitted light and absorb the assay reagents to create better bioimaging analysis.

3.2   Biocompatibility

Biocompatibility is the ability of a foreign body to coexist with tissue without generating minor alterations (Dugan et al. 2013). Although cellulose is not biodegradable inside the body (*in vivo*), studies have indicated that oxidized cellulose is possibly degradable (Luo et al. 2013). The MCC and NCC's surface that is coated with many reactive hydroxyl groups and other functional groups available for various types of chemical modification is an essential aspect of MCC and NCC to be biocompatible *in vivo* as well as *in vitro*. The presence of functional groups also allows various labels, medicinal chemicals and targeting vectors to be applied on their surface. Drugs with functional groups that would not ordinarily bind to nanocellulose, such as nonionized or hydrophobic drugs, could be loaded and released through surface modification (Abdul Khalil et al. 2015). Acid hydrolyzed NCC form a widely dispersed colloidal suspension in an aqueous medium contributed from their strongly negative surface charge, which can extend its circulation in the bloodstream and minimize non-specific interactions with plasma proteins.

NCC was generated from acid hydrolysis of cotton fibers was tested for its biocompatibility with murine macrophages, and breast adenocarcinoma cell lines have shown a good interaction with the cell lines. The results indicated good compatibility and suitability of NCC as a candidate for nanosystem development for targeted drug delivery in breast cancer and theragnostic applications. To employ the NCC's biocompatibility further, NCCs have been applied in bone tissue engineering for the fabrication of various artificial bones. NCC extracted from wood pulp using sulfuric acid hydrolysis has been used to fabricate NCCs/hydroxyapatite composite to imitate bone scaffold. The protein stabilization test (bovine serum albumin) was performed on the artificial bone scaffold to test its biocompatibility. The scaffold was found to be able to stabilize the protein and slow its inactivation recorded (Huang et al. 2018), indicating a strong affinity between NCCs composite and bones' collagen and mineral. The good biocompatibility demonstrated by the composite scaffold indicated that it is a suitable choice for bone scaffolding.

3.3   Antimicrobial Properties

Antimicrobial and antibacterial properties can be added to nanocellulose materials through functionalization and/or inclusion of antimicrobial/antibacterial agents which may be added to nanocellulose materials. In general, cellulose must be processed, changed and conjugated with the enzyme before it can be used. Nanocellulose could be combined with inorganic antibacterial agents, such as silver particles and derivatives and organic antimicrobial agents, such as lysozyme, natural antimicrobial agents found in human secretions, to produce nanocellulose conjugate (Jebali et al. 2013).

The antimicrobial activity of nanocellulose conjugated with enzymes and antibacteria has shown good antifungal and antibacterial effects against standard strains of bacteria. The cellulose crystals used in the study were isolated from cotton fiber and hydrolyzed using an acid mixture (Lin and Dufresne, 2014). Its evaluation of the antimicrobial susceptibility towards Candida albicans, Aspergillus niger, Staphylococcus aureus and *Escherichia coli* was conducted according to the Clinical and Laboratory Standards Institute. Beside from several studies on the antibacterial properties of nanocellulose conjugated with antimicrobial agents, several concerns, such as the duration of antimicrobial action, remain under investigation. Some other studies of organic NC-based antimicrobial summarized in Table 10.5.

Table 10.5 NCC conjugated with antimicrobial agents

Agents	Composite Form	Synthesis Method	Microbial Growth Inhibition
Chitosan	Film	Flax CNC incorporated in CS film solution by following the solution casting method	*E. coli, E. faecalis, L. monocytogenes, P. aeruginosa, S. aureus*
Chitosan (PVP)	Film	Solution casting method	*P. aeruginosa, S. aureus*
Chitosan, PCL, grape seed extract	Film	Casting method for preparation of CS films with GSE and NC, the addition of PCL was achieved with coating and compression molding method for PCL	*E. coli, L. monocytogenes*
Nisin (PLA)	Film	PLA-CNC films were treated with nisin by adsorption/diffusion coating method	*L. monocytogenes*
Curcumin	Film	CNF suspended in PVA the solution, and then curcumin was added	*B. coagulans, C. Albicans, E. coli, P. mirabilis, S. aureus, Streptococcus* sp.

Adapted from (Kupnik et al. 2020)

4 SAFETY EVALUATION

In an earlier article, Ikuo Horii (2016) stipulated that to obtain approval from both Investigational New Drugs and New Drug Applications, the studies of toxicity were still playing a vital role in the development of pharmaceutical areas, as well as in the subpart of safety evaluation. Additionally, there were steps which were allocated from exploration of drugs to scientific development as well as post-marketing: (1) Study of toxicology screening and early phase drug development, (2) study of toxic exploration for scientific compound exclusion, (3) study of drug safety evaluation for human primary clinical study, (4) study of NDA's toxicology, and the last (5) study of the validation of post-marketing safety evaluation. Paracelsus, who was the founder of toxicology described that 'All substances are poisons; there is none which is not a poison. The right dose differentiates a poison from a remedy.' Thus, safety evaluation was extremely essential in making the first phase of the 'mixture' in drug discovery perfect.

4.1 Toxicology Evaluations of NCC and MCC *in vitro* and *in vivo*

In the development of the pharmaceutical industry, preclinical safety testing of new drugs is a vital pathway before the drug was delivered to patients, which is classified in a few sets, such as *in vivo* and *in vitro* tests. Jones and Grainger (2009) proposed that there was a need to have toxicity risk assessment because nanoparticles may possess the chance of putting human health in jeopardy, although they have brought a vast quantity of advantages to our life. This was due to the unique characteristics that they owned, for example like tiny in size and high reactivity, which may lead them to become poisonous to human cells (Pereira et al. 2013).

In vivo toxicity test was proposed to identify the toxicology effects of both MCC and NCC within a living organism such as laboratory animals (rats), while *in vitro* toxicity test was applied to investigate the toxic risks of crystalline cellulose by growing cells or tissues under controlled environments. *In vitro* assays were popular in nanomaterials investigation sites. Moreover, the human epithelial cells were frequently utilized in this assay as the phenotype of such cells was parallel with *in vivo* assay, such as the presence of mucin that surrounded the cell membrane (Jones and Grainger 2009). In addition, based on some researches proved that both MCC and NCC did not show any cytotoxicity effects in both *in vitro* and *in vivo* toxicity tests, including animal feeding, human Dermal Fibroblasts (hDF) treatment, CaCo-2 cell viability test and series epithelial cell lines exposure (Dong et al. 2012; Ferraz et al. 2012; Pelissari et al. 2014).

Principle of safety evaluation in medicinal drug

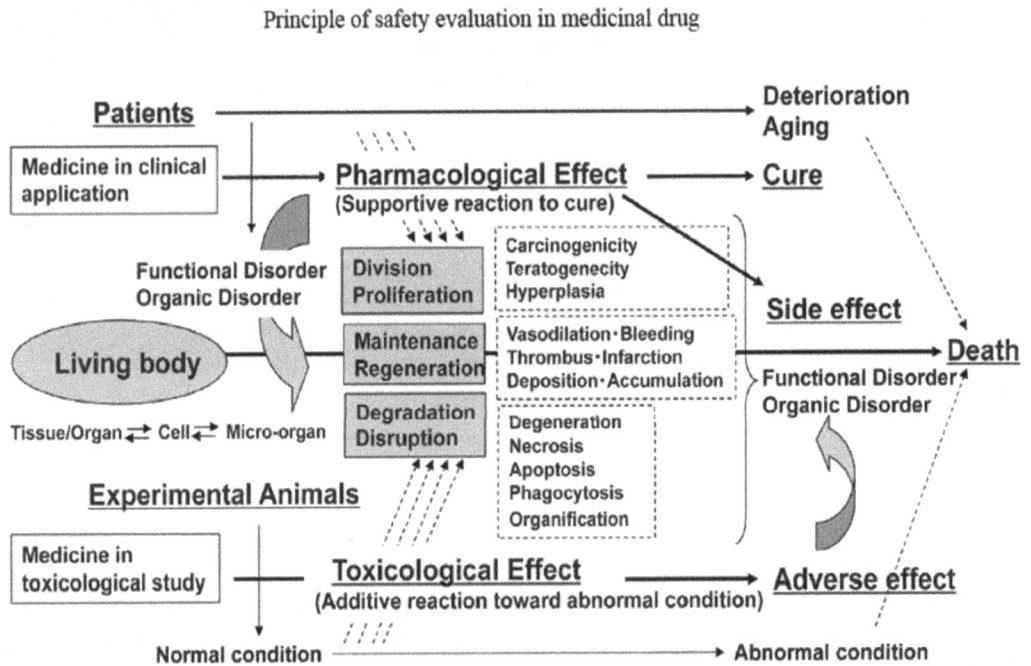

Figure 10.5   Pharmacological and Toxicological Effect.

Nevertheless, it has been stated that there were abundant new *in vitro* assays that have been appeared in the past few years in order to help in sorting out the particle which contained the risk of being cytotoxic, while the usage of particular *in vivo* tests was replaced in a few studies indirectly. This was due to the commitment in supporting the analysis of potential drugs and complying with the 3Rs legislation, which summarized the standards of replacement, reduction, and refinement on animal experimentation (Russell and Burch 1959).

To determine the cytotoxicity and toxicity risk of both MCC and NCC (as shown in Table 10.6), both *in vivo* and *in vitro* assays were conducted on cell lines and animals, respectively. In the area of cell studies (*in vitro* assay), the treated *Gelidium amansii* cellulose nanocrystal (TGa CNC) did not show any cytotoxicity signals on Human keratinocyte HaCaT, Beas-2B, and Raw 264.7 cells through MTT assay. In addition, TGa CNC with a concentration range between 25 to 250 µg/mL also showed non-cytotoxicity effects in human lung epithelial Beas-2B and human skin cells. In the Western Blotting assay, results showed that UVB-induced COX-2 (cyclooxygenase-2) was successfully restrained by TGa CNC, this outcome showed that NCC that treated with *Gelidium amansii* possessed the ability in controlling inflammation and cancer (Ha et al. 2018).

Although NCC did not show cytotoxicity effect in low concentration, yet it may induce cell viability to reduce when high concentration (5000 μg/mL) of NCC (CNF) was treated to cell line

Table 10.6 *In vivo* and *in vitro* tests from MCC and NCC

In vivo and In vitro assay				
Cellulosic Material	**Mammalian Cells**	**Method**	**Remarkable Results**	**References**
CNF solutions with concentrations (50,100,500,1000,2000 and 5000 μg/mL) were incubated and tested with yellow tetrazolium MTT.	Human colon carcinoma CaCo-2 cell line	MTT assay	• Produced CNFs did not show a toxicity effect on the CaCo-2 cells for studied concentrations lower than 1000 mg/mL. • Cell viability reduced (74.59 to 73.13%) when exposed to 5000 μg/mL samples.	Tibolla et al. 2019
The cell's line was incubated and treated with MTT solution	Human keratinocyte HaCaT, Beas-2B, and Raw 264.7 cells		• No cytotoxicity on the three types of cell lines.	So et al. 2021
Cells were treated with TGa CNC for one hours and exposed to UVB.	—	Western Blot assay	• UVB-induced COX-2 was suppressed by TGa CNC.	
200uL of TGa CNC was fed to the mice on the first day. The same amount was fed to the mice on the second day but exposed with UVB.	Five-week-old male ICR ice		• There was anti-inflammation effect on mice skin.	
A wound was created on the mouse model. The wound was covered with three types of dressing: 1. traditional bandage (control group), 2. pure gelatine film, 3. crystalline cellulose biocomposite film	Experimental mice model	*In vivo* wound healing	• No significant weight loss or fever was found during the total healing process. • The wound surface was found almost totally healed. • Crystalline cellulose biocomposite film showed a faster healing rate than gelatine and bandage.	Bhowmik et al. 2017
The drug concentration released was determined by UV-Vis spectrophotometer at 232 nm.	—	*In vitro* drug dissolution	• MCC-F released 83.48% of the drug. • NCC-F released 92.27% of drug in a duration of 5 minutes.	AnjiReddy and Karpagam, 2020
Rats were fed with a 5 mg/kg dose of drug. Blood samples were taking after 3,10,15,30 minutes and 1 hour of feeding.	Three months old Albino Wister rats	High-Performance Liquid Chromatography (HPLC) method	• The plasma concentration of NCC-F is the highest compared to MCC-F and marketed tablet formulation. • NCC-F possessed excellent drug absorption.	

(Tibolla et al. 2019). In an earlier article, Pereira et al. (2013) stipulated that cell viability may be dropped when the content of CNF increased due to the occurrence of apoptosis. Furthermore, there were no toxicity signs appeared on animal models as reported by previous researchers about the safety level of NCC in animals (Bhowmik et al. 2017; AnjiReddy and Karpagam 2020; So et al. 2021).

5   PHARMACEUTICAL APPLICATION OF MCC AND NCC

Recently, the production of MCC as a pharmaceutical usage has become a big concern due to its excellent compatibility under the least compression. In addition, MCC also owns high binding capability, which also allows it to formulate stable and fast break down tablets. In addition, it also plays roles as filler and binders in tablet preparation through the direct compression method (Setu et al. 2014).

Lipinski, 2002, stipulated that drug nanocrystal had been the focal point to be applied in the pharmaceutical industry for the past 20 years. However, most of the pharmaceutical products are made by the most stable form of crystalline nanoparticles that may encounter some issues, including weak solubility, which contributes to low bioavailability. This problem is caused by the abundant amount of hydroxyl group located on the surface of NCC, which led to the half-life of NCC is longer than other hydrophobic particles in a circulatory system due to its hydrophilic properties. Thus, this situation stops the progress of phagocytosis by preventing the absorption of opsonin protein (Roman et al. 2009). George and Sabapathi (2015) stated that there are several modification methods such as sulfonation, oxidation and esterification that can be carried out by alternating the functional groups of NCC in order to bind and link the hydrophobic drugs to the surface of NCC (Lam et al. 2012).

Table 10.7   Pharmaceutical Applications from MCC and NCC

Crystalline Cellulose	Application	References
MCC	• Reinforcing agent • Tablet excipient	Cataldi et al. 2015, Hindi 2016
NCC	• Drug dispersant • Tablet excipient • Drug disintegrant • Tablet fabrication • Transdermal drug delivery system • Oral drug delivery system • Hydrogel system • Cell culture	Wang et al. 2015a; Wang et al. 2015b; Dan et al. 2016; Emara et al. 2016; Song et al. 2016; Gopinath et al. 2018; Kamel et al. 2020b
MCC and NCC	• Antibacterial and wound healing applications • Strength agent	Alavi 2019; Chen et al. 2020

Based on Table 10.7, NCC was used as the potential drug disintegrant in the pharmaceutical industry (Wang et al. 2015a). This can be proven by comparing the time needed for calcium carbonate tablets with different ratios of NCC to decompose in distilled water or artificial gastric fluid. The tablet without the presence of NCC took 2 hours to finish the whole disintegration process, while 2 and half minutes was spent by the tablet, which contained a 10:3 ratio of calcium carbonate and NCC. This showed that NCC possessed the potential of high-quality disintegrant which the disintegration time of tablets reduced when the ratio of NCC used was increased. Moreover, both MCC and NCC were also popular to be applied as tablet excipients in the pharmaceutical area (Hindi 2016, Emara et al. 2016) due to their high disintegration rate and outstanding dry binding characteristic. However, the tablet's dissolution rate may be influenced by the properties of the drugs contained. NCC application in fabricating tablets was proposed to optimize the chances for tablets to reach the requested requirements, such as hardness and dissolution rate. It reduced the side effect that contributed to the process of compression. In an earlier article, Moon et al. (2011) stipulated that NCC owned the tensile strength of 7.5–7.7 Gpa range, which was stronger than Kevlar-49 fiber (3.5 Gpa), carbon fiber (1.5–5.5 Gpa), and even steel wire (4.1 Gpa). This can be concluded that NCC was able to enhance the hardness of tablets by stabilizing the extra frictions formed during tablet compression. Due to the presence of short microfibrils and the nature of crystallinity, MCC was selected as the suitable reinforcing agent. It fulfilled the requirements of achieving good interaction between matrix and reinforcement compounds, which was a large surface area. In addition, other outstanding properties such as sustainable sources and biocompatible were also made MCC vital in strengthening the mechanical and supporting elements from biomass (Yang et al. 2014, Wang et al. 2015b).

In the element of drug released control, hydrogel which is formulated from crystalline cellulose played a vital role in providing an optimum environment to undergo cell culture scaffold (Bhattacharya et al. 2012). This is because hydrogel was low in viscosity, which offered excellent injectability when mixing cells with it. It is also able to yield 3D multicellular spheroids and strengthen cell adhesion which supported the division of the cultured cells (Kamel et al. 2020b). Besides, the hydrogel was used to avoid the sudden burst when drug content was released. The presence of pores on hydrogels allowed drug access into the matrix of the particular gel and released the drug at a rate that relied on the diffusion properties of the molecules (Hoare and Kohane, 2008).

In the drug delivery system (Song et al. 2016) and Transdermal Drug Delivery System (TDDS) (Kamel et al. 2020b), the bonding of water-repelled drugs such as paclitaxel was strengthened due to the modification of the surface when NCC was combined with cetyltrimethylammonium bromide, which caused the prolonged release of the drug for two days (Taheri and Mohammadi, 2015). Due to the capability of altering the chemical surface of the drugs, NCC was selected to be the drug reservoir as the main component to carry out TDDS. In addition, the size of NCC (1–100 nm diameter and 10–1000 nm length) was 40–400 times smaller than the size of stratum corneum cell (40 um diameter and 0.5 μm thick), which was the skin cells located at the epidermis layer. Thus, the permeability of drugs was improved by applying NCC in TDDS due to its smaller particle size (Subedi et al. 2010).

Wiegand and his team (2015) stated that NCC played a role in transporting antimicrobial agents and scaffold wound filling due to its high stability and biocompatibility. The network complex structure of NCC made the microorganisms hard to reach into the internal environment. Moreover, the structure of NCC provided the functions of moisturization and gaseous exchanges, which helped reduce inflammation and pain sensation and speed up the recovery of skin damage. Although both MCC and NCC were non-toxic, NCC was targeted to be the potential ingredients in pharmaceutical industry development as it possessed higher porosity compared to MCC which had greater fluid penetration degree and disintegration rate (Jackson et al. 2011).

In the field of pharmaceuticals, Lin and Dufresne (2014) proposed cellulose had created a great contribution, including tablet development. Yet, the ongoing exploration of new types of cellulose is still proceeding in the advanced drug-loading systems. In the drug delivery system, tablets present in a solid dosage form achieved the highest applications due to their satisfying chemical stability, accurate dose and high compliance from patients (Rodrigues and Emeje, 2012). Disintegrant is a type of excipients that are used in tablets production. It is functioned to strengthen the solubility and dissolution of the drugs or Active Pharmaceutical Ingredients (API) in solution by separating the tablets into pieces (Niwa and Hiraishi 2013; Desai et al. 2016). MCC was one of the pure celluloses which possess a high crystallinity degree that provides supposition for polymers (Coelho et al. 2018). It was valuable in the pharmaceutical industry as it was able to suppress the after effect of drugs as well as aided in the accessibility and stability of tablets via reacted with API (Pishnamazi et al. 2019). Moreover, Thoorens and his team (2014) testified that the flowability and compressibility of MCC were performed in a granulated form. This property has made it be highly beneficial in tablets formulation through direct compression way.

In an earlier article, it was studied that NCC, which is also called cellulose whiskers, was present in a colloidal form in an aqueous state due to its crystalline and non-crystalline regions in its complex structure (Zaman et al. 2013; Lin and Dufresne, 2014). A few journals stipulated that NCC possesses a higher crystallinity degree than MCC after its amorphous region was eliminated through the acid hydrolysis process. Their structure and properties were still preserved as nano-typed materials although the amorphous part was detached (Cha et al. 2013; Lin and Dufresne, 2014). Although MCC was a popular component that was applied in the pharmaceutical industry, Gohel et al. (2007) stipulated the low porosity of MCC affects the penetration of the fluid to the tablets which may directly alter the disintegration of tablets. On the contrary, NCC has a higher porosity than MCC due to its smaller size, which can bind significant amounts of ionizable water antibiotics (Jackson et al. 2011). Additionally, NCC has also proved that it is toxic to cells through the Human Brain Microvascular Endothelial Cells (HBMEC) evaluation (Roman et al. 2009). These statements strongly support that NCC possesses the potential in specific drug delivery applications due to the designated uptake progress for drug targeting.

6   CONCLUSION

In recent years, the trend of sustainable development in crystalline cellulose-based products was expanded in different industries, including the pharmaceutical industry. In this study, a few origins of agricultural waste were mentioned as the potential sources in formulating micro (MCC) and nano (NCC) crystalline cellulose. Some aspects are considered to have proved that both MCC and NCC are the potential excipients or fillers that could contribute greatly to the pharmaceutical industry. Moreover, both MCC and NCC are also non-toxic and non-inflammatory when tested on cell lines and animal models in low concentrations. However, some studies investigated that MCC and NCC are toxic when present in high concentrations. There is still insufficient information about the relationship between concentration and toxicity of MCC and NCC. Thus, this chapter recommends that more research and studies should be carried out the toxicity evaluation on crystalline cellulose that possess the potential to be the new alternative approaches in pharmaceutical.

ACKNOWLEDGMENT

Funding from the Fundamental Research Grant Scheme (FRGS/1/2019/TK10/UKM/02/1) by the Malaysian Ministry of Education and the Ministry of Higher Education of Malaysia is gratefully acknowledged for the completion of our original work.

■ References

Abdul Khalil, H.P.S., A.H. Bhat and A.F. Ireana Yusra. 2012. Green composites from sustainable cellulose nanofibrils: A review. Carbohydrate Polymers 87: 963–979. https://doi.org/10.1016/j.carbpol.2011.08.078.

Abdul Khalil, H.P., A.H. Bhat, A. Abu Bakar, P.M. Tahir, I.S. Zaidul and M. Jawaid. 2015. Cellulosic nanocomposites from natural fibers for medical applications: A review. pp. 475–511. In: Handbook of Polymer Nanocomposites. Processing, Performance and Application: Volume C: Polymer Nanocomposites of Cellulose Nanoparticles. https://doi.org/10.1007/978-3-642-45232-1_72.

Abdullah, N.A., M.H. Sainorudin, M.S.A. Rani, M. Mohammad, N.H.A. Kadir and N. Asim. 2021a. Structure and thermal properties of microcrystalline cellulose extracted from coconut husk fiber. Polimery/ Polymers 66(3): 187–192. https://doi.org/10.14314/POLIMERY.2021.3.4.

Abdullah, A., M. Rani, M. Mohammad, M.H. Sainorudin, N. Asim, Z. Yaakob, et al. 2021b. Nanocellulose from agricultural waste as an emerging material for nanotechnology applications–An overview. Polimery 66: 157–168.

Ahmad, Z., N.N. Roziaizan, R. Rahman, A.F. Mohamad and W.I.N.W. Ismail. 2016. Isolation and Characterization of Microcrystalline Cellulose (MCC) from Rice Husk (RH). The 3rd International Conference on Civil and Environmental Engineering for Sustainability (IConCEES 2015). 47: 05013.

Alavi, M. 2019. Modifications of microcrystalline cellulose (MCC), nanofibrillated cellulose (NFC), and nanocrystalline cellulose (NCC) for antimicrobial and wound healing applications. E-Polymers 19: 103–119.

AnjiReddy, K. and S. Karpagam. 2020. Micro and nanocrystalline cellulose based oral dispersible film; preparation and evaluation of *in vitro/in vivo* rapid release studies for donepezil. Brazilian Journal of Pharmaceutical Sciences 56: e17797.

Antunes, F.A.F., A.K. Chandel, L.P. Brumano, R. Terán Hilares, G.F.D. Peres, L.E.S. Ayabe, et al. 2018. A novel process intensification strategy for second-generation ethanol production from sugarcane bagasse in fluidized bed reactor. Renew Energy, SI: Waste Biomass to Biofuel 124: 189–196.

Araki, J., M. Wada, S. Kuga and T. Okano. 1999. Influence of surface charge on viscosity behavior of cellulose microcrystal suspension. Journal of Wood Science 45: 258–261. https://doi.org/10.1007/BF01177736.

Azubuike, C.P., J. Odulaja and A.O. Okhamafe. 2012. Physicotechnical, spectroscopic and thermogravimetric properties of powdered cellulose and microcrystalline cellulose derived from groundnut shells. Journal of Excipients and Food Chemistry 3: 106–115.

Bahloul, A., Z. Kassab, M. El Bouchti, H. Hannache, A.E.K. Qaiss, M. Oumam, et al. 2021. Micro- and nano-structures of cellulose from eggplant plant (*Solanum Melongena* L) agricultural residue. Carbohydrate Polymers 253: 117311. https://doi.org/10.1016/j.carbpol.2020.117311.

Bhattacharya, M., M.M. Malinen, P. Lauren, Y.-R. Lou, S.W. Kuisma, L. Kanninen, et al. 2012. Nanofibrillar cellulose hydrogel promotes three-dimensional liver cell culture. pp. 291–298. *In*: W.E. Hennink and J.F.J. Engbersen (eds). The 12th edition of the European Symposium on Controlled Drug Delivery, (ESCDD2012) April 4–6, 2012, Egmond aan Zee, The Netherlands. Journal of Controlled Release 164: 291–298.

Bhowmik, S., J.M.M. Islam, T. Debnath, M.Y. Miah, S. Bhattacharjee and M.A. Khan. 2017. Reinforcement of gelatin-based nanofilled polymer biocomposite by crystalline cellulose from cotton for advanced wound dressing applications. Polymers 9: 222.

Brinchi, L., F. Cotana, E. Fortunati and J.M. Kenny. 2013. Production of nanocrystalline cellulose from lignocellulosic biomass: Technology and applications. Carbohydrate Polymers 94: 154–169.

Cataldi, A., F. Deflorian and A. Pegoretti. 2015. Poly 2-ethyl-2-oxazoline/microcrystalline cellulose composites for cultural heritage conservation: Mechanical characterization in dry and wet state and application as lining adhesives of canvas. International Journal of Adhesion and Adhesives 62: 92–10.

Cha, C., S.R. Shin, N. Annabi, M.R. Dokmeci and A. Khademhosseini. 2013. Carbon-based nanomaterials: multifunctional materials for biomedical engineering. ACS Nano 7: 2891–2897. doi: 10.1021/nn401196a

Chen, Q., Y. Shi, G. Chen and M. Cai. 2020. Enhanced mechanical and hydrophobic properties of composite cassava starch films with stearic acid modified MCC (microcrystalline cellulose)/NCC (nanocellulose) as strength agent. International Journal of Biological Macromolecules 142: 846–854.

Coelho, C.C.S., M. Michelin, M.A. Cerqueira, C. Gonçalves, R.V. Tonon, L.M. Pastrana, et al. 2018. Cellulose nanocrystals from grape pomace: Production, properties and cytotoxicity assessment. Carbohydrate Polymers 192: 327–336. https://doi.org/10.1016/j.carbpol.2018.03.023.

Dan, J., Y. Ma, P. Yue, Y. Xie, Q. Zheng, P. Hu, et al. 2016. Microcrystalline cellulose-carboxymethyl cellulose sodium as an effective dispersant for drug nanocrystals: A case study. Carbohydrate Polymers 136: 499–506.

Deepa, B., E. Abraham, B.M. Cherian, A. Bismarck, J.J. Blaker, L.A. Pothan, et al. 2011. Structure, morphology and thermal characteristics of banana nano fibers obtained by steam explosion. Bioresouse Technology 102: 1988–1997.

Desai, M.S., A.M. Seekatz, N.M. Koropatkin, N. Kamada, C.A. Hickey, M. Wolter, et al. 2016. A dietary fiber-deprived gut microbiota degrades the colonic mucus barrier and enhances pathogen susceptibility. Cell 167(5): 1339–1353.e21. doi: 10.1016/j.cell.2016.10.043.

Dong, S., A.A. Hirani, K.R. Colacino, Y.W. Lee and M. Roman. 2012. Cytotoxicity and cellular uptake of cellulose nanocrystals. Nano Life 02(03): 1241006. https://doi.org/10.1142/s1793984412410061.

Domingues, R.M.A., M.E. Gomes and R.L. Reis. 2014. The potential of cellulose nanocrystals in tissue engineering strategies. Biomacromolecules 15: 2327–2346.

DOSM. 2020. Press Release: Selected Agricultural Indicators, Malaysia, Available online 17072021. https://www.dosm.gov.my/v1/index.php?r=column/cthemeByCat&cat=72&bul_id=RXVKUVJ5TitHM0cwYWxlOHcxU3dKdz09&menu_id=Z0VTZGU1UHBUT1VJMFlpaXRRR0xpdz09

Du, H., W. Liu, M. Zhang, C. Si, X. Zhang and B. Li. 2019. Cellulose nanocrystals and cellulose nanofibrils based hydrogels for biomedical applications. Carbohydrate Polymers 209: 130–144.

Dufresne, A. 2013. Nanocellulose: From Nature To High Performance Tailored Materials. Walter De Gruyter, Berlin, Boston.

Dugan, J.M., J.E. Gough and S.J. Eichhorn. 2013. Bacterial cellulose scaffolds and cellulose nanowhiskers for tissue engineering. Nanomedicine 8(2): 287–298.

Dungani, R., M. Karina, Subyakto, A. Sulaeman, D. Hermawan and A. Hadiyane. 2016. Agricultural waste fibers towards sustainability and advanced utilization: A review. Asian Journal of Plant Sciences 15: 42–55.

Eichhorn, S.J., A. Dufresne, M. Aranguren, N.E. Marcovich, J.R. Capadona, S.J. Rowan, et al. 2010. Review: Current International Research into Cellulose Nanofibres and Nanocomposites. Journal of Materials Science 45: 1–33. https://doi.org/10.1007/s10853-009-3874-0.

Emara, L., A. El-Ashmawy, N. Taha, K. El-Shaffei, E.-S. Mahdey and H. El-kholly. 2016. Nano-crystalline cellulose as a novel tablet excipient for improving solubility and dissolution of meloxicam. Journal of Applied Pharmaceutical Science 6(2): 032–043.

Evangelista, D.K., Z. Rico, E. Vellezas and K. Badajos. 2017. Isolation and characterization of potential microcrystalline cellulose (MCC) present in coconut (*Cocos nucifera*) fibers. available at doi: 10.13140/RG.2.2.19581.87520/1.

Ferraz, N., D.O. Carlsson, J. Hong, R. Larsson, B. Fellström, L. Nyholm, et al. 2012. HaeFmocompatibility and ion exchange capability of nanocellulose polypyrrole membranes intended for blood purification. Journal of the Royal Society Interface 9: 1943–1955. http://doi.org/10.1098/rsif.2012.0019.

Feynman, R.P. 1960. There's plenty of room at the bottom. Engineering and Science 23(5): 22–36.

Fortunati, E., D. Puglia, M. Monti, L. Peponi, C. Santulli, J.M. Kenny, et al. 2013. Extraction of Cellulose Nanocrystals from Phormium tenax Fibres. Journal of Polymers and the Environment 21, 319–328.

Fug, L., J. Zhang and G. Yang. 2013. Present status and applications of bacterial cellulose-based materials for skin tissue repair. Carbohydrate Polymers 92: 1432–1442. https://doi.org/10.1016/j.carbpol.2012.10.071.

George, J. and S. Sabapathi. 2015. Cellulose nanocrystals: synthesis, functional properties, and applications. Nanotechnology, Science and Applications 8: 45–54.

Giorgio P. and R. Patrizia. 2003. The safety of pharmaceutical excipients. Il Farmaco 58: 541–550.

Gohel, M.C., R.K. Parikh, B.K. Brahmbhatt and A.R. Shah. 2007. Improving the tablet characteristics and dissolution profile of ibuprofen by using a novel coprocessed superdisintegrant: A technical note. AAPS PharmSciTech 8: E94–E99.

Gopinath, V., S. Saravanan, A.R. Al-Maleki, M. Ramesh and J. Vadivelu. 2018. A review of natural polysaccharides for drug delivery applications: Special focus on cellulose, starch and glycogen. Biomedicine & Pharmacotherapy 107: 96–108.

Guo, X., R.-K. Chang and M.A. Hussain. 2009. Ion-exchange resins as drug delivery carriers. Journal of Pharmaceutical Sciences 98: 3886–3902.

Ha, S.J., J. Lee, J. Park, Y.H. Kim, N.H. Lee, Y.E. Kim, et al. 2018. Syringic acid prevents skin carcinogenesis via regulation of NoX and EGFR signaling. Biochemical Pharmacology 154: 435–445.

Habibi, Y., A.L. Lucian and J.R. Orlando. 2010. Cellulose nanocrystals: Chemistry, self-assembly, and applications. Chemical Reviews 110(6): 3479–3500. DOI: 10.1021/cr900339w

Hanif, Z., F.R. Ahmed, S.W. Shin, Y.K. Kim and S.H. Um. 2014. Size- and dose-dependent toxicity of cellulose nanocrystals (CNC) on human fibroblasts and colon adenocarcinoma. Colloids and Surfaces B: Biointerfaces 119: 162–165. https://doi.org/10.1016/j.colsurfb.2014.04.018.

Hanani, A.S.N., A. Zuliahani, W.I. Nawawi, N. Razif and A.R. Rozyanty. 2017. The effect of various acids on properties of microcrystalline cellulose (MCC) extracted from rice husk (RH). IOP Conference Series: Materials Science and Engineering 204: 012025.

Hindi, S. 2016. Microcrystalline cellulose: Its processing and pharmaceutical specifications. BioCrystal Journal 1: 26–38.

Hoare, T.R. and D.S. Kohane. 2008. Hydrogels in drug delivery: Progress and challenges. Polymer 49: 1993–2007.

Huang, C., N. Hao, S. Bhagia, M. Li, X. Meng, Y. Pu, et al. 2018. Porous artificial bone scaffold synthesized from a facile *in situ* hydroxyapatite coating and crosslinking reaction of crystalline nanocellulose. Materialia 4: 237–246. https://doi.org/10.1016/j.mtla.2018.09.008.

Ikuo, H. 2016. The principle of safety evaluation in medicinal drug—how can toxicology contribute to drug discovery and development as a multidisciplinary science? The Journal of Toxicological Sciences 41: SP49–SP67.

Imlimthan, S., A. Correia, P. Figueiredo, K. Lintinen, V. Balasubramanian, A.J. Airaksinen, et al. 2020. Systematic *in vitro* Biocompatibility Studies of Multimodal Cellulose Nanocrystal 2020 and Lignin Nanoparticles. Journal of Biomedical Materials Research Part A 108: 770–783. https://doi.org/10.1002/jbm.a.36856.

Jackson, J.K., K. Letchford, B.Z. Wasserman, L. Ye, W.Y. Hamad, H.M. Burt, et al. 2011. The use of nanocrystalline cellulose for the binding and controlled release of drugs. International Journal of Nanomedicine 6: 321–330.

Jebali, A., S. Hekmatimoghaddam, A. Behzadi, I. Rezapor, B.H. Mohammadi, T. Jasemizad, et al. 2013. Antimicrobial activity of nanocellulose conjugated with allicin and lysozyme. Cellulose 20: 2897–2907. https://doi.org/10.1007/s10570-013-0084-3.

Jia, X., R. Xu, W. Shen, M. Xie, M. Abid, S. Jabbar, et al. 2015. Stabilizing oil-in-water emulsion with amorphous cellulose. Food Hydrocolloids 43: 275–282.

Johar, N., I. Ahmad and A. Dufresne. 2012. Extraction, preparation and characterization of cellulose fibres and nanocrystals from rice husk. Industrial Crops and Products 37: 93–99.

Jones, C.F. and D.W. Grainger. 2009. *In vitro* assessments of nanomaterial toxicity. Advanced Drug Delivery Reviews 61: 438–456..

Kamel, R., N.A. El-Wakil, A. Dufresne and N.A. Elkasabgy. 2020a. Nanocellulose: From an agricultural waste to a valuable pharmaceutical ingredient. International Journal of Biological Macromolecules 163: 1579–1590.

Kamel, R., N.A. El-Wakil, A.A. Abdelkhalek and N.A. Elkasabgy. 2020b. Nanofibrillated cellulose/cyclodextrin based 3D scaffolds loaded with raloxifene hydrochloride for bone regeneration. International Journal of Biological Macromolecules 156: 704–716.

Kargarzadeh, H., I. Ahmad, I. Abdullah, A. Dufresne, S.Y. Zainudin and R.M. Sheltami. 2012. Effects of hydrolysis conditions on the morphology, crystallinity, and thermal stability of cellulose nanocrystals extracted from kenaf bast fibers. Cellulose 19: 855–866. https://doi.org/10.1007/s10570-012-9684-6.

Karthik, T. and R. Murugan. 2013. Characterization and analysis of ligno-cellulosic seed fiber from pergularia daemia plant for textile applications. Fibers and Polymers 14: 465–472.

Kassab, Z., M. El Achaby, Y. Tamraoui, H. Sehaqui, R. Bouhfid and A.E.K. Qaiss. 2019. Sunflower oil cake-derived cellulose nanocrystals: extraction, physico-chemical characteristics and potential application. International Journal of Biological Macromolecules 136: 241–252. https://doi.org/10.1016/j.ijbiomac.2019.06.049.

Kassab, Z., Y. Abdellaoui, M.H. Salim, R. Bouhfid, A.E.K. Qaiss and M. El Achaby. 2020a. Micro- and nano-celluloses derived from hemp stalks and their effect as polymer reinforcing materials. Carbohydrate Polymers 245: 116506. https://doi.org/10.1016/j.carbpol.2020.116506.

Kassab, Z., I. Kassem, H. Hannache, R. Bouhfid, A.E.K. Qaiss and M. El Achaby. 2020b. Tomato plant residue as new renewable source for cellulose production: extraction of cellulose nanocrystals with different surface functionalities. Cellulose 27: 4287–4303. https://doi.org/10.1007/s10570-020-03097-7.

Kassab, Z., Y. Abdellaoui, M.H. Salim and M. El Achaby. 2020c. Cellulosic materials from pea (*Pisum Sativum*) and broad beans (*Vicia Faba*) pods agro-industrial residues. Materials Letters 280: 128539. https://doi.org/10.1016/j.matlet.2020.128539.

Klemm, D., F. Kramer, S. Moritz, T. Lindström, M. Ankerfors, D. Gray, et al. 2011. Nanocelluloses: A new family of nature-based materials. Angewandte Chemie International Edition 50: 5438–5466.

Kupnik, K., M. Primožič, V. Kokol and M. Leitgeb. 2020. Nanocellulose in drug delivery and antimicrobially active materials. Polymers 12: 2825. https://doi.org/10.3390/polym12122825.

Kusmono, R.F. Listyanda, M.W. Wildan and M.N. Ilman. 2020. Preparation and characterization of cellulose nanocrystal extracted from ramie fibers by sulfuric acid hydrolysis. Heliyon 6(11): e05486.

Lam, E., K.B. Male, J.H. Chong, A.C.W. Leung and J.H.T. Luong. 2012. Applications of functionalized and nanoparticle-modified nanocrystalline cellulose. Trends Biotechnol. 30: 283–290.

Lin, N. and A. Dufresne. 2014. Nanocellulose in biomedicine: Current status and future prospect. European Polymer Journal 59: 302–325. https://doi.org/10.1016/j.eurpolymj.2014.07.025.

Lipinski, A.C. 2002. Physicochemical properties and the discovery of orally active drugs: Technical and people issues. Molecular Informatics: Confronting Complexity, May 13th–16th 2002, Bozen, Proceedings of the Beilstein-Institut Workshop.

Liu, D., X. Chen, Y. Yue, M. Chen and Q. Wu. 2011. Structure and rheology of nanocrystalline cellulose. Carbohydrate Polymers 84: 316–322.

Liu, Y., A. Liu, S.A. Ibrahim, H. Yang and W. Huang. 2018. Isolation and characterization of microcrystalline cellulose from pomelo peel. International Journal of Biological Macromolecules 111: 717–721. https://doi.org/10.1016/j.ijbiomac.2018.01.098.

Luo, H., G. Xiong, D. Hu, K. Ren, F. Yao, Y. Zhu, et al. 2013. Characterization of TEMPO-oxidized bacterial cellulose scaffolds for tissue engineering applications. Materials Chemistry and Physics 143: 373–379. https://doi.org/10.1016/j.matchemphys.2013.09.012.

Mishra, D., P. Khare, M.R. Das, S. Mohanty, D.B. Kule and P.A. Kumar. 2018. Characterization of crystalline cellulose extracted from distilled waste of cymbopogon winterianus. Cellulose Chemistry and Technology 52: 9–17.

Moon, R.J., A. Martini, J. Nairn, J. Simonsen and J. Youngblood. 2011. Cellulose nanomaterials review: Structure, properties and nanocomposites. Chemical Society Review 40: 3941–3994.

Nagalakshmaiah, M., N.E. Kissi, G. Mortha and A. Dufresne. 2016. Structural investigation of cellulose nanocrystals extracted from chili leftover and their reinforcement in cariflex-OR rubber latex. Carbohydrate Polymers 136(1): 945–954.

Ndika, E.V., U.S. Chidozie and U.K. Ikechukwu. 2019. Chemical modification of cellulose from palm kernel de-oiled cake to microcrystalline cellulose and its evaluation as a pharmaceutical excipient. African Journal of Applied Chemistry 13: 49–57.

Niwa, M. and Y. Hiraishi 2013. Quantitative analysis of visible surface defect risk in tablets during film coating using terahertz pulsed imaging. International Journal of Pharmaceutics 461(1–2): 342–350. doi: 10.1016/j.ijpharm.2013.11.051.

Nordin, N.A., O. Sulaiman, R. Hashim and M.H.M. Kassim. 2017. Oil palm frond waste for the production of cellulose nanocrystals. Journal of Physical Science 28(2): 115–126. https://doi.org/10.21315/jps2017.28.2.8.

O'Brien, P.J., W. Irwin, D. Diaz, E. Howard-Cofield, C.M. Krejsa, M.R. Slaughter, et al. 2006. High concordance of drug-induced human hepatotoxicity with *in vitro* cytotoxicity measured in a novel cell-based model using high content screening. Archives of Toxicology 80: 580–604.

Owolabi, A.F., M.K.M. Haafiz, M.S. Hossain, M.H. Hussin and M.R.N. Fazita. 2017. Influence of alkaline hydrogen peroxide pre-hydrolysis on the isolation of microcrystalline cellulose from oil palm fronds. International Journal of Biological Macromolecules 95: 1228–1234. https://doi.org/10.1016/j.ijbiomac.2016.11.016.

Patra, J.K., G. Das, L.F. Fraceto, E.V.R. Campos, M. del P. Rodriguez-Torres, L.S. Acosta-Torres, et al. 2018. Nano based drug delivery systems: Recent developments and future prospects. Journal of Nanobiotechnology 16: 71. https://doi.org/10.1186/s12951-018-0392-8.

Pelissari, F.M., P.J.d. Sobral and F.C. Menegalli. 2014. Isolation and characterization of cellulose nanofibers from banana peels. Cellulose 21: 417–432. https://doi.org/10.1007/s10570-013-0138-6.

Peng, B.L., N. Dhar, H.L. Liu and K.C. Tam. 2011. Chemistry and applications of nanocrystalline cellulose and its derivatives: A nanotechnology perspective. Canadian Journal of Chemical Engineering 89: 1191–1206.

Pereira, M.M., N.R.B. Raposo, R. Brayner, E.M. Teixeira, V. Oliveira, C.C.R. Quintão, et al. 2013. Cytotoxicity and expression of genes involved in the cellular stress response and apoptosis in mammalian fibroblast exposed to cotton cellulose nanofibers. Nanotechnology 24: 075103.

Petersen, N. and P. Gatenholm. 2011. Bacterial cellulose-based materials and medical devices: current state and perspectives. Applied Microbiology and Biotechnology 91: 1277–1286. https://doi.org/10.1007/s00253-011-3432-y.

Pishnamazi, M., H. Hafizi, S. Shirazian, M. Culebras, G.M. Walker and M.N. Collins. 2019. Design of controlled release system for paracetamol based on modified lignin. Polymers 11(6): 1059. DOI: 10.3390/polym11061059.

Prado, K.S. and M.A.S. Spinacé. 2019. Isolation and characterization of cellulose nanocrystals from pineapple crown waste and their potential uses. International Journal of Biological Macromolecules, 122: 410–416. https://doi.org/10.1016/j.ijbiomac.2018.10.187.

Rasheed, M., M. Jawaid, Z. Karim and L.C. Abdullah. 2020. Morphological, physiochemical and thermal properties of microcrystalline cellulose (MCC) extracted from bamboo fiber. Molecules 25: 2824.

Raza, Z.A., M. Aslam, A. Azeem and H.S. Maqsood. 2019. Development and characterization of nano-crystalline cellulose incorporated poly(Lactic Acid) composite films. Materials Science and Engineering Technology 50: 64-73. https://doi.org/10.1002/mawe.201800081.

Robles, E., I. Urruzola, J. Labidi and L. Serrano. 2015. Surface-modified nano-cellulose as reinforcement in poly(lactic acid) to conform new composites. Industrial Crops and Products 71: 44–53.

Rodrigues, A. and M. Emeje. 2012. Recent applications of starch derivatives in nanodrug delivery. Carbohydrate Polymers 87(2): 987–994. https://doi.org/10.1016/j.carbpol.2011.09.044.

Rojas, J., A. Lopez, S. Guisao and C. Ortiz. 2011. Evaluation of several microcrystalline celluloses obtained from agricultural by-products. Journal of Advanced Pharmaceutical Technology & Research 2: 144–150.

Roman, M. and W.T. Winter. 2004. Effect of sulfate groups from sulfuric acid hydrolysis on the thermal degradation behavior of bacterial cellulose. Biomacromolecules 5: 1671–1677.

Roman, M., S. Dong, A. Hirani and Y.W. Lee. 2009. Cellulose nanocrystals for drug delivery. pp. 81–91. In: K.J. Edgar, T. Heinze and C.M. Buchanan (eds). Polysaccharide Materials: Performance by Design, ACS Symposium Series. American Chemical Society. https://doi.org/10.1021/bk-2009-1017.ch004.

Russell, W.M.S. and R.L. Burch. 1959. The Principles of Humane Experimental Technique. Methuen, London. ISBN 0900767782.

Sacui, I.A., R.C. Nieuwendaal, D.J. Burnett, S.J. Stranick, M. Jorfi, C. Weder, et al. 2014. Comparison of the properties of cellulose nanocrystals and cellulose nanofibrils isolated from bacteria, tunicate, and wood processed using acid, enzymatic, mechanical, and oxidative methods. ACS Applied Materials Interfaces 6: 6127–6138.

Semachai, T., P. Chandranupap and P. Chandranupap. 2018. Preparation of microcrystalline cellulose from water hyacinth reinforced polylactic acid biocomposite. MATEC Web of Conferences 187: 02003.

Setu, M.N.I., M.Y. Mia, N.J. Lubna and A.A. Chowdhury. 2014. Preparation of microcrystalline cellulose from cotton and its evaluation as direct compressible excipient in the formulation of naproxen tablets, Dhaka University. Journal of Pharmaceutical Sciences 13(2): 187–192.

Shaikh, S., A. Birdi, S. Qutubuddin, E. Lakatosh and H. Baskaran. 2007. Controlled release in transdermal pressure sensitive adhesives using organosilicate nanocomposites. Annals of Biomedical Engineering 35: 2130–2137.

Shanmugam, N., R.D. Nagarkar and M. Kurhade. 2015. Microcrystalline cellulose powder from banana pseudostem fibres using bio-chemical route. Indian Journal of Natural Products and Resources 6: 42–50.

Shao, X., J. Wang, Z. Liu, N. Hu, M. Liu and Y. Xu. 2020. Preparation and characterization of porous microcrystalline cellulose from corncob. Industrial Crops and Products 151: 112457. https://doi.org/10.1016/j.indcrop.2020.112457.

Sharma, A., M. Thakur, M. Bhattacharya, T. Mandal and S. Goswami. 2019. Commercial application of cellulose nano-composites – A review. Biotechnology Reports 21: e00316.

Singh, M., S. Kanawjia, A. Giri and Y. Khetra. 2015. Moisture sorption characteristics of shredded mozzarella cheese. Journal of Food Processing and Preservation 39: 521–529. https://doi.org/10.1111/jfpp.12257.

Siqueira, G., S. Tapin-Lingua, J. Bras, D. da Silva Perez and A. Dufresne. 2011. Mechanical properties of natural rubber nanocomposites reinforced with cellulosic nanoparticles obtained from combined mechanical shearing, and enzymatic and acid hydrolysis of sisal fibers. Cellulose 18: 57–65.

Sluiter, J.B., H. Chum, A.C. Gomes, R.P.A. Tavares, V. Azevedo, M.T.B. Pimenta, et al. 2016. Evaluation of brazilian sugarcane bagasse characterization: An interlaboratory comparison study. Journal of AOC International 99: 579–585.

So, B.R., H.J. Yeo, J.J. Lee, Y.H. Jung and S.K. Jung. 2021. Cellulose nanocrystal preparation from Gelidium amansii and analysis of its anti-inflammatory effect on the skin *in vitro* and *in vivo*. Carbohydrate Polymers 254: 117315.

Song, Y.K., I.M.L. Chew, T.S.Y. Choong and K.W. Tan. 2016. NanoCrystalline cellulose, an environmental friendly nanoparticle for pharmaceutical application—A quick study. MATEC Web of Conferences 60: 01006.

Sosiati, H., D.A.Wijayanti, K. Triyana and B. Kamiel. 2017. Morphology and crystallinity of sisal nanocellulose after sonication. *In*: AIP Conference Proceedings 2017. https://doi.org/10.1063/1.4999859.

Subedi, R.K., S.Y. Oh, M.-K. Chun and H.-K. Choi. 2010. Recent advances in transdermal drug delivery. Archives of Pharmacal Research 33: 339–351.

Taheri, A. and M. Mohammadi. 2015. The use of cellulose nanocrystals for potential application in topical delivery of hydroquinone. Chemical Biology & Drug Design 86: 102–106.

Tang, Y., S. Yang, N. Zhang and J. Zhang. 2014. Preparation and characterization of nanocrystalline cellulose via low-intensity ultrasonic-assisted sulfuric acid hydrolysis. Cellulose 21: 335–346. https://doi.org/10.1007/s10570-013-0158-2.

Tibolla, H., F.M. Pelissari, J.T. Martins, A.A. Vicente and F.C. Menegalli. 2018. Cellulose nanofibers produced from banana peel by chemical and mechanical treatments: Characterization and cytotoxicity assessment. Food Hydrocolloids 75: 192–201.

Tibolla, H., F.M. Pelissari, J.T. Martins, E.M. Lanzoni, A.A. Vicente, F.C. Menegalli, et al. 2019. Banana starch nanocomposite with cellulose nanofibers isolated from banana peel by enzymatic treatment: *In vitro* cytotoxicity assessment. Carbohydrate Polymers 207: 169–179.

Thoorens, G., F. Krier, B. Leclercq, B. Carlin and B. Evrard. 2014. Microcrystalline cellulose, a direct compression binder in a quality by design environment—A review. International Journal of Pharmaceutics 473: 64–72.

Ventura-Cruz, S. and A. Tecante. 2019. Extraction and characterization of cellulose nanofibers from Rose stems (Rosa spp.). Carbohydrate Polymers 220(15): 53–59.

Villanova, J.C.O., E. Ayres, S.M. Carvalho, P.S. Patrício, F.V. Pereira and R.L. Oréfice. 2011. Pharmaceutical acrylic beads obtained by suspension polymerization containing cellulose nanowhiskers as excipient for drug delivery. European Journal of Pharmaceutical Sciences 42: 406–415. https://doi.org/10.1016/j.ejps.2011.01.005.

Wang, C., H. Huang, M. Jia, S. Jin, W. Zhao and R. Cha. 2015a. Formulation and evaluation of nanocrystalline cellulose as a potential disintegrant. Carbohydrate Polymers 130: 275–279.

Wang, M., X.-W. Han, L. Liu, X.-F. Zeng, H.-K. Zou, J.-X. Wang, et al. 2015b. Transparent aqueous Mg (OH)$_2$ nanodispersion for transparent and flexible polymer film with enhanced flame-retardant property. Industrial & Engineering Chemistry Research 54: 12805–12812.

Wiegand, C., S. Moritz, N. Hessler, D. Kralisch, F. Wesarg, F.A. Müller, et al. 2015. Antimicrobial functionalization of bacterial nanocellulose by loading with polihexanide and povidone-iodine. Journal of Materials Science: Materials in Medicine 26: Article number 245.

Xiang, L.Y., M.A.P. Mohammed and A.S. Baharuddin. 2016. Characterization of microcrystalline cellulose from oil palm fibres for food applications. Carbohydrate Polymers 148: 11–20.

Xu, J., E.F. Krietemeyer, V.M. Boddu, S.X. Liu and W.-C. Liu. 2018. Production and characterisation of cellulose nanofibril (CNF) from agricultural waste corn stover. Carbohydrate Polymers 192: 202–207.

Yang, S., Y. Tang, J. Wang, F. Kong and J. Zhang. 2014. Surface treatment of cellulosic paper with starch-based composites reinforced with nanocrystalline cellulose. Industrial and Engineering Chemistry Research 53: 13980–13988.

Yu, H.-Y., D.-Z., Zhang, F.-F. Lu, and J. Yao. 2016. New approach for single-step extraction of carboxylated cellulose nanocrystals for their use as adsorbents and flocculants. ACS Sustainable Chemistry & Engineering 4(5): 2632–2643. https://doi.org/10.1021/acssuschemeng.6b00126.

Zaman, M., H. Liu, H. Xiao, F. Chibante and Y. Ni. 2013. Hydrophilic modification of polyester fabric by applying nanocrystalline cellulose containing surface finish. Carbohydrate Polymers 91: 560–567.

Zheng, D., Y. Zhang, Y. Guo and J. Yue. 2019. Isolation and characterization of nanocellulose with a novel shape from walnut (*Juglans Regia* L.) shell agricultural waste. Polymers 11: 1130.

Zoppe, J.O., V. Ruottinen, J. Ruotsalainen, S. Rönkkö, L.-S. Johansson, A. Hinkkanen, et al. 2014. Synthesis of cellulose nanocrystals carrying tyrosine sulfate mimetic ligands and inhibition of alphavirus infection. Biomacromolecules 15: 1534–1542.

Chapter 11

Recycling and Sustainability of Nanocomposites

Salma Shaik and Matthew J. Franchetti*

The University of Toledo, 2801 W. Bancroft St, Toledo, OH–43606, USA

salma.shaik@rockets.utoledo.edu; matthew.franchetti@utoledo.edu

1 INTRODUCTION

Since the advent of nanotechnology in the early 1980s (Nano 101, n.d), has revolutionized the scientific and industrial advancement by opening a whole new domain for research and innovations of new products and processes. Due to their superior mechanical, thermal, electrical properties and more importantly due to the fact that nanomaterials can be tailored to obtain desired properties, they have made their mark in almost all fields ranging from engineering to agriculture, medicine and personal care. The market size value for nanomaterials is poised to grow from USD 9.6 billion in 2020 to USD 22.9 billion by 2027 at a predicted annual growth of 13.1% (MAR 2020). Unfortunately, the unprecedented growth of any new technology or industrial sector leads to the rise in consumerism by the creation of more goods and ultimately results in more waste. It was estimated that out of 260,000–309,000 metric tons of nanomaterials (more specifically ENMs which are man-made with desired properties for specific applications) that were produced in 2010, almost 63–91% ended up in landfills and the rest in soils, water bodies and atmosphere (Keller et al. 2013). Researchers have categorized ENM waste into five types namely pristine, dissolved, transformed, matrix embedded and product embedded as shown in Fig. 11.1 to study the flow of ENM waste across different channels (Adam et al. 2018).

As one has witnessed with the growth of plastics' industry, any progress or technological advancement, if not dealt with a sustainable approach with efficient waste management systems in place, will inevitably create a huge burden on the already depleting natural resources and will irreparably damage the fragile environment. Superior characteristics such as high electrical and thermal conductivity with a great potential for weight decrease (Gardiner 2014) helped carbon nanotubes (CNTs) dominate the market in 2019 accounting for almost 26.9% of the total market share (MAR 2020). It is expected that the fast-paced growth of Carbon Fiber Reinforced

---

*Corresponding Author

Composites (CFRC) will increase the CFRC waste at 20 kt per annum by 2025 which will unavoidably burden the existing waste management systems (Karuppannan and Karki 2020). In some cases, even though the matrix is made up of recyclable polymers, if the nano-fillers are carbon or other niche fibers it will render the whole composite non-recyclable or in some cases not profitable to be recycled (Petrakli et al. 2020). CNTs exhibit strong affinities to pollutants and could carry contaminants facilitating the accumulation of pollutants *in vivo* (Li et al. 2015). It has been further reported that carbon nano fibers can have toxic effects on the human body through intracellular metabolic pathways, oxidative stress and physical membrane damage causing rupture (Roberto and Christofoletti 2019; Petrakli et al. 2020). Given the rapid growth of nanotechnology and the subsequent increase in the production and manufacturing of nanocomposites, NPs will inevitably escape into the environment during manufacturing, transportation, use and disposal. Some of the major product categories containing nanomaterials are listed in Table 11.1 which also contribute to the release of most of the ENMs into environment.

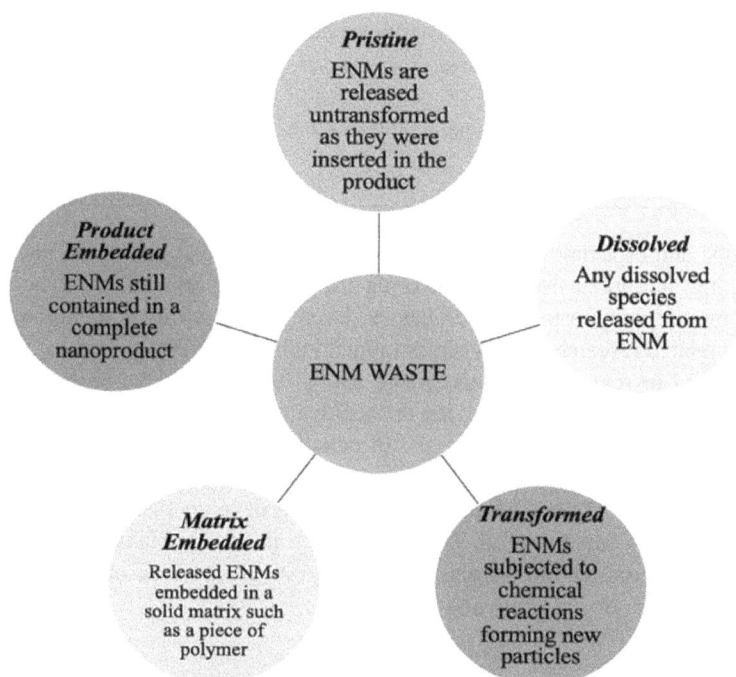

Figure 11.1   Types of ENM Waste (Adapted from Adam et al. 2018).

The exponential growth of nanotechnology and applications of NPs in various industries is giving rise to concerns about their potential hazardous impacts on the health and environment (Mirshafiee et al. 2018). The aerospace industry is one of the biggest consumers of nanomaterials and it is estimated that around 6000–8000 commercial aircrafts will be approaching their EoL by 2030 (Zhang et al. 2020). Non-recyclable waste is a significant issue facing global communities as landfills are filling up at a faster pace and there is a sense of urgency to go green in order to combat global warming (TEC 2019). Hence, economically viable waste management and recycling technologies should be set in place to efficiently manage waste from nanomaterials used in such products. As part of its circular economy plan, the European Commission aims to increase recycling rates and to reduce the amount of urban waste going to landfills to 10% by 2030 (SWD 2019). More stringent laws and regulations for the design of sustainable NPs, efficient waste management and recycling practices can be expected to be passed soon (Nasrollahzadeh et al. 2020a).

Table 11.1 Major product categories with ENM composites (Adapted from Caballero-Guzman et al. 2015)

Product category	Product examples	ENM
Electronics	Keyboards, mice, hair dryers, refrigerators, mobile phones, housings, semiconductor devices	silicon-based, iron-based, titanium-based, carbon-based
Personal care products	Sunscreens, cosmetics	titanium-based, zinc-based
Paints	Wall paints	titanium-based, zinc-based, silver-based, CNT
Textiles	Shirts, blankets, socks, sportswear, gloves	silver-based
Coatings	Anti-microbial and photocatalytic coatings, door locks, kitchenware, watch chains	silver-based, CNT
Energy and environment	Batteries, solar cells	CNT; silicon-based, titanium-based
Automotive	Dashboards, fuel system components	CNT
Catalysts	Catalytic additives in wastewater treatment	silicon-based, aluminum-based, titanium-based, zinc-based
Packaging	Plastic packaging, food packaging	silicon-based, zinc-based, silver-based
Construction and Demolition	Paints, glass	titanium-based, zinc-based

Governments, and especially consumers, are becoming more aware of the undeniable reality of global warming and are increasingly favoring those companies that are striving to obtain a greener image by adopting sustainable processes. Even though millions of dollars are spent on nanotechnology-related research every year, not much money or attention is given to develop tools for the safe disposal or recycling of nano waste. Only about 4% of the federal nanotechnology research budget in the U.S. goes into studying the health and environmental effects of nanomaterials (Faunce and Kolodziejczyk 2017). The uptake of nanocomposites might be challenged by a lack of efficient and sustainable recycling infrastructure (Gardiner 2014). Hence there is a need to understand the impact of NPs on the environment and to develop sustainable and cost-effective processed and products (Taynton et al. 2016; Hajian et al. 2018). Making sustainability one of the core objectives in product and process design is no longer a moral obligation but a required responsibility on the part of academia and industry.

2 CURRENT TREATMENT PROCESSES OF WASTE CONTAINING NANOPARTICLES

2.1 Sources of ENM Waste

NPs can be released into the atmosphere either directly through human activities or indirectly due to a consequence of human action (Fig. 11.2). Non-biodegradable waste is detrimental to all life forms and creates a plethora of environmental problems. As reported in a sustainability analysis study performed using waste to product ratio, the current nanomaterials production methods produce up to 1000 times more waste (at around 100,000 kg waste per kg product) when compared to the production of pharmaceuticals and fine chemicals (Patwardhan et al. 2018). Increasing risk to health and the environment may hamper the growth unless the manufacturing methods of nanocomposites are designed to be more sustainable and eco-friendlier. With the increasing use of NPs in the medicinal, food, pharmaceutical and agricultural industries, there is an emphasis on the production of eco-friendly, non-toxic and environmentally benign NPs (Patra and Baek 2014). Since physiochemical methods to fabricate nanomaterials can be hazardous to the environment, there is a growing need to design sustainable, greener, non-toxic and inexpensive syntheses methods for nanocomposites (NPTELHRD 2014; Nasrollahzadeh et al. 2020b).

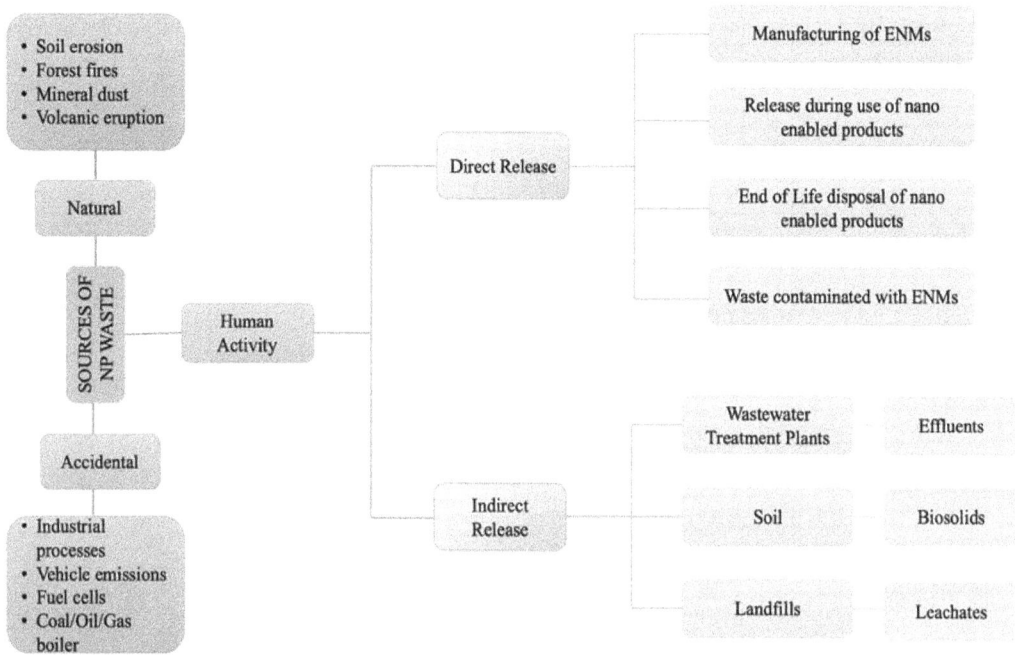

Figure 11.2 Sources of nanoparticles into environment. (Adapted from WCN 2018; Boldrin et al. 2014; Bundschuh et al. 2018; Smita et al. 2012).

2.2 ENM Releases from Waste Management Systems

Detachment, sorting and separation of NPs from nanocomposites or nano additive composites for recycling/repurposing are difficult due to their small size (Li et al. 2015; Petrakli et al. 2020). It is also challenging to track and monitor the fate of NPs both in the environment and the human body. As a result, there are no specific standards that relate to the safe disposal or recycling of a large variety of nanomaterials (Faunce and Kolodziejczyk 2017). Hence, the waste management and recycling of NPs are considered in terms of the entire composite or product. Recycling, incineration, landfilling and wastewater treatment (WWT) are currently the four waste treatment operations that receive and handle the waste containing NPs (Nanotechnologies and Waste 2015). These waste management processes also represent the channels through which the NPs can further leak into the surrounding environment as shown in Table 11.2 (Gottschalk and Nowack 2011).

Nanocomposites that are discarded can let NPs enter landfill leachate, but it is unlikely that a significant amount will be carried to the atmosphere because the degradation occurring in a landfill is a slow process. The presence of pH and other substances influence any potential structural changes of NPs and their binding to other substances present in leachate (WCN 2018). Though some studies (Siddique 2013; Zuin et al. 2013) report that advanced membrane liners may limit the dispersion of NPs from landfills, the efficacy of liners in restricting the NPs from leaching to the environment and the leaking of landfill gas has not yet been studied providing conclusive results (OECD 2016). Recycling of Construction and Demolition (C&D) waste is mostly done by crushing, shredding and milling activities which can release NPs into the air and leachate when C&D waste is used as an aggregate (Boldrin et al. 2015). Lack of proper occupational controls during various thermal processes such as heating, welding and pyrolysis involving NPs and transportation, sorting, shredding and grinding of nanomaterials' waste could also release ultrafine dust containing free NPs (WCN 2018). Also, the behavior of NPs during dismantling, shredding, and thermal processes of nanocomposites in recycling facilities is relatively unknown (Gottschalk and Nowack 2011). NPs that are still trapped in the composite matrix will continue to be present in recycled products and it is estimated that less than 10% of engineered nanomaterials (ENMs) will be looped back in the production and manufacturing chain (Caballero-Guzman et al. 2015).

Table 11.2 Leakage Routes of NPs from Waste Management Processes (adapted from OECD, 2016)

Waste treatment process	Source of NP waste	Possible leakage route
Recycling	– Municipal Solid Waste – Construction and Demolition (C and D) waste – EoL products	– Embedded in secondary materials – Airborne particles emissions – Leachate when C&D waste is used as aggregate
Incineration	– Municipal Solid Waste – Sludge and biosolids from WWT plants	– Fuel gas emissions – Fly ash and bottom ash to landfills, storage facilities and industrial applications
Landfilling	– Municipal Solid Waste – Sludge and biosolids from WWT plants	– Landfill gas and surface emissions – Leachate to leachate and WWT facilities
Wastewater Treatment	– Household drainage – Commercial and industrial sewage – Landfill leachate	– Emissions to surface water – Wastewater sludge to incinerators, landfills, and agricultural applications

NPs could be destroyed or liberated, if the matrix is combustible, during incineration if temperatures are sufficiently high although complete destruction cannot be guaranteed (Mueller et al. 2013; Vejerano et al. 2014 (HCN 2011). The fate of NPs during incineration is not well described because a complex set of factors influence their behavior during incineration and sampling and analysis of waste containing NPs is difficult (Sotiriou et al. 2016; Wiesner and Plata 2012). NPs that are not completely combusted end up in bottom ash or in the liquid effluents. The ashes can be sent to a landfill (Hjelmar and van der Sloot, 2010) which could release NPs to landfill leachate or recycled in construction work by using in the road sub-base. On the other hand, wastewater treatment plants (WWTPs) receive the liquid effluents from incineration plants which may cause the NPs to mix with the output sludge. This may release further NPs to the environment if such sludge is not properly managed (WCN 2018). NPs such as nano-titanium oxide (nTio), nano-silver (nAg) nano-cerium oxide (nCeo) or nano-copper (nCu) are some of the most commonly found in municipal WWT plants. Around 80% of the commonly found nanomaterials such as nano-titanium oxide (nTio), nano-silver (nAg) nano-cerium oxide (nCeo) and nano-copper (nCu) were able to be captured and diverted into solid sludge through transformation, bacterial aggregation, biological polymer adsorption and sedimentation in aerobic processes in pilot WWT plants (Kiser et al. 2009; Kiser et al. 2010; Kaegi et al. 2011; Ganesh et al. 2010; Wang et al. 2012; Gomez-Rivera et al. 2012). This allows the residuals to remain in the form of NPs and appear in surface water while waste incinerators and fuel gas treatment systems could divert a significant share of NPs into fly ash bottom ash (Tied et al. 2010; Kim et al. 2010).

2.3   Life Cycle Stages of ENMs

Researchers have relied on modeling approaches based on mass balances covering the life-cycle stages of ENMs to estimate the quantities of wastes generated from EoL nanocomposites

(Boldrin et al. 2014; Walser and Gottschalk 2014; Caballero-Guzman et al. 2015; Adam and Nowack 2017). But these models have limited model validation and quantitative assessment due to the scarcity of reliable data (Heggelund 2017). Increased availability of data would greatly benefit such modeling and characterization studies of waste-containing NPs which will be indispensable to develop efficient waste management systems. As a result, it is not possible to assess the effectiveness of current disposal and handling processes of waste-containing NPs in mitigating the harmful effects on human health and the environment. Currently, NP waste is disposed off along with other waste without any special precautions or treatments in place (Nanotechnologies and Waste 2015; OECD 2016). Figure 11.3 summarizes the major life cycle stages of ENMs starting from ENM production to end of life disposal.

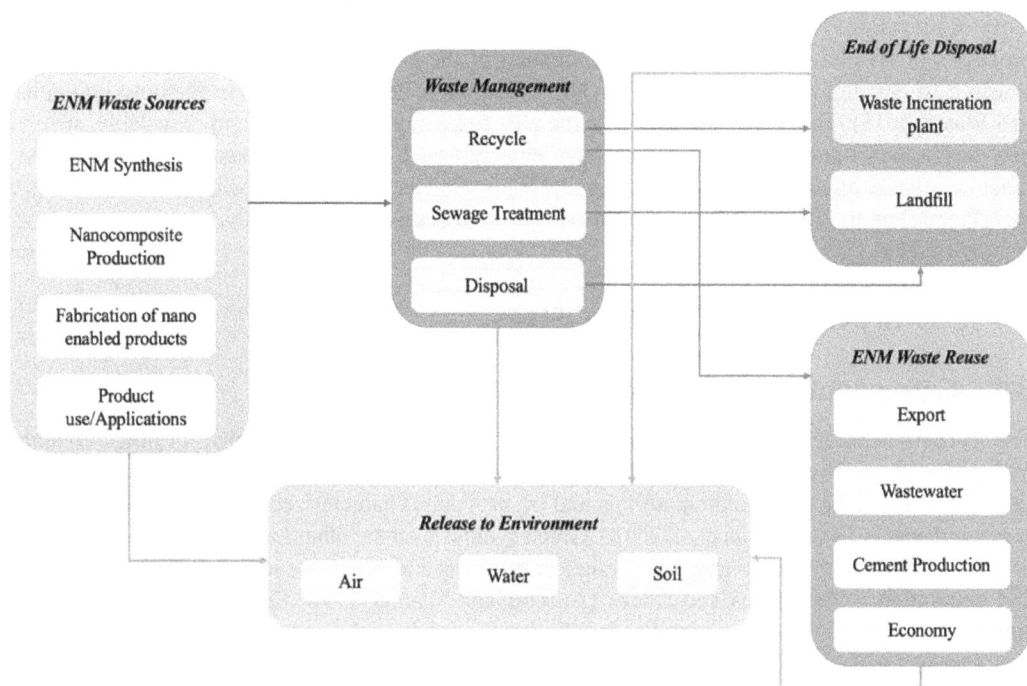

Figure 11.3   Life cycle stages of ENMs. (Adapted from Sun et al. 2014; Caballero-Guzman et al. 2015; Kuenen et al. 2021).

The life cycle of ENMs starts with their synthesis and incorporation into products using various manufacturing processes and these nano-enabled products are then subsequently used by consumers. During the consumption stage along with all the earlier synthesis and manufacturing stages, ENMs find their way into various waste handling streams such as incineration, landfill and recycling. Here, the recycling stream for ENMs implies the recycling of nano-enabled products such as plastics, textiles, electronics, etc. but not of the ENMs themselves. During each of the life cycle stages, ENMs may be released into the eco-sphere contaminating air, water and soil (Kuenen et al. 2021). Once the ENMs enter the recycling system following different separation and sorting techniques, the ENMs can further travel to different destinations such as exporting or in wastewater treatment, cement kilns, etc. Since the nanomaterials are included in composites to enhance performance and since the recycling techniques are directed at recycling of the composites, nanomaterials are usually sent to the incineration and landfill and not in the material sent for further processing as a new product (DEPA 2014).

A major drawback of these studies is that most of them have relied on laboratory experiments or modeling results to understand the fate of some NPs in different waste streams but very few have investigated actual waste treatment facilities (OECD 2016). There is also a scarcity of reliable studies looking into the extent to which the nanomaterials are released or recovered from different products and processes during various waste treatment processes (Nanotechnologies and Waste 2015). Estimates of manufacturing waste and ENM contaminated waste could be provided by the manufacturers of nanomaterials. This is rarely done and is purely voluntary due to confidentiality reasons. Further, manufacturers are not subjected to any reporting requirements about the quantities of waste containing nanomaterials in many countries (WCN 2018). Due to this, there is a shortage of reliable data to draw conclusions or estimates about the quantities of NPs waste produced (Keller et al. 2013; Boldrin et al. 2014). Since nanocomposites contain reinforcements with at least one dimension in nanoscale and with only around 0.5 to 5% of the amount added by weight, it becomes inherently challenging to collect and segregate the small number of NPs for reuse or recycling (AZONANO 2007; Maham et al. 2020). So far, recycling and waste treatment processes of NPs have been studied in the context of the entire composite. But there is growing research to develop processes to recover NPs from waste for targeted reuse (Nanotechnologies and Waste, 2015). To specifically fabricate the NPs to be more sustainable and recyclable, several new technologies and research avenues have emerged such as magnetic nanoparticles (MNPs), green synthesis of nanomaterials, biodegradable or bio-based nanocomposites (biocomposites), etc. (Petrakli et al. 2020) have been explored and are discussed subsequently.

3   SUSTAINABLE NANOCOMPOSITES

3.1   Green Synthesis of Nanoparticles/Green Chemistry

The current approaches to manufacture nanomaterials such as lithography, milling, etching, hydrothermal or sol-gel synthesis could result in unsustainable manufacturing processes because they either use or produce toxic materials and by-products, characterized by high consumption of water and energy (Patwardhan et al. 2018). Green synthesis on the other hand follows the principles of green chemistry that encourage the design of products and processes aimed at reducing the use and generation of hazardous substances (Anastas and Warner 1998; Singh et al. 2018). Green synthesis of NPs involves incorporating renewable materials in the composite mix which requires low energy for their extraction and production (Koronis and Silva 2019). Nanocomposites that are manufactured through green synthesis processes are called "Green Nanocomposites". Some of the main advantages of green synthesis are (a) eliminating the need for toxic chemicals and nonrenewable resources (b) ability to be used at large scale production (c) high energy and high-pressure experimental conditions are not required leading to energy savings and (d) overall cost of synthesis process is reduced since the biological components themselves act as reducing and capping agents (Singh et al. 2018; Abdelbary and Abdelfattah 2020). Biosynthesis of NPs uses a bottom-up approach characterized by the choice of solvent medium, choice of an eco-friendly reducing agent and finally, the choice of a nontoxic material as a capping agent to stabilize the synthesized NPs (Mohanpuria et al. 2008; Singh et al. 2011; Patra and Baek 2014).

Figure 11.4 outlines some of the major advantages of green synthesis and common nanofillers used in bio nanocomposites. Among the numerous natural chemical substances provided by nature, plant extracts are the most studied category (Lu and Ozcan 2015). Synthesis of metal NPs, bioceramics used in electronics and medical applications respectively, using plant extracts and natural biopolymers as reducing agents has been particularly successful (Shamim and Sharma 2013; Omanovic-Miklicanin et al. 2020). Biogenic, sustainable and inexpensive nanomaterials which can be produced in an energy-efficient manner are playing significant roles in decontamination protocols for drinking and industrial wastewaters (Gautam et al. 2019). Given the abundance of carbon-rich natural sources such as vegetable oil, sugars and biopolymers, their use as precursors for carbon nanomaterials has been rapidly increasing in the last few years (Titirici et al. 2015).

The development of green chemistry to synthesize nanomaterials has become a key component of the nanotechnology future (Lu and Ozcan, 2015).

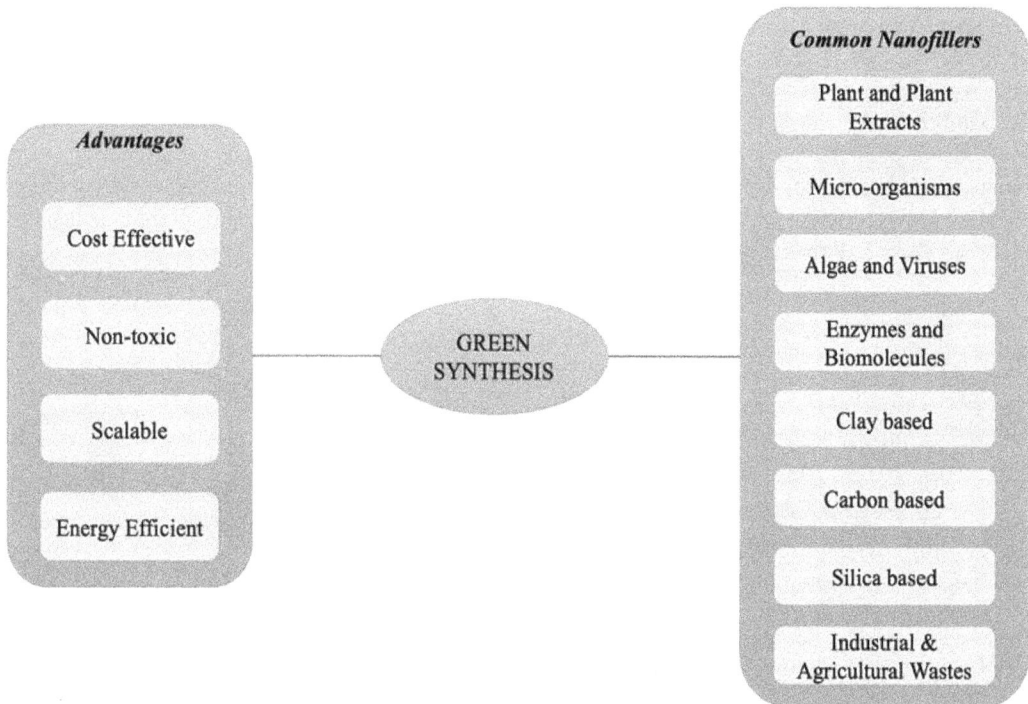

Figure 11.4   Advantages of Green Synthesis and Common Green Nanofillers. (Adapted from Parveen et al. 2016; Singh et al. 2018; Kausar, 2020).

In addition to using natural starting and processing materials for NPs, research is also focused on green processing by developing environmentally friendly and sustainable processes (Lu and Ozcan 2015). In a quest to find alternatives to organic solvents, water and supercritical carbon dioxide are investigated (Sui et al. 2012), and hydrothermal approaches using water as the reaction medium is gaining popularity (Lu and Ozcan 2015). Flow chemical processes have been developed for the improvement and cost-effectiveness of green chemistry (Deadman et al. 2013). Alternative heat sources for the synthesis methods such as Microwave (MW)-Assisted Synthesis (Nadagouda et al. 2011), Focused Sunlight (Vinayan et al. 2013), Ultrasonic-Assisted Synthesis, Laser Ablation Synthesis, etc. (Nasrollahzadeh et al. 2020b) are being investigated, but their cost-effectiveness when compared to conventional heat sources and any potential challenges when scaling to commercial production should be thoroughly evaluated (Lu and Ozcan 2015; Nasrollahzadeh et al. 2020b).

3.2 Biocomposites

Growing ecological and environmental concerns has led to active research for the development of sustainable, bio-based, degradable and renewable reinforcing materials from natural sources for applications in various fields such as energy, electronics, cosmetics, food packaging, environmental remediation, genetic engineering, construction, automotive and biomedical sciences (Carus et al. 2015; Saba et al. 2019). The research in the synthesis of wood-plastic composites and natural fiber composites constitutes about 10–15% of the total European composite market (Loureiro and Esteves 2019). When compared to glass fibers, natural fibers can have increased tensile strength with low filler content due to their low density and high fiber diameter. This contributes to low consumption of fuel and coupled with their natural advantages of biodegradability, renewability, nontoxicity and abundancy, natural fibers are proving to be ideal for the composites automotive industry (Nagalakshmaiah et al. 2019). The degradability of biodegradable nanomaterials under biological conditions in the body makes them promising candidates in sophisticated medical applications such as in controlled drug delivery systems, as agents in magnetic resonance imaging and for magnetic-induced tumor treatment via hyperthermia (Kafrouni and Savadogo 2016; Wiwanitkit 2019; Liu et al. 2020). Bio nanocomposites are generally produced using common methods through which some composites are produced as listed in Fig. 11.5. Recently, new methods such as electrospinning and nanofiber direct dispersion techniques are also gaining traction. However, some methods such as solution mixing and sol-gel process remain at academic study levels only (Ates et al. 2020).

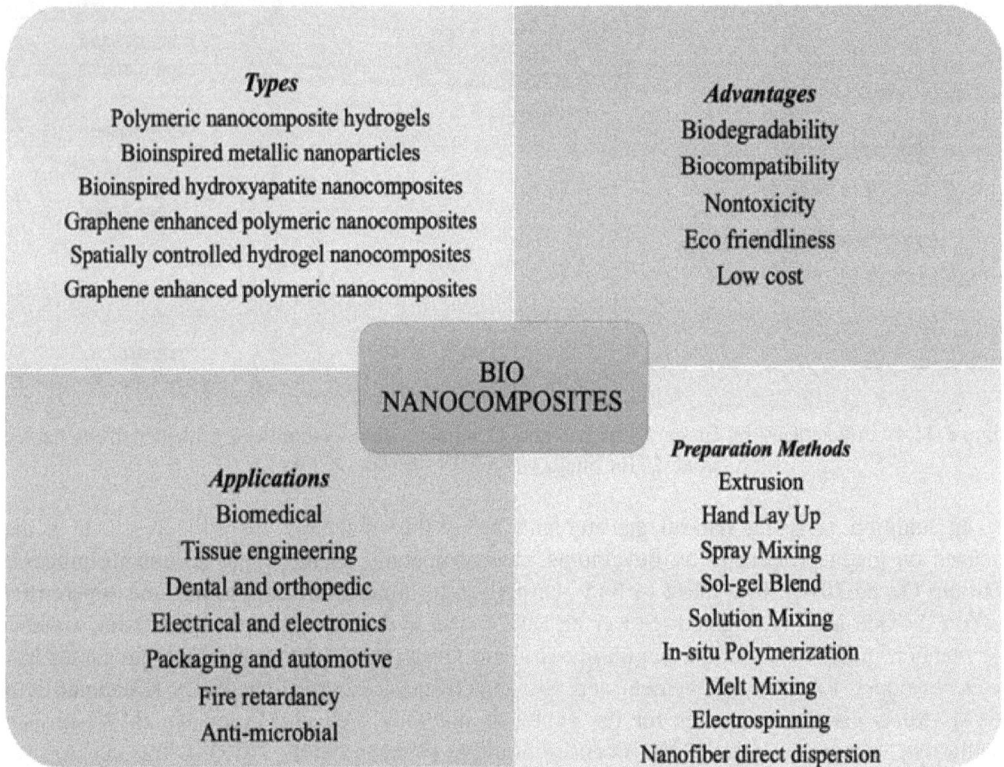

Types
Polymeric nanocomposite hydrogels
Bioinspired metallic nanoparticles
Bioinspired hydroxyapatite nanocomposites
Graphene enhanced polymeric nanocomposites
Spatially controlled hydrogel nanocomposites
Graphene enhanced polymeric nanocomposites

Advantages
Biodegradability
Biocompatibility
Nontoxicity
Eco friendliness
Low cost

BIO NANOCOMPOSITES

Applications
Biomedical
Tissue engineering
Dental and orthopedic
Electrical and electronics
Packaging and automotive
Fire retardancy
Anti-microbial

Preparation Methods
Extrusion
Hand Lay Up
Spray Mixing
Sol-gel Blend
Solution Mixing
In-situ Polymerization
Melt Mixing
Electrospinning
Nanofiber direct dispersion

Figure 11.5 Bio Nanocomposites—Types, Applications, Advantages, and Preparation Methods. (Adapted from Mishra et al. 2019; Ates et al. 2020).

Natural vegetable fibers, hollow cellulose fibrils held together by lignin and hemicelluloses matrix, are perfect examples of composites designed by nature (Loureiro and Esteves 2019). Cellulose is the most abundant natural polymer and has been leading innovation and development of biodegradable composites containing nanosized cellulose fibers or crystals produced by bacteria or derived from plants (Lu and Ozcan 2015). There is a renewed interest in natural fibers in automotive industries due to their excellent acoustic properties, thermal insulation, low cost and growing requirements to pursue environmentally friendly practices by using materials from renewable sources (Loureiro and Esteves 2019). Cellulose nanofibers with excellent features such as low-density, renewability, biodegradability, nontoxicity, low-cost and robust mechanical properties are increasingly used as organic alternatives to clay-based particulates (Nanda et al. 2002; Orts et al. 2005; Wilson et al. 2017; Saba et al. 2019). Research has shown that they can compete with components made of conventional materials (Singh et al. 2017). Cellulose

Nanocrystals (CNCs) and Cellulose Nanofibrils (CNFs) are the two promising classifications of nanocellulose with the potential to serve as a platform for the next generation of green nanomaterials (Eichhorn et al. 2010; Saba et al. 2019; Nagalakshmaiah et al. 2019). Chitin and chitosan are other important biocompatible nanofillers that possess antimicrobial properties as well as the ability to absorb heavy metal ions and can be employed for synthesizing metallic NPs (Nasrollahzadeh et al. 2020b). Chitin and bamboo nanofibers are blended without the use of a solvent to make biodegradable green nanocomposites which are beneficial for food packaging (Hai et al. 2019). Organoclays are one of the attractive and promising hybrid organic/inorganic nanomaterials using clay, a naturally occurring mineral composed primarily of fine-grained minerals. These are generally used for polymers and polymer-based composite modification (Saba et al. 2019). Exhibiting improved balance of stiffness and toughness, excellent mechanical and barrier properties, the inexpensive reinforcement fillers of organic nano clay and nano clay-based polymer coatings are the best candidates for applications in construction, aerospace, automobile, and marine industries (Kim et al. 2015).

3.3 Sustainable Polymer Matrix Nanocomposites (PMNCs)

PMNCs are one of the most widely researched nanocomposites and environmentally friendly polymer nanocomposites that hold promise to be the next generation of sustainable, recyclable, and eco-friendly materials combining performance and environmental compatibility. Polylactide (PLA)-based nanocomposites utilize PLA, one of the most popular (bio)degradable polymers, as matrices in the composites (Musiol et al. 2019). Due to their desirable property of biodegradation, natural nanofillers such as nano clays, nanocellulose, jute, oil palm and coir nanofiller are being increasingly incorporated as reinforcing agents in polymer composites to form the green nanocomposites (Musiol et al. 2019; Saba et al. 2019). Researchers have shown that conductive composites of reduced graphene oxide-silver particles with a (bio)degradable matrix, such as PCL, can be used in a wide range of bioelectronic applications (Kumar et al. 2016) while 3D printing technology can benefit from biodegradable thermoplastic polymers (Wlodarczyk et al. 2018). Biodegradable PMNCs are also finding extensive adaptation in biomedical applications such as implantable electronics (Liu et al. 2019) while bio-resins made from vegetable oils could offer a sustainable alternative to petroleum-based thermoset resins (Rwahwire et al. 2019). Natural biopolymer-based nanocomposite packaging materials have great potential for active food packaging applications owing to their bio-functional properties (Shankar et al. 2018). The susceptibility of biodegradable polymers to organic recycling at the end of their life allows rational utilization of these materials in accordance with new trends in waste management reducing the environmental impact of polymeric waste (Musiol et al. 2019). Until recently, very little research work was focused on the recycling of nanofillers when compared to the recycling of polymer matrices since the latter is cheaper and less energy-consuming. Nevertheless, since bio nanofillers are abundant in nature and are renewable, they present a cost-effective and eco-friendly alternative to traditional filler material (Majka et al. 2016; Cestari et al. 2019).

3.4 Limitations of Bio-based Nanocomposites

Despite possessing many favorable characteristics such as biodegradability, renewability, low-toxicity, high modulus, mechanical strength, lightweight and low density (Saba et al. 2019), bio-based nanomaterials syntheses suffer from high costs and complex synthesis processes among others which significantly limit their industrial uptake (Hernandez 2015). Translating different novel biodegradable NP designs into a clinical setting poses a huge challenge along with other limitations such as lack of knowledge about their clinical safety, the requirement of composition purity and long-term stability of payload as well as challenges with low drug encapsulation efficiency and exorbitant costs in scaled-up production (Lee et al. 2016; Flores et al. 2019; Su and Kang 2020). Such limitations have encouraged researchers to understand the molecular behavior of biomolecules and a golden mean between synthetic and bio-based methods called "Bioinspiration" has been identified. This strategy involves identifying the chemical principles underlying the biological routes so that synthetic molecules can be designed with the desired motifs (Patwardhan et al. 2018). The bioinspiration approach presents a win-win situation because the synthetic molecules retain the green nature of the biological synthesis while still allowing control of key nanomaterial properties such as particle size, crystallinity and porosity at the laboratory scale. Bioinspiration synthesis helps to substantially reduce the time and energy usage while enabling the synthesis of high-value non-hazardous materials for desired applications (Walsh and Knecht, 2017; Patwardhan et al. 2018).

3.5   Recyclable Magnetic Nanocomposites for Catalysis

Nanocomposite materials have gained considerable attention due to their excellent catalytic ability under mild conditions and are fast emerging to be sustainable alternatives to hazardous aromatic nitro compounds (Cyganowski 2021). Though nanoscale metal particles demonstrate remarkable catalytic activity (Lu et al. 2005), their practical application to environmentally friendly processes are limited by the inability of metal nanocatalysts to be separated from the reaction solution for recycling (Zhang et al. 2011). Catalytic recycling and reusability, especially of nanocatalysts, are important factors that affect the commercialization success of catalysts (Veisi et al. 2019; Li et al. 2020). To achieve this, researchers have studied immobilizing metal NPs on a suitable magnetic support matrix, which can be reversibly recovered and redispersed by applying an external magnetic field (Stevens et al. 2005; Jiang et al. 2008; Tsang et al. 2004; Kurtan et al. 2016; Wang et al. 2020). At the end of the catalytic reaction, these magnetic nanocatalysts can be efficiently removed from the reaction mixtures with an external magnet and can be reused for multiple reactions until the catalyst is deactivated (Shang et al. 2016; Maham et al. 2017). Some of the major applications of MNPs are in biomedical, environmental and chemical fields (Hajizadeh and Maleki 2018; Foroutan et al. 2020).

   Due to strong magnetic properties, superparamagnetic behavior, low toxicity, low cost and ease of preparation, magnetite (Fe_3O_4) and ferrite nanoparticles are widely used as these can be quickly and easily separated from the reaction medium using a magnet (Xiong et al. 2013; Wang and Astruc, 2014; Mahdavinasab et al. 2019; Elkodous et al. 2019; Islam et al. 2020; Sharma et al. 2020). Fe_3O_4 based nanocomposites have proven to be a highly efficient, renewable, and eco-friendly heterogeneous class of catalysts suitable for industrial applications (Tang et al. 2014). In recent years, the incorporation of magnetic components into TiO_2 nanoparticle-based catalysts (Pang et al. 2011; Wang et al. 2012; Chen et al. 2017) for environmental remediation (Wilson et al. 2017; Ali et al. 2019; Popli, 2020) has also been extensively studied. Analyzing the extent to which the magnetic nanocatalysts can be recycled is an important parameter for measuring the catalytic activity, stability and efficiency of magnetic nanocomposites used for catalytic operations (Nasseh et al. 2019). Studies have demonstrated that they can be recycled and reused for photocatalysis multiple times without substantial degradation in catalytic activity (Tseng et al. 2018). At the end of the catalytic reaction, the magnetic nanocomposites are separated from the reaction medium with the help of an external magnetic field and can sometimes be purified with water and ethanol solution before reuse. Reusability and recycling efficiencies of some of the magnetic nano photocatalysts found in the literature are shown in Table 11.2. The 'Recycling' column represents the tools or procedures with which the composites were retrieved and purified (where applicable) to be reused again. 'Cycles' represents the total number of catalytic runs or cycles up to which the nanocomposites were separated from the media and successfully used in the subsequent reaction without any degradation in performance.

Table 11.3 Recycling efficiencies of magnetic nanocomposites used in catalysis

Sl. No	Magnetic Nanocomposite	Recycling	Cycles	References
1.	Fe_3O_4-CdTe	external magnet	10	Guo et al. 2020
2.	Fe_3O_4@APTES@ PAMAM-Ag	external magnet	5	Kurtan and Baykal, 2014
3.	Pd/Fe_3O_4	external magnet; washed with warmish water and ethanol	7	Veisi et al. 2019
4.	$MnFe_2O_4$@GO@ Chitosan/Cu	external magnet; washed with EtOH (5 mL) several times, dried under vacuum	9	Mahdavinasab et al. 2019
5.	rMGO-Au	external magnet	6	Fathalipour et al. 2019
6.	Ag/SiO_2–$CoFe_2O_4$	external magnet; washed with 10 mL deionized water	7	Chen et al. 2015
7.	$CNTs/Fe_3O_4$	external magnetic field; methanol-wash	5	Li et al. 2015
8.	TiO_2/graphene oxide	external magnet	10	Zhang et al. 2018
9.	$Bi/Bi_{25}FeO_{40}$-C	external magnet; rinsed	5	Li et al. 2020
10.	Fe_3O_4@C nanocomposite	external magnet	4	Kurtan et al. 2016
11.	$Fe_3O_4/La(OH)_3$	external magnet	10	Ahmed and Lo, 2020
12.	glutathione@magnetite	external magnet	6	Sharma et al. 2020
13.	Fe_3O_4@TAS	external magnet; washed with 0.01 M HCl solution	5	Alqadami et al. 2017
14.	$Bi_4O_5Br_2/SrFe_{12}O_{19}$	external magnetic field	10	Wang et al. 2020
15.	Ag/ZrO_2	external magnet	5	Maham et al. 2020
16.	GO-Fe_3O_4@NPVP-Ag	external magnetic field	5	Li et al. 2018
17.	Graphene oxide/$CuFe_2O_4$	external magnet	10	Rouhani et al. 2018
18.	Fe_3O_4@SiO_2@TiO_2	external magnet	6	Wang et al. 2012
19.	rGO-PD-MCNT	external magnetic field; washed with ethanol and water for MB and 4-NP	15	Islam et al. 2020

The use of silver, TiO_2 and ZnO and other NPs in wastewater treatment has skyrocketed in the last few decades owing to their excellent photocatalytic abilities (Singh et al. 2019; Li et al. 2020). This increased adaptation has highlighted the potential toxicity and biosafety issues with NPs (Nasrollahzadeh et al. 2020a). NPs used as purification agents should be reclaimed from the water after treatment so that additional contaminants are not introduced into the water systems. The reclaimed NPs can then be reused multiple times before the end of their life. Hence recyclability, reusability and non-toxicity are essential characteristics for NPs so that they remain a cost-effective, viable, reusable and an environmentally friendly solution for WWT (Xiong et al. 2013; Yong et al. 2018; Nasseh et al. 2019; Tran et al. 2020).

4 LIFE CYCLE AND RISK ASSESSMENTS

Several regulatory bodies recommend proper evaluation and monitoring of the generation of hazardous substances due to manufacturing and processing of nanomaterials to manage the risks posed by nano-enabled products on health and the environment (Nanomaterials Market, n.d). Information on potential emissions during the various life cycle stages of ENMs such as manufacturing, integration of ENMs into products, transport, usage and disposal is critical to estimating the environmental concentration and exposure of ENMs (Keller et al. 2013). In recent years there has been growing interest in studying the toxicity of engineered nanoparticles (ENPs) especially using holistic tools such as life-cycle analysis and risk assessment to better evaluate environmental and safety impacts of nano-enabled products (Patwardhan et al. 2018). Figure 11.6 outlines the possible pathways through with ENPs could be released into the environment leading to human and wildlife exposure.

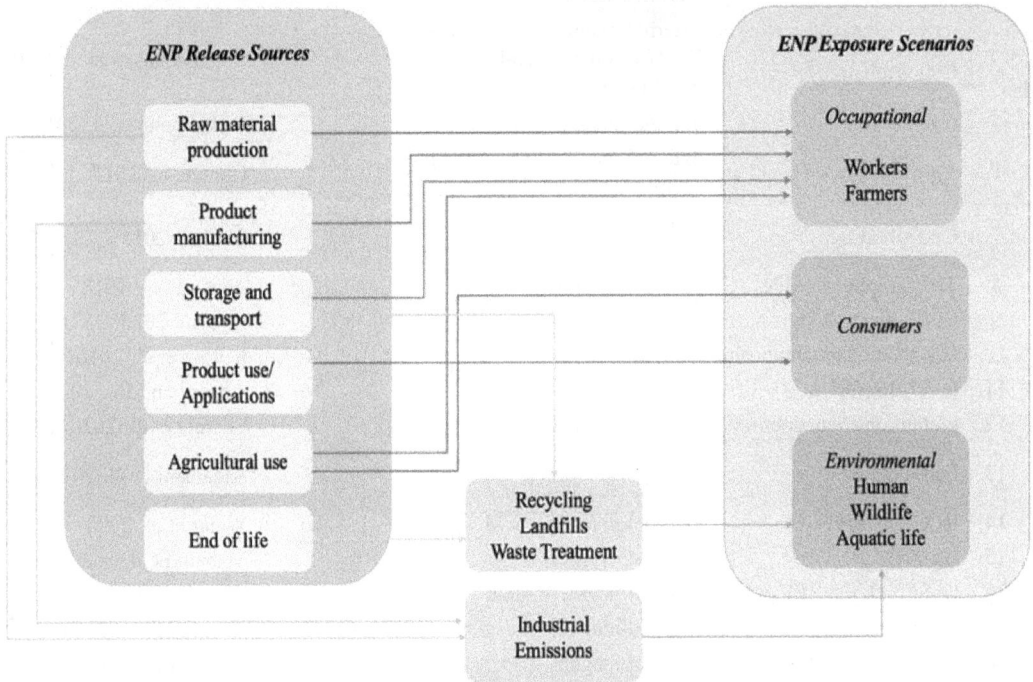

Figure 11.6   ENP Exposure Pathways.
(Adapted from Hristozov and Malsch, 2009; Mirshafiee et al. 2018).

4.1   Life Cycle Assessment (LCA)

LCA is a methodological framework that considers the entire life cycle of a product or system to assess its environmental impacts at every stage of its life starting from raw material extraction to manufacturing, distribution, consumption and EoL disposal (Muralikrishna and Manickam 2017). As such, material production, manufacturing, use and end-of-life are the four main stages of a product's life cycle (Petrakli et al. 2020). Since the ENMs and their engineered nanoparticles (ENPs) have raised concerns about their environmental impacts in the recent past (Miseljic and Olsen 2014), LCA is becoming a key requirement of a product's environmental requirement (Hogan 2015). Performing LCA as the first step of a product development life cycle right after its inception, will give the flexibility of adopting the most sustainable, cost-effective and environmentally friendly route for the development of nanocomposites. The results of a comprehensive LCA conducted by Petrakli et al. (2020) to assess and evaluate the environmental impacts of (nano)enhanced Carbon Fiber-Reinforced Polymer (CFRP) prototypes showed that extraction and production of materials contribute to more than 65 and 70% to the climate change impact followed by manufacture stage at around 30% (Petrakli et al. 2020). Such analysis would be indispensable for researchers and engineers to identify the potential hot spots to pay more attention to find more green and sustainable solutions to manufacturing nanocomposites.

4.2 Health, Safety and Environment (HSE) Risk Assessment

The European Commission's regulatory board on NPs concluded that certain nanomaterials used for specific applications are toxic and emphasizes on conducting a risk evaluation on a case-by-case basis (MAR 2020). Other regulatory bodies also recommend to thoroughly evaluate and address the health and environmental risks posed by any potential hazardous substances that are generated during the manufacturing, processing, consumption and disposal of nanomaterials (Nanomaterials Market, n.d; Faunce and Kolodziejczyk 2017). Keller et al. (2013) recommended estimating the likelihood of ENM exposure to humans and ecosystem receptors by considering potential emissions during manufacturing, incorporation of ENMs into intermediate and final products, use of a product containing ENMs, disposal or recycling. Though efforts are being made to estimate the quantity of nanomaterials released into the environment based on the usage of products that contain nanomaterials, nanomaterial concentration in these products, etc., there is a lack of information on the content of nanomaterials in wastes and the emissions of nanomaterials from disposal processes (Mirshafiee et al. 2018). Nevertheless, the amount of toxicity data on the nanomaterials that are proven to be hazardous should aid in carrying out exploratory research focusing on risk evaluations and building iterative models as and when more data becomes available (Nanotechnologies and Waste 2015).

Due to the shortage of reliable toxicity and ecotoxicity data of nanomaterials and nanomaterial composites, a complete exposure/release assessment might be a challenging task (Silva et al. 2019). But efforts can be made based on the five-step risk assessment approach provided by Health Safety Executive from the United Kingdom (HSE-UK) (Risk assessment, n.d).

The main steps in this approach are (a) identify the hazard (b) decide who might be harmed and how (c) evaluate the risks and decide on precautions (d) record your findings and implement them (e) review your assessment and update if necessary. Though this approach might be useful to get a baseline idea, it should be followed with a precautionary view since some adjustments to the analysis due to uncertainties and lack of information might be warranted (Silva et al. 2019). The amount of daily human exposure to nanomaterials may be negligible but their prolonged accumulation into the human body and in wildlife organisms could be a cause for concern (Sun et al. 2016; Mirshafiee et al. 2017). Hence it is essential to explore safer design approaches that can reduce or mitigate any adverse impacts the inadvertent nanomaterial exposures can have on human health and the environment (Mirshafiee et al. 2018; Silva et al. 2019). The integration of life cycle analysis and risk assessment tools can help in better evaluation of the environmental and safety impacts of nanomaterials to find ways to minimize waste and toxicity of nanocomposites (Patwardhan et al. 2018) since toxicity can be a barrier to the commercialization of nanotechnology (MAR, 2020).

5 CHALLENGES

There is a lot of uncertainty around the fate of ENMs in different waste treatment plants and this requires further investigation to quantify and to develop a reliable database that can be used for future research (OECD 2016). It is important to understand the properties of specific nano wastes before developing effective disposal practices since a single procedure for disposal may not suffice for the broad range of existing nanomaterials (Kolodziejczyk 2016). Biocomposites where both phases (reinforced polymer and matrix) are derived from renewable and completely biodegradable natural sources will improve manufacturing speed and recycling with enhanced environmental compatibility, but their production remains a challenge (Nagalakshmaiah et al. 2019). The application of LCA of ENMs is not well developed due to limited data on the potential release of ENP during the life cycle and the fate and effect of released ENPs (Miseljic and Olsen 2014). The adaptation of green nanotechnology often requires entirely new processes, so a cost-benefit analysis considering the engineering and scale-up challenges is required before moving it into production (Lu and Ozcan 2015). There is a need for complementary studies of polymer-based nanocomposites with emphasis on both the material properties and biodegradability (Musiol et al. 2019). A major challenge plaguing the fabrication of biocomposites is the difficulty in procuring biopolymers from natural resources as matrices (Nagalakshmaiah et al. 2019). Though cellulose, gelatin, chitosan and plant-based oils are effective biopolymers, they are scarcer and entail a tedious and costly production process (Nagalakshmaiah et al. 2019). Cost is definitely the biggest deciding factor for companies to adopt a new environmentally friendly technology. So, it is not enough to just devise a green product or process, but it is equally crucial to make it cost-effective and/or result in cost savings for the company and easy to adapt.

5.1   Future Directions

A promising development is that international bodies such as ISO are working to develop standards related to safe disposal and/or recycling of nano waste. Experts are proposing that at least 10% of the nanotechnology budget should be spent on making the field more sustainable and safer (Faunce and Kolodziejczyk 2017). Only a few studies have investigated the presence of NMs' emissions in Municipal Solid Waste (MSW) incineration and there is an urgent need to investigate the fate of different nanomaterials in waste treatment routes (Petrakli et al. 2020). Though green synthesis techniques have attracted research focus lately, the challenges in scaling them up to real-time production still remain. Hence future research could be directed toward translating laboratory-based work to industrial production scale by conducting informative LCA and HSE assessments (Patwardhan et al. 2018; Singh et al. 2018; Silva et al. 2019). Active collaboration between academic researchers and the private sector will continue to be a key to successful commercialization (Lu and Ozcan 2015). Another promising research area is to determine how to separate and recycle the small CNTs in a water purification process so that they donot carry contaminants (Li et al. 2015). Nanomaterials have significantly enhanced the overall performances of biodegradable polymers, but the properties of electric conductivity, flexibility and biocompatibility still need to be improved (Liu et al. 2019). The applications of green synthesized nanocatalysts in environmental remediation applications are encouraging but issues pertaining to toxicity, biosafety should be systematically evaluated to develop low-cost green nanocatalysts (Nasrollahzadeh et al. 2020a).

6   CONCLUSION

The extent to which nanomaterials have encompassed all industries and our lives through their inclusion in industrial and everyday products alike will inevitably lead to more waste generation. But the important discussion on managing the waste containing NPs either through source reduction or EoL recycling is still very scarce. There is an urgent need to conduct rigorous evaluations of potential health, environmental and societal risks of nanocomposites, establish better recycling and waste management practices along with progressing the development of sustainable synthetic approaches for nanocomposites. Such endeavors are essential so that the growth of the nanocomposites industry is not hampered by the negative effects and it grows into a viable, environmentally friendly and cost-effective technology.

■ References

Abdelbary, S. and H. Abdelfattah. 2020. Modern Trends in Uses of Different Wastes to Produce Nanoparticles and Their Environmental Applications, Nanotechnology, and the Environment, Mousumi Sen, IntechOpen. DOI: 10.5772/intechopen.93315.

Adam, V. and B. Nowack. 2017. European country-specific probabilistic assessment of nanomaterial flows towards landfilling, incineration, and recycling. Environmental Science: Nano 4: 1961–1973. https://doi.org/10.1039/C7EN00487G

Adam, V., A. Caballero-Guzman and B. Nowack. 2018. Considering the forms of released engineered nanomaterials in probabilistic material flow analysis. Environmental Pollution 243 Part A: 17–27. https://doi.org/10.1016/j.envpol.2018.07.108.

Ahmed, S. and I.M.C. Lo. 2020. Phosphate removal from river water using a highly efficient magnetically recyclable Fe_3O_4/La $(OH)_3$ nanocomposite. Chemosphere 261: 128118. doi:10.1016/j.chemosphere.2020.128118.

Ali, S., Z. Li, S. Chen, A. Zada, I. Khan, I. Khan, et al. 2019. Synthesis of activated carbon-supported TiO_2-based nano-photocatalysts with well recycling for efficiently degrading high-concentration pollutants. Catalysis Today 335: 557–564. doi:10.1016/j.cattod.2019.03.044.

Alqadami, A.A., M. Naushad, M.A. Abdalla, T. Ahamad, Z.A. ALOthman, S.M. Alshehri, et al. 2017. Efficient removal of toxic metal ions from wastewater using a recyclable nanocomposite: A study of adsorption parameters and interaction mechanism. Journal of Cleaner Production 156: 426–436. doi:10.1016/j.jclepro.2017.04.085.

Anastas, P.T. and J.C. Warner. 1998. Green Chemistry: Theory & Practice. Oxford University Press. Oxford.

Ates, B., S. Koytepe, A. Ulu, C. Gurses and V.K. Thakur. 2020. Chemistry, structures, and advanced applications of nanocomposites from biorenewable resources. Chemical Reviews 120(17): 9304–9362. doi:10.1021/acs.chemrev.9b00553.

AZONANO. 2007. Nanocomposites – An Overview of Properties, Applications and Definition. Available: https://www.azonano.com/article.aspx?ArticleID=1832#:~:text=Nanocomposites%20are%20materials%20that%20incorporate,and%20electrical%20or%20thermal%20conductivity

Boldrin, A., S.F. Hansen, A. Baun, N.I. Hartmann and T.F. Astrup. 2014. Environmental exposure assessment framework for nanoparticles in solid waste. Journal of Nanoparticle Research 16(6): 2394. https://doi.org/10.1007/s11051-014-2394-2.

Boldrin, A., L. Heggelund and S.F. Hansen. 2015. Report on methods for characterizing of the composition and physical properties of NOAA-containing waste. Deliverable report 3.5 for the EU-FP7 project "SUN – Sustainable Nanotechnologies", Grant Agreement Number 604305.

Bundschuh, M., J. Filser, S. Lüderwald, M.S. McKee, G. Metreveli, G.E. Schaumann, et al. 2018. Nanoparticles in the environment: Where do we come from, where do we go to? Environmental sciences Europe 30(1): 6. https://doi.org/10.1186/s12302-018-0132-6.

Caballero-Guzman, A., T. Sun and B. Nowack. 2015. Flows of engineered nanomaterials through the recycling process in Switzerland. Waste Management 36: 33–43. https://doi.org/10.1016/j.wasman.2014.11.006.

Carus, M., A. Eder, L. Dammer, H. Korte, L. Scholz and R. Essel. 2015. NCF market study: European and global markets 2012 and future trends in automotive and Construction. Nova Institute for Ecology and Innovation. http://bio-based.eu/downloads/wood-plastic-composites-wpc-and-natural-fibre-composites-nfc-european-and-global-markets-2012-and-future-trends-in-automotive-and-construction-3/

Cestari, S., F. Daniela, D. Rodrigues, L. Mendes. 2019. Recycling processes and issues in natural fiber-reinforced polymer composites. pp. 285–299. In: G. Koronis and A. Silva (eds). Green Composites for Automotive Applications, A volume in Woodhead Publishing Series in Composites Science and Engineering. Woodhead Publishing, UK. doi:10.1016/b978-0-08-102177-4.00012-4.

Chen, J., Z.-H. Lu, Y. Wang, X. Chen and L. Zhang. 2015. Magnetically recyclable Ag/SiO_2–$CoFe_2O_4$ nanocomposite as a highly active and reusable catalyst for H_2 production. International Journal of Hydrogen Energy 40(14): 4777–4785. doi:10.1016/j.ijhydene.2015.02.054.

Chen, C.C., D. Jaihindh, S.H. Hu, and Y.P. Fu. 2017. Magnetic recyclable photocatalysts of Ni-Cu-Zn ferrite@SiO_2@TiO_2@Ag and their photocatalytic activities. Journal of Photochemistry and Photobiology, A 334: 74–85. doi:10.1016/j.jphotochem.2016.11.005.

Cyganowski, P. 2021. Fully recyclable gold-based nanocomposite catalysts with enhanced reusability for catalytic hydrogenation of p-nitrophenol. Colloids and Surfaces A: Physicochemical and Engineering Aspects 612: 125995. https://doi.org/10.1016/j.colsurfa.2020.125995.

Deadman, B., C. Battilocchio, E. Sliwinski, L. Steven. 2013. A prototype device for evaporation in batch and flow chemical processes. Green Chemistry 15(8): 2050. doi:10.1039/c3gc40967h

DEPA [The Danish Environmental Protection Agency]. 2014. Nanomaterials in waste. Danish Ministry of the Environment. Environmental Project No. 1608. ISBN: 978-87-93283-10-7.

Eichhorn, S.J., A. Dufresne, M.I. Aranguren, N.E. Marcovich, J.B. Capadona, S.J. Rowan. 2010. Review: Current international research into cellulose nanofibres and nanocomposites. Journal of Materials Science 45. doi: 10.1007/s10853-009-3874-0.

Elkodous, M.A., G.S. El-Sayyad, A.E. Mohamed, K. Pal, N. Asthana, F.G. de Souza Junior, et al. 2019. Layer-by-layer preparation and characterization of recyclable nanocomposite ($Co_xNi_{1-x}Fe_2O_4$; X=0.9/ SiO_2/TiO_2). Journal of Materials Science: Materials in Electronics 30: 8312–8328. doi:10.1007/s10854-019-01149-8.

Fathalipour, S., B. Ataei and F. Janati. 2019. Aqueous suspension of biocompatible reduced graphene oxide- Au NPs composite as an effective recyclable catalyst in a Betti reaction. Materials Science and Engineering: C 97: 356–366. https://doi.org/10.1016/j.msec.2018.12.048

Faunce, T. and B. Kolodziejczyk. 2017. Nanowaste: need for disposal and recycling standards. G20 Insights. https://www.g20-insights.org/policy_briefs/nanowaste-need-disposal-recycling-standards/.

Flores, A.M., J. Ye, K.U. Jarr, N. Hosseini-Nassab, B.R. Smith and N.J. Leeper. 2019. Nanoparticle therapy for vascular diseases. Arteriosclerosis, Thrombosis, and Vascular Biology 39(4): 635–646. https://doi.org/10.1161/ATVBAHA.118.311569.

Foroutan, R., R. Mohammadi, F. MousaKhanloo, S. Sahebi, B. Ramavandi, P.S. Kumar, et al. 2020. Performance of montmorillonite/graphene oxide/$CoFe_2O_4$ as a magnetic and recyclable nanocomposite for cleaning methyl violet dye-laden wastewater. Advanced Powder Technology 31(9): 3993–4004 doi:10.1016/j.apt.2020.08.001.

Ganesh, R., J. Smeraldi, T. Hosseini, L. Khatib, B.H. Olson and D. Rosso. 2010. Evaluation of nanocopper removal and toxicity in municipal wastewaters. Environmental Science & Technology 44(20): 7808–7813. https://doi.org/10.1021/es101355k

Gardiner, G. 2014. Recycled carbon fiber update: Closing the CFRP lifecycle loop. Composites World 20: 28–33.

Gautam, P.K., A. Singh, K. Misra, A.K. Sahoo and S.K. Samanta. 2019. Synthesis and applications of biogenic nanomaterials in drinking and wastewater treatment. Journal of Environmental Management 231: 734–748. doi: 10.1016/j.jenvman.2018.10.104.

Gomez-Rivera, F., J.A. Field, D. Brown and R. Sierra-Alvarez. 2012. Fate of cerium dioxide (CeO_2) nanoparticles in municipal wastewater during activated sludge treatment. Bioresource Technology 108: 300–304. https://doi.org/10.1016/j.biortech.2011.12.113.

Gottschalk, F. and B. Nowack. 2011. The release of engineered nanomaterials to the environment. Journal of Environmental Monitoring 13(5): 1145–1155. https://doi.org/10.1039/c0em00547a.

Guo, Y., H. Liu, D. Chen, J. Qu, and J. Yang. 2020. High recycling Fe_3O_4-CdTe nanocomposites for the detection of organophosphorothioate pesticide chlorpyrifos. Green Energy & Environment 7(2): 229–235. https://doi.org/10.1016/j.gee.2020.09.001.

Hai, L., E.S. Choi, L. Zhai, P.S. Panicker and J. Kim. 2019. Green nanocomposite made with chitin and bamboo nanofibers and its mechanical, thermal, and biodegradable properties for food packaging. International Journal of Biological Macromolecules 144: 491–499.

Hajian, A., Q. Fu and L. Berglund. 2018. Recyclable and superelastic aerogels based on carbon nanotubes and carboxymethyl cellulose. Composites Science and Technology 159: 1–10. 10.1016/j.compscitech. 2018. 01.002.

Hajizadeh, Z. and A. Maleki. 2018. Poly(ethylene imine)-modified magnetic halloysite nanotubes: A novel, efficient and recyclable catalyst for the synthesis of dihydropyrano[2,3-c]pyrazole derivatives. Molecular Catalysis 460: 87–93. doi.org/10.1016/j.mcat.2018.09.018.

Health Council of Netherlands [HCN]. 2011. Nanomaterials in waste. Publication No. 2011/14E.

Heggelund, L.R. 2017. Characterization of waste from nanoenabled products: Occurrence, distribution, fate and nanoparticle release. PhD Thesis. Department of Environmental Engineering. Technical University of Denmark (DTU).

Hernandez, R. 2015. Continuous manufacturing: a changing processing paradigm. BioPharm International 28:(4) 20–27.

Hogan, H. 2015. Everything Old is New Again. Composites Manufacturing Magazine. http://compositesmanufacturingmagazine.com/2015/02/composites-recycling-set-to-expand/

Hristozov, D. and I. Malsch. 2009. Hazards and risks of engineered nanoparticles for the environment and human health. Sustainability 1(4): 1161–1194. https://doi.org/10.3390/su1041161.

Hjelmar, O. and H.A. van der Sloot. 2010. Landfilling: Mineral waste landfills. pp. 755–771 In: Christensen, T.H. (ed.). Solid Waste Tech and Management. John Wiley & Sons, Chichester.

Islam, Md. R., Md. Ferdous, M.I. Sujan, X. Mao, H. Zeng and Md. S. Azam. 2020. Recyclable Ag-decorated highly carbonaceous magnetic nanocomposites for the removal of organic pollutants. Journal of Colloid and Interface Science 562: 52–62. https://doi.org/10.1016/j.jcis.2019.11.119.

Jiang, Y., J. Jiang, Q. Gao, M. Ruan, H. Yu and L. Qi. 2008. A novel nanoscale catalyst system composed of nanosized Pd catalysts immobilized on $Fe_3O_4@SiO_2$–PAMAM. Nanotechnology 19(7): 075714. doi:10.1088/0957-4484/19/7/075714.

Kafrouni, L. and O. Savadogo. 2016. Recent progress on magnetic nanoparticles for magnetic hyperthermia. Progress in Biomaterials 5(3–4): 147–160. https://doi.org/10.1007/s40204-016-0054-6.

Kaegi, R., A. Voegelin, B. Sinnet, S. Zuleeg, H. Hagendorfer, M. Burkhardt, et al. 2011. Behavior of metallic silver nanoparticles in a pilot wastewater treatment plant. Environmental Science & Technology 45(9): 3902–3908. https://doi.org/10.1021/es1041892.

Karuppannan, S.G. and T. Karki. 2020. A review on the recycling of waste carbon fibre/glass fibre-reinforced composites: Fibre recovery, properties, and life-cycle analysis. SN Applied Sciences 2: 1–21.

Kausar, A. 2020. Progress in green nanocomposites for high-performance applications. Materials Research Innovations 25: 53–65. doi:10.1080/14328917.2020.1728489.

Keller, A.A., S. McFerran A. Lazareva and S. Suh. 2013. Global life cycle releases of engineered nanomaterials. Journal of Nanoparticle Research 15: 1692. https://doi.org/10.1007/s11051-013-1692-4.

Kim, B., C.S. Park, M. Murayama and M.F. Hochella. 2010. Discovery and characterization of silver sulfide nanoparticles in final sewage sludge products. Environmental Science & Technology 44(19): 7509–7514. https://doi.org/10.1021/es101565j

Kim, B., G.H. Kim and M.H. Lee. 2015. Application of biodegradable polymer to construction materials. International Journal of Engineering and Innovative Technology 4(9): 1–7.

Kiser, M.A., P. Westerhoff, T. Benn, Y. Wang, J. Pérez-Rivera and K. Hristovski. 2009. Titanium nanomaterial removal and release from wastewater treatment plants. Environmental Science & Technology 43(17): 6757–6763. https://doi.org/10.1021/es901102n.

Kiser, M.A., H. Ryu, H. Jang, K. Hristovski and P. Westerhoff. 2010. Biosorption of nanoparticles to heterotrophic wastewater biomass. Water Research 44(14): 4105–4114. doi:10.1016/j.watres.2010.05.036

Kolodziejczyk, B. 2016. Nanotechnology, nanowaste and their effects on ecosystems: A need for efficient monitoring, disposal and recycling. United Nations Sustainable Development. Available: https://sdgs.un.org/documents/brief-gsdr-nanotechnology-nanowaste-and-21498

Koronis, G. and A. Silva. 2019. Eco-impact assessment of a hood made of a ramie reinforced composite. pp. 99–114. In: G. Koronis and A. Silva (eds). Green Composites for Automotive Applications, A volume in Woodhead Publishing Series in Composites Science and Engineering. Woodhead Publishing, UK. doi:10.1016/b978-0-08-102177-4.00005-7.

Kuenen, J., V. Pomar-Portillo, A. Vilchez, A. Visschedijk, H.D. van der Gon, S. Vasquez-Campos, et al. 2021. Inventory of country-specific emissions of engineered nanomaterials throughout the life cycle. Environmental Science Nano 7: 3824–3839. https://doi.org/10.1039/D0EN00422G.

Kumar, S., S. Raj, S. Jain and K. Chatterje. 2016. Multifunctional biodegradable polymer nanocomposite incorporating graphene-silver hybrid for biomedical applications. Mater Des 108(C): 319–332. https://doi.org/10.1016/j.matdes.2016.06.107.

Kurtan, U. and A. Baykal. 2014. Fabrication and characterization of Fe_3O_4 @APTES@PAMAM-Ag highly active and recyclable magnetic nanocatalyst: Catalytic reduction of 4-nitrophenol. Materials Research Bulletin 60: 79–87. doi:10.1016/j.materresbull.2014.08.016.

Kurtan, U., M. Amir and A. Baykal. 2016. Fe_3O_4@Nico-Ag magnetically recyclable nanocatalyst for azo dyes reduction. Applied Surface Science 363: 66–73. doi:101016/japsusc201511214.

Lee, B.K., Y. Yun and K. Park. 2016. PLA micro- and nanoparticles. Advanced Drug Delivery Reviews. 107: 176–191. https://doi.org/10.1016/j.addr.2016.05.020.

Li, S., Y. Gong, Y. Yang, C. He, L. Hu, L. Zhu, et al. 2015. Recyclable CNTs/Fe_3O_4 magnetic nanocomposites as adsorbents to remove bisphenol A from water and their regeneration. Chemical Engineering Journal 260: 231–239.

Li, Q., C. Yong, W. Cao, X. Wang, L. Wang, J. Zhou, et al. 2018. Fabrication of charge reversible graphene oxide-based nanocomposite with multiple antibacterial modes and magnetic recyclability. Journal of Colloid and Interface Science 511: 285–295. https://doi.org/10.1016/j.jcis.2017.10.002.

Li, Y., Y. Wang, H. Lu and X. Li. 2020. Preparation of $CoFe_2O_4$–P4VP@Ag NPs as effective and recyclable catalysts for the degradation of organic pollutants with $NaBH_4$ in water. International Journal of Hydrogen Energy 45(32): 16080–16093. https://doi.org/10.1016/j.ijhydene.2020.04.002.

Liu, H., R. Jian, H. Chen, X. Tian, C. Sun, J. Zhu, et al. 2019. Application of biodegradable and biocompatible nanocomposites in electronics: Current status and future directions. Nanomaterials (Basel) 9(7): 950. doi:10.3390/nano9070950. PMID: 31261962; PMCID: PMC6669760.

Liu, X., Y. Zhang, Y. Wang, W. Zhu, G. Li, X. Ma, et al. 2020. Comprehensive understanding of magnetic hyperthermia for improving antitumor therapeutic efficacy. Theranostics 10(8): 3793–3815. https://doi.org/10.7150/thno.40805.

Loureiro, N.C. and J.L. Esteves. 2019. Green composites in automotive interior parts. pp. 81–97. In: G. Koronis and A. Silva (eds). Green Composites for Automotive Applications, A volume in Woodhead Publishing Series in Composites Science and Engineering. Woodhead Publishing, UK. doi:10.1016/b978-0-08-102177-4.00004-5.

Lu, Q., B. Yang, L. Zhuang and J. Lu. 2005. Anodic activation of PtRu/C catalysts for methanol oxidation. The Journal of Physical Chemistry, B 109(5): 1715–1722. https://doi.org/10.1021/jp0461652.

Lu, Y. and S. Ozcan. 2015. Green nanomaterials: On track for a sustainable future. Nano Today 10(4): 417–420. doi:10.1016/j.nantod.2015.04.010.

Maham, M., M. Nasrollahzadeh, S.M. Sajadi and M. Nekoei. 2017. Biosynthesis of Ag/reduced graphene oxide/Fe_3O_4 using *Lotus garcinii* leaf extract and its application as a recyclable nanocatalyst for the reduction of 4-nitrophenol and organic dyes. Journal of Colloid and Interface Science 497: 33–42. doi:10.1016/j.jcis. 2017.02.064.

Maham, M., M. Nasrollahzadeh and S. Mohammad Sajadi. 2020. Facile synthesis of Ag/ZrO_2 nanocomposite as a recyclable catalyst for the treatment of environmental pollutants. Composites Part B. 107783. doi:10.1016/j.compositesb.2020.107783.

Mahdavinasab, M., M. Hamzehloueian and Y. Sarrafi. 2019. Preparation and application of magnetic chitosan/graphene oxide composite supported copper as a recyclable heterogeneous nanocatalyst in the synthesis of triazoles. International Journal of Biological Macromolecules 138: 764–772. https://doi.org/10.1016/j.ijbiomac.2019.07.013.

Majka, T.M., O. Bartyzel, K.N. Raftopoulos, J. Pagacz, A. Leszczynska and K. Pielichowski. 2016. Recycling of polypropylene/montmorillonite nanocomposites by pyrolysis. Journal of Analytical and Applied Pyrolysis 119: 1–7. doi:10.1016/j.jaap.2016.04.005.

MAR [Market Analysis Report]. 2020. Nanomaterials Market Size, Share & Trends Analysis Report By Product (Carbon Nanotubes, Titanium Dioxide), By Application (Medical, Electronics, Paints & Coatings), By Region, And Segment Forecasts, 2020–2027. Grand View Research. Available: https://www.grandviewresearch.com/industry-analysis/nanotechnology-and-nanomaterials-market

Mirshafiee, V., W. Jiang, B. Sun, X. Wang and T. Xia. 2017. Facilitating translational nanomedicine via predictive safety assessment. Molecular Therapy: The Journal of the American Society of Gene Therapy 25(7): 1522–1530. https://doi.org/10.1016/j.ymthe.2017.03.011.

Mirshafiee, V., O.J. Osborne, B. Sun and T. Xia. 2018. Safety concerns of industrial engineered nanomaterials. pp. 1063–1072. In: C.M. Hussain (ed.). Handbook of Nanomaterials for Industrial Applications—A volume in Micro and Nano Technologies. Elsevier, Cambridge: U.S. doi:10.1016/b978-0-12-813351-4.00062-6.

Miseljic, M. and S.I. Olsen. 2014. Life-cycle assessment of engineered nanomaterials: A literature review of assessment status. Journal of Nanoparticle Research 16: 1–33. https://doi.org/10.1007/s11051-014-2427-x

Mishra, S., S. Sharma, M.N. Javed, F.H. Pottoo, M. Abul Barkat, Harshita et al. 2019. Bioinspired nanocomposites: Applications in disease diagnosis and treatment. Pharmaceutical Nanotechnology 07. doi:10.2174/2211738507666190425121509.

Mohanpuria, P., N.K. Rana and S.K. Yadav. 2008. Biosynthesis of nanoparticles: Technological concepts and future applications. Journal of Nanoparticle Research 10: 507–517. https://doi.org/10.1007/s11051-007-9275-x

Mueller, N.C., J. Buha, J. Wang, A. Ulrich and B. Nowack. 2013. Modeling the flows of engineered nanomaterials during waste handling. Environmental Science: Processes & Impacts 15: 251–259. doi: https://doi.org/10.1039/C2EM30761H

Muralikrishna, I.V. and V. Manickam. 2017. Life cycle assessment. pp. 57–75. *In*: I.V. Muralikrishna and V. Manickam (eds). Environmental Management: Science and Engineering for Industry. Butterworth-Heinemann. https://doi.org/10.1016/B978-0-12-811989-1.00005-1.

Musiol, M., S. Jurczyk, W. Sikorska and J. Rydz. 2019. (Bio)degradable polymer nanocomposites for environmental protection. pp. 1–27. *In*: C. Hussain and S. Thomas (eds). Handbook of Polymer and Ceramic Nanotechnology. Springer, Cham. https://doi.org/10.1007/978-3-030-10614-0_42-1.

Nadagouda, M.N., T.F. Speth and R.S. Varma. 2011. Microwave-assisted green synthesis of silver nanostructures. Accounts of Chemical Research 44(7): 469–478. doi:10.1021/ar1001457.

Nagalakshmaiah, M., S. Afrin, R.P. Malladi, S. Elkoun, M. Robert and M.A. Ansari, et al. 2019. Biocomposites: Present trends and challenges for the future. pp. 197–215. *In*: G. Koronis and A. Silva (eds). Green Composites for Automotive Applications, A volume in Woodhead Publishing Series in Composites Science and Engineering. Woodhead Publishing, UK. doi:10.1016/b978-0-08-102177-4.00009-4.

Nanda, A., G.M. Dutta and A.K. Banthia. 2002. Sulfonated polybutadiene ionomer templates nanonickel composite. Materials Letters 52(3): 203–205. https://doi.org/10.1016/S0167-577X(01)00394-9.

Nanomaterials Market. (n.d.). Nanomaterials Market Size, Industry Analysis Report, Regional Outlook, Application Development Potential, Price Trend, Competitive Market Share & Forecast, 2020–2026. Global Markets Insights. https://www.gminsights.com/industry-analysis/nanomaterials-market

Nanotechnologies and Waste. 2015. Report by the Ministry of the Environment. https://oekopol.de/src/files/06-FD3_nanomaterials-and-waste_report_bf.pdf

Nano 101. (n.d.). Nanotechnology Timeline. National Nanotechnology Initiative. https://www.nano.gov/timeline

Nasrollahzadeh, M., M. Sajjadi, S. Iravani and R.S. Varma. 2020a. Green-synthesized nanocatalysts and nanomaterials for water treatment: current challenges and future perspectives. Journal of Hazardous Materials 401: 123401. https://doi.org/10.1016/j.jhazmat.2020.123401.

Nasrollahzadeh, M., M. Sajjadi, S. Iravani, and R. Varma. 2020b. Trimetallic nanoparticles: greener synthesis and their applications. Nanomaterials (Basel, Switzerland) 10(9): 1784. https://doi.org/10.3390/nano10091784.

Nasseh, N., L. Taghavi, B. Barikbin, M. Nasseri and A. Allahresani. 2019. $FeNi_3/SiO_2$ magnetic nanocomposite as an efficient and recyclable heterogeneous fenton-like catalyst for the oxidation of metronidazole in neutral environments: Adsorption and degradation studies. Composites, Part B. 166: 328–340.

NPTELHRD. 2014. Mod-03 Lec-27 Nanocomposites - I. Nano structured materials-synthesis, properties, self-assembly and applications. YouTube. Available: https://www.youtube.com/watch?v=WQdLDMLrYIA

OECD. 2016. Nanomaterials in Waste Streams: Current Knowledge on Risks and Impacts. OECD Publishing, Paris, doi: https://doi.org/10.1787/9789264249752-en.

Omanovic-Miklicanin, E., A. Badnjevic and A. Kazlagic. 2020. Nanocomposites: A brief review. Health and Technology 10: 51–59. https://doi.org/10.1007/s12553-019-00380-x

Orts, W.J., J. Shey and S.H. Imam, G.M. Glenn, M.E. Guttman and J.-F. Revol. 2005. Application of cellulose microfibrils in polymer nanocomposites. Journal of Polymers and the Environment 13: 301–306. https://doi.org/10.1007/s10924-005-5514-3.

Pang, H., Y. Li, L. Guan, Q. Lu and F. Gao. 2011. TiO_2/Ni nanocomposites: Biocompatible and recyclable magnetic photocatalysts. Catalysis Communications 12(7): 611–615. https://doi.org/10.1016/j.catcom.2010.12.015.

Parveen, K., V. Banse and L. Ledwani. 2016. Green synthesis of nanoparticles: Their advantages and disadvantages. AIP Conference Proceedings 1724: 020048. doi:10.1063/1.4945168.

Patra, J.K. and K.–H. Baek. 2014. Green nanobiotechnology: Factors affecting synthesis and characterization techniques. Journal of Nanomaterials 2014: 417305. https://doi.org/10.1155/2014/417305.

Patwardhan, S.V., Joseph R.H. Manning and M. Chiacchia. 2018. Bioinspired synthesis as a potential green method for the preparation of nanomaterials: Opportunities and challenges. Current Opinion in Green and Sustainable Chemistry 12: 110–116. doi: https://doi.org/10.1016/j.cogsc.2018.08.004.

Petrakli, F., A. Gkika, A. Bonou, P. Karayannis, E.P. Koumoulos, D. Semitekolos, et al. 2020. End-of-life recycling options of (Nano)enhanced CFRP composite prototypes waste—A life cycle perspective. Polymers 12(9): 2129. https://doi.org/10.3390/polym12092129.

Popli, S. 2020. Use of nano particles in wastewater treatment Dr. Snehal Popli G.H. Patel College of Engineering. Mechanical FDP. https://www.youtube.com/watch?v=TS7pCm9-JZE

Risk Assessment. (n.d). Managing risks and risk assessment at work. Health and Safety Execute (HSE). https://www.hse.gov.uk/simple-health-safety/risk/risk-assessment-template-and-examples.htm

Roberto, M.M. and C.A. Christofoletti. 2019. How to Assess Nanomaterial Toxicity? An Environmental and Human Health Approach, Nanomaterials - Toxicity, Human Health and Environment, Simona Clichici, Adriana Filip and Gustavo M. do Nascimento, IntechOpen. DOI: 10.5772/intechopen.88970. https://www.intechopen.com/chapters/68905.

Rouhani, S., A. Rostami, A. Salimi and O. Pourshiani. 2018. Graphene oxide/$CuFe_2O_4$ nanocomposite as a novel scaffold for the immobilization of laccase and its application as a recyclable nanobiocatalyst for the green synthesis of arylsulfonyl benzenediols. Biochemical Engineering Journal 133: 1–11. doi:10.1016/j.bej.2018.01.004.

Rwahwire, S., B. Tomkova, A.P. Periyasamy and B.M. Kale. 2019. Green thermoset reinforced biocomposites. pp. 61–80. In: G. Koronis and A. Silva (eds). Green Composites for Automotive Applications, A volume in Woodhead Publishing Series in Composites Science and Engineering. Woodhead Publishing, UK. doi:10.1016/b978-0-08-102177-4.00003-3.

Saba, N., M. Jawaid and M. Asim. 2019. Nanocomposites with nanofibers and fillers from renewable resources. pp. 145–170. In: G. Koronis and A. Silva (eds). Green Composites for Automotive Applications, A volume in Woodhead Publishing Series in Composites Science and Engineering. Woodhead Publishing, UK. doi:10.1016/b978-0-08-102177-4.00007-0.

Salah, A. and A. Hadeer. 2020. Modern Trends in Uses of Different Wastes to Produce Nanoparticles and Their Environmental Applications, Nanotechnology and the Environment, Mousumi Sen, IntechOpen, DOI: 10.5772/intechopen.93315. https://www.intechopen.com/chapters/72931.

Shamim, N. and V.K. Sharma (eds). 2013. Sustainable Nanotechnology and the Environment: Advances and Achievements. American Chemical Society, Washington: DC. pp. 11–39.

Shang, M., W. Wang, H. Zou and G. Ren. 2016. Coating Fe_3O_4 spheres with polypyrrole-Pd composites and their application as recyclable catalysts. Synthetic Metals 221: 142–148. https://doi.org/10.1016/j.synthmet. 2016.08.015.

Shankar, S., R. Pangeni, J.W. Park and J.-W. Rhim. 2018. Preparation of sulfur nanoparticles and their antibacterial activity and cytotoxic effect. Materials Science and Engineering: C 92: 508–517. doi:10.1016/j.msec.2018. 07.015.

Sharma, M., K. Chaudhary, M. Kumari, P. Yadav, K. Sachdev, V. Chandra Janu, et al. 2020. Highly efficient, economic, and recyclable glutathione decorated magnetically separable nanocomposite for uranium (VI) adsorption from aqueous solution. Materials Today Chemistry. doi:10.1016/j.mtchem.2020.100379.

Siddique, S.N. 2013. Simulation of Mobility and Retention of Selected Engineered Nanoparticles Beneath Landfills. Msc dissertation. School of Graduate and Postdoctoral Studies, The University of Western Ontario London, Ontario, Canada.

Singh, M., S. Manikandan and A.K. Kumaraguru. 2011. Nanoparticles: A new technology with wide applications. Research Journal of Nanoscience and Nanotechnology 1: 1–11. 10.3923/rjnn.2011.1.11.

Singh, A.A., S. Afrin and Z. Karim. 2017. Green composites: Versatile material for future. pp. 29–44. *In*: M. Jawaid, S.M. Sapuan and O.Y. Alothman (eds). Green Biocomposites: Manufacturing and Properties. Series (GREEN): Green Energy and Technology. Springer, Cham. https://doi.org/10.1007/978-3-319-49382-4_2.

Singh, J., T. Dutta and K.H. Kim. 2018. 'Green' synthesis of metals & their oxide nanoparticles: applications for environmental remediation. Journal of Nanobiotechnology 16: 84. https://doi.org/10.1186/s12951-018-0408-4.

Singh S., V. Kumar, R. Romero, K. Sharma and J. Singh. 2019. Applications of nanoparticles in wastewater treatment. pp. 395–418. *In*: R. Prasad, V. Kumar, M. Kumar and D. Choudhary (eds). Nanobiotechnology in Bioformulations (Nanotechnology in the Life Sciences). Springer, Cham. https://doi.org/10.1007/978-3-030-17061-5_17.

Silva, G.L., C. Viana, D. Domingues and F. Vieira. 2019. Risk assessment and health, safety, and environmental management of carbon nanomaterials. pp. 31–52. *In*: S. Clichici, G.A. Filip and G.M. Do Nascimento (eds). Nanomaterials – Toxicity, Human Health and Environment. IntechOpen, London: U.K.

Smita, S., S.K. Gupta, A. Bartonova, M. Dusinska, A.C. Gutleb and Q. Rahman. 2012. Nanoparticles in the environment: assessment using causal diagram approach. Environmental Health 11: S13. https://doi.org/10.1186/1476-069X-11-S1-S13.

Sotiriou, G.A., D. Singh, F. Zhang, M.C.G. Chalbot, E. Spielman-Sun, L. Hoering, et al. 2016. Thermal decomposition of nano-enabled thermoplastics: Possible environmental health and safety implications. Journal of Hazardous Materials 305: 87–95. https://doi.org/10.1016/j.jhazmat.2015.11.001.

Stevens, P.D., J. Fan, H.M.R. Gardimalla, M. Yen and Y. Gao. 2005. Superparamagnetic nanoparticle-supported catalysis of suzuki cross-coupling reactions. Organic Letters 7(11): 2085–2088. doi:10.1021/ol050218w

Su, S. and P.M. Kang. 2020. Systemic review of biodegradable nanomaterials in nanomedicine. Nanomaterials 10: 656. doi:10.3390/nano10040656.

Sui, Z., Q. Meng, X. Zhang, R. Ma and B. Cao. 2012. Green synthesis of carbon nanotube-graphene hybrid aerogels and their use as versatile agents for water purification. Journal of Materials Chemistry 22(18): 8767–8771.

Sun, T.Y., F. Gottschalk, K. Hungerbuhler and B. Nowack. 2014. Comprehensive probabilistic modelling of environmental emissions of engineered nanomaterials. Environmental Pollution 185: 69–76. doi:10.1016/j.envpol.2013.10.004.

Sun, B., X. Wang, Y.-P. Liao, Z. Ji, C.H. Chang, S. Pokhrel, et al. 2016. Repetitive dosing of fumed silica leads to profibrogenic effects through unique structure–activity relationships and biopersistence in the lung. ACS Nano 10(8): 8054–8066. doi:10.1021/acsnano.6b04143.

SWD. 2019. Report from the Economic and Social Committee and the Committee of the Regions on the Implementation of the Circular Economy Action Plan. European Commission. https://ec.europa.eu/environment/circular-economy/pdf/report_implementation_circular_economy_action_plan.pdf

Tang, M., S. Zhang, X. Li, X. Pang and H. Qiu. 2014. Fabrication of magnetically recyclable Fe_3O_4@Cu nanocomposites with high catalytic performance for the reduction of organic dyes and 4-nitrophenol. Materials Chemistry and Physics 148(3): 639–647. doi: 10.1016/j.matchemphys.2014.08.029.

Taynton, P., H. Ni, C. Zhu, K. Yu, S. Loob, Y. Jin, et al. 2016. Repairable woven carbon fiber composites with full recyclability enabled by malleable polyimine networks. Advanced Materials 28(15): 2904–2909. doi:10.1002/adma.201505245.

TEC [The European Commission]. 2019. The European Green Deal COM/2019/640 Final. https://eur-lex.europa.eu/legal-content/EN/TXT/?uri=COM%3A2019%3A640%3AFIN

Tiede, K., A.B.A. Boxall, X. Wang, D. Gore, D. Tiede, M. Baxter, et al. 2010. Application of hydrodynamic chromatography-ICP-MS to investigate the fate of silver nanoparticles in activated sludge. Journal of Analytical Atomic Spectrometry 25(7): 1149. https://www.semanticscholar.org/paper/Application-of-hydrodynamic-chromatography-ICP-MS-Tiede-Boxall/984e0cd7e8927a1c6db70bd339cf6ab917c7902d

Titirici, M., R.J. White, N. Brun, V. Budarin, D. Su, F. Del Monte, et al. 2015. Sustainable carbon materials. Chemical Society Reviews 44: 250–290. doi: https://doi.org/10.1039/C4CS00232F

Tran, T.V., T.T. Phan, D.T. Nguyen, T.T. Nguyen, D. Nguyen, D.N. Vo, et al. 2020. Recyclable Fe_3O_4@C nanocomposite as potential adsorbent for a wide range of organic dyes and simulated hospital effluents. Environmental Technology & Innovation 20: 101122.

Tsang, S.C., V. Caps, I. Paraskevas, D. Chadwick and D. Thompsett. 2004. Magnetically separable, carbon-supported nanocatalysts for the manufacture of fine chemicals. Angewandte Chemie International Edition 43(42): 5645–5649. doi:10.1002/anie.200460552.

Tseng, W.J., Y.C. Chuang and Y.A. Chen. 2018. Mesoporous Fe_3O_4@Ag@TiO_2 nanocomposite particles for magnetically recyclable photocatalysis and bactericide. Advanced Powder Technology 29(3): 664–671. doi:10.1016/j.apt.2017.12.008.

Veisi, H., S. Razeghi, P. Mohammadi and S. Hemmati. 2019. Silver nanoparticles decorated on thiol modified magnetite nanoparticles (Fe_3O_4/SiO_2-Pr-S-Ag) as a recyclable nanocatalyst for degradation of organic dyes. Materials Science and Engineering C 97: 624–631. doi: 10.1016/j.msec.2018.12.076.

Vejerano, E.P., E.C. Leon, A.L. Holder and L.C. Marr. 2014. Characterization of particle emissions and fate of nanomaterials during incineration. Environmental Science Nano 1: 133. https://doi.org/10.1039/c3en00080j

Vinayan, B.P., R. Nagar and S. Ramaprabhu. 2013. Solar light assisted green synthesis of palladium nanoparticle decorated nitrogen doped graphene for hydrogen storage application. Journal of Materials Chemistry A 1(37): 11192. doi:10.1039/c3ta12016c

Walser, T. and F. Gottschalk. 2014. Stochastic fate analysis of engineered nanoparticles in incineration plants. Journal of Cleaner Production 80: 241–251. https://doi.org/10.1016/j.jclepro.2014.05.085.

Walsh, T.R. and M.R. Knecht. 2017. Biointerface structural effects on the properties and applications of bioinspired peptide-based nanomaterials. Chemical Reviews 117: 12641–12704. https://doi.org/10.1021/acs.chemrev.7b00139.

Wang, Y., P. Westerhoff and K.D. Hristovski. 2012. Fate and biological effects of silver, titanium dioxide, and C60 (fullerene) nanomaterials during simulated wastewater treatment processes. Journal of Hazardous Materials 201–202: 16–22. https://doi.org/10.1016/j.jhazmat.2011.10.086.

Wang, D. and D. Astruc. 2014. Fast-growing field of magnetically recyclable nanocatalysts. Chemical Reviews 114: 6949–6985, doi: https://doi.org/10.1021/cr500134h.

Wang, H., L. Xu, X. Wu and M. Zhang. 2020. Eco-friendly synthesis of a novel magnetic $Bi_4O_5Br_2$/$SrFe_{12}O_{19}$ nanocomposite with excellent photocatalytic activity and recyclable performance. Ceramics International 47(6): 8300–8307. https://doi.org/10.1016/j.ceramint.2020.11.191.

WCN [Waste Containing Nanomaterials]. 2018. Report on issues related to waste containing nanomaterials and options for further work under the Basel Convention. Basel Convention. United Nations.

Wiesner, M.R. and D.L. Plata. 2012. Environmental, health and safety issues: Incinerator filters nanoparticles. Nature Nanotechnology 7: 487–488. https://doi.org/10.1038/nnano.2012.133.

Wilson, R., J. Joy, G. George and V. Anuraj. 2017. Nanocellulose: A novel support for water purification. pp. 456–476. In: Kharisov CMHB (ed.). Advanced Environmental Analysis: Applications of Nanomaterials, Vol. 1. Royal Society of Chemistry, Cambridge.

Wiwanitkit, V. 2019. Biodegradable nanoparticles for drug delivery and targeting. pp. 167–181. In: Y.V. Pathak (ed.). Surface Modification of Nanoparticles for Targeted Drug Delivery. Springer, Cham. https://doi.org/10.1007/978-3-030-06115-9_9.

Wlodarczyk, J., W. Sikorska and J. Rydz. 2018. 3D processing of PHA containing (bio)degradable materials. pp. 121–168. In: M. Koller (ed.). Current Advances in Biopolymer Processing and Characterization. Biomaterials – Properties, Production, and Devices Series. Nova Science Publishers, New York.

Xiong, P., J. Zhu and X. Wang. 2013. Cadmium sulfide–ferrite nanocomposite as a magnetically recyclable photocatalyst with enhanced visible-light-driven photocatalytic activity and photostability. Industrial and Engineering Chemistry Research 52(48): 17126–17133. doi:10.1021/ie402437k

Yong, C., X. Chen, Q. Xiang, Q. Li and X. Xing. 2018. Recyclable magnetite-silver heterodimer nanocomposites with durable antibacterial performance. Bioactive Materials 3(1): 80–86. https://doi.org/10.1016/j.bioactmat.2017.05.008.

Zhang, X., W. Jiang, Y. Zhou, S. Xuan, C. Peng, L. Zong, et al. 2011. Magnetic recyclable Ag catalysts with a hierarchical nanostructure. Nanotechnology 22(37): 375701. doi:10.1088/0957-4484/22/37/375701.

Zhang, L., X. Hu, C. Wang and Y. Tai. 2018. Water-dispersible and recyclable magnetic TiO_2/graphene nanocomposites in wastewater treatment. Materials Letters 231: 80–83. doi:10.1016/j.matlet.2018.08.011

Zhang, J., V.S. Chevali, H. Wang and C.-H. Wang. 2020. Current status of carbon fibre and carbon fibre composites recycling. Composites Part B: Engineering 193: 108053. https://doi.org/10.1016/j.compositesb.2020.108053.

Zuin, S., A. Massari, S. Motellier, S. Golanski and Y. Sicard. 2013. Nanowaste Management. NanoHouse Dissemination report. http://www.basel.int/Implementation/Wastecontainingnanomaterials/Followupto OEWG11/tabid/7964/ctl/Download/mid/21778/Default.aspx?id=9&ObjID=21266.

Index

For Product Safety Concerns and Information please contact our EU
representative GPSR@taylorandfrancis.com
Taylor & Francis Verlag GmbH, Kaufingerstraße 24, 80331 München, Germany